Brown Theory Group

Lecture Notes in Computer Science 4000

Commenced Publication in 1973
Founding and Former Series Editors:
Gerhard Goos, Juris Hartmanis, and Jan van Leeuwen

Sławomir Stańczak Marcin Wiczanowski
Holger Boche

Resource Allocation in Wireless Networks

Theory and Algorithms

Springer

Authors

Sławomir Stańczak
Fraunhofer German-Sino Lab for Mobile Communications
Einsteinufer 37, 10587 Berlin, Germany
E-mail: stanczak@hhi.fhg.de

Marcin Wiczanowski
Heinrich-Hertz Chair, Faculty of EECS Technical University of Berlin
Einsteinufer 25, 10587 Berlin, Germany
E-mail: marcin.wiczanowski@tu-berlin.de

Holger Boche
Heinrich-Hertz Chair, Faculty of EECS Technical University of Berlin
Einsteinufer 25, and Fraunhofer German-Sino Lab for Mobile Communications and
Fraunhofer-Institute for Telecommunications, Heinrich-Hertz-Institut (HHI)
Einsteinufer 37, 10587 Berlin, Germany
E-mail: boche@hhi.fhg.de

Library of Congress Control Number: 2006934732

CR Subject Classification (1998): C.2, F, G.2.2, D.4.4

LNCS Sublibrary: SL 1 – Theoretical Computer Science and General Issues

ISSN 0302-9743
ISBN-10 3-540-46248-1 Springer Berlin Heidelberg New York
ISBN-13 978-3-540-46248-4 Springer Berlin Heidelberg New York

Springer is a part of Springer Science+Business Media

springer.com

Printed in Germany

Typesetting: Camera-ready by author, data conversion by Markus Richter, Heidelberg
Printed on acid-free paper SPIN: 11818762 06/3142 5 4 3 2 1 0

To our families

Der König hat viele Gnade für meine geringen Dienste, und das Publikum viel Nachsicht für die unbedeutenden Versuche meiner Feder; ich wünschte, dass ich einigermaßen etwas zu der Verbesserung des Geschmackes in meinem Lande, zur Ausbreitung der Wissenschaften beitragen könnte. Denn sie sind's allein, die uns mit anderen Nationen verbinden, sie sind's, die aus den entferntsten Geistern Freunde machen, und die angenehmste Vereinigung unter denen selbst erhalten, die leider durch Staatsverhältnisse öfters getrennt werden.

Clavigo, Johann Wolfgang von Goethe

Nader wiele łaski raczy mi okazywać król za moje skromne służby, publiczność zaś zbyt jest wyrozumiała dla niepozornych płodów mego pióra, byłbym wszelako szczęśliwy mogąc się nieco przyczynić do urobienia literackiego smaku w moim kraju, do rozprzestrzenienia nauk. Jedynie to bowiem może nas zbliżyć do innych nacji, tylko dzięki temu zdobywamy w najdalszych stronach przyjaciół pośród przodujących umysłów, przyczyniając się do utrwalenia najcenniejszych więzów, które niestety jakże często zrywają interesa państwowe.

Clavigo, Johann Wolfgang von Goethe
(Tłumaczenie: Wanda Markowska)

Preface

The wireless industry is in the midst of a fundamental shift from providing voice-only services to offering customers an array of multimedia services, including a wide variety of audio, video and data communications capabilities. Future wireless networks will be integrated into every aspect of daily life, and therefore could affect our life in a magnitude similar to that of the Internet and cellular phones. However, the emerging applications and directions require fundamental understanding on how to design and control wireless networks that lies far beyond what the currently existing theory can provide. We are deeply convinced that mathematics is the key technology to cope with central technical problems in the design of wireless networks since the complexity of the problem simply precludes the use of engineering common sense alone to identify good solutions.

The main objective of this book is to provide tools for better understanding the fundamental tradeoffs and interdependencies in wireless networks, with the goal of designing resource allocation strategies that exploit these interdependencies to achieve significant performance gains. The book consists of three largely independent parts: theory, applications and appendices. The first part ends with some bibliographical comments and the second part starts with a short introduction to the problem of resource allocation in wireless networks. Below we briefly summarize the content of each part.

Theory: Chapters 1 and 2 deal with some fundamental problems in the theory of nonnegative matrices and provide a theoretical basis for the resource allocation problem addressed in the second part of the book. It should be emphasized that our intent is *not* to provide a thorough treatment of this wide subject. Instead, we focus on problems that naturally appear in the design of resource allocation strategies for wireless networks. When developing such strategies, different characterizations of the Perron root of nonnegative irreducible matrices turn out to be vital to better understanding of fundamental tradeoffs between diverse optimization objectives. Our main attention will be directed to the Perron root of nonnegative irreducible matrices whose entries continuously depend on some parameter vector. In this case, the Perron root can be viewed as a map from a convex parameter set into the set of positive reals. The book is concerned with the properties of this map and, in particular, with the question under which conditions it is a convex function of the parameter vector. With few exceptions, we focus on a special structure of matrix-valued functions that is particularly relevant to applications in wireless networks. We provide necessary and sufficient conditions for the Perron root to be a convex function of the parameter vector as well as address a closely related problem of convexity of the so-called feasibility set. Chapter 2 is devoted to some properties of a positive solution to a system of linear equations with nonnegative coefficients. Applications that involve systems of linear equations with nonnegative coefficients are numerous, ranging from the physical and engineering sciences to other mathematical areas like graph theory and optimization. Such systems also occur in the power control problem for power-constrained wireless networks.

Applications: The second part of the book (Chaps. 4-6) deals with the problem of resource allocation in wireless networks. Roughly speaking, the objective is to maximize the sum of utilities of link rates for best-effort (elastic) traffic. This is equivalent to the problem of joint power control and link scheduling, which has been extensively investigated in the literature and is known to be notoriously difficult to solve, even in a centralized manner. Although the book provides some interesting insights into this problem, the main focus will be on the power control part. In particular, a class of utility functions is identified for which the power control problem can be converted into an equivalent convex optimization problem. The convexity property is a key ingredient in the development of powerful and efficient power control algorithms.

Appendices: The main purpose of the appendices is to make the book more understandable to readers who are not familiar with some basic concepts and results from linear algebra and convex analysis. The treatment is very superficial and formal proofs are presented only for the most important results such as the Perron–Frobenius theorem. Moreover, the presentation is limited to results used somewhere in the book. However, we hope that this collection of basic results will help some readers to better understand the material covered by the book. Finally, the presentation introduces the notation and terminology used throughout the book.

Acknowledgments: The work of Holger Boche and Sławomir Stańczak was supported in part by the *Bundesministerium für Bildung und Forschung (BMBF)* under grants 01BU150 (Hyeff), 01BU350 (3GET) and 01BU566 (ScaleNet). Marcin Wiczanowski was supported by the *Deutsche Forschungsgemeinschaft (DFG)* under grant BO1734/7-1. The authors also acknowledge support from Alcatel SEL Forschungszentrum in Stuttgart, and Siemens CT in München, as well as valuable suggestions and comments from colleagues.

And finally, we would like to thank our families for their patience, support and understanding. This book is dedicated to you.

Berlin, June 2006

Sławomir Stańczak
Marcin Wiczanowski
Holger Boche

Contents

List of Figures

List of Symbols

$a, b, c, \alpha, \beta, \mu, \ldots$	Scalars over $\mathbb{R}$ or $\mathbb{C}$
$\mathbf{A}, \mathbf{B}, \mathbf{X}, \mathbf{Y} \ldots$	Matrices; Sect. A.2
$\mathbf{A} \leq \mathbf{B}$	Partial ordering; Sect. A.2
$\mathbf{A}^{-1}$	Matrix inverse; Sect. A.3
$\mathbf{A}^T$	Transpose matrix; Definition A.3
$\mathrm{A}_K(\mathbf{X})$	Eq. (1.8)
$\mathrm{A} \times \mathrm{B}$	Cartesian product
A	Sect. 5.2.1
$\mathbf{A} \circ \mathbf{B}$	Hadamard product; Sect. A.2
B_K	Sect. 1.6.2
$\overline{\mathrm{B}}_K$	Sect. 1.6.2
B	Sects. 4.3 and 5.2.1
$\mathbb{C}$	Sect. A.1
$\mathrm{cl}(\mathrm{A})$	Closure
C	Eq. (5.11)
$\tilde{\mathrm{C}}$	Eq. (5.15)
$\mathrm{diag}(\mathbf{u})$	Diagonal matrix; Sect. A.2
$\det(\mathbf{A})$	Matrix determinant; Sect. A.3
δ_l	The Kronecker delta
$\mathbf{e}_i$	Sect. A.1
$\mathrm{E}_K(\mathbf{X})$	Sect. 1.6.2
$\mathrm{E}_K^+(\mathbf{X})$	Sect. 1.6.2
$\boldsymbol{\eta}(\mathbf{p})$	Eq. (6.30)
F	Eq. (1.53) and Eq. (2.5)
$\partial\mathrm{F}$	Eq. (1.55)
F^c	Eq. (1.60)

$\mathrm{F}(P_{\mathrm{t}})$	Eq. (2.9)
$\mathrm{F}(P_1, \ldots, P_K)$	Eq. (2.11)
$\mathrm{F}_k(\alpha)$	Eq. (2.12)
$\mathrm{F}(P_{\mathrm{t}}; P_1, \ldots, P_K)$	Eq. (2.13)
$\partial\mathrm{F}(P_{\mathrm{t}})$	Definition 2.15
$\partial\mathrm{F}(P_1, \ldots, P_K)$	Definition 2.15
$\mathrm{F}^c(P_{\mathrm{t}})$	Sect. 2.4
F_γ	Eq. (5.32)
$\mathrm{F}_\gamma(\mathrm{P})$	Eq. (5.31)
$\partial\mathrm{F}_\gamma(\mathrm{P})$	Eq. (5.34)
$f'(x), x \in \mathbb{R}$	The first derivative; Sect. B.1
$f''(x), x \in \mathbb{R}$	The second derivative; Sect. B.1
$F(\mathbf{p})$	Eq. (6.2)
$F_e(\mathbf{s})$	Eq. (6.12)
$\boldsymbol{\Gamma}(\boldsymbol{\omega})$	Eqs. (1.48) and (5.12)
$g_k(\mathbf{p})$	(6.21)
$h_k(\mathbf{s})$	Eq. (6.14)
$\mathbf{I}$	Identity matrix; Sect. A.2
$I_k(\mathbf{p})$	Eq. (6.5)
K	Sect. 4.1
$\mathsf{K}(n)$	Sect. 4.1
$\lambda_p(\boldsymbol{\omega})$	Sect. 1.3.1
$\mathrm{LC}_K(\Omega)$	Definition 1.34
$\mathrm{lc}(\Omega)$	Sect. 2.3
L	Sect. 4.1
$\mathbb{N}$	Natural numbers
$\mathbb{N}_0$	Nonnegative integers
N	Sect. 4.1
$\mathrm{N}_K = \mathbb{R}_+^{K \times K}$	Definition A.18
N_K^+	Sect. 1.6.2
$\mathrm{N}_K(\Omega)$	Definition 1.32
$\mathrm{N}_{K,\boldsymbol{\Gamma}}(\Omega)$	Eq. (2.8)
$\Vert\mathbf{u}\Vert$	Vector norms; Sect. A.1
$\Vert\mathbf{X}\Vert$	Matrix norms; Sect. A.2
$\nabla_k f(\mathbf{x})$	Definition B.9
$\nabla f(\mathbf{x})$	Eq. (B.3)
$\nabla^2 f(\mathbf{x})$	Definition B.12

$\mathbf{1}$	Sect. A.1
$\Omega \subset \mathbb{R}^K$	Eq. (1.45)
Π_K	Sect. 1.1
Π_K^+	Sect. 1.1
$\mathrm{P}_K = \mathbb{R}_{++}^{K\times K}$	Definition A.18
$\mathrm{P}_K(\Omega)$	Sect.
$\boldsymbol{\omega}$	Sect. 1.3.1
$\mathbf{p}(\mathbf{X})$	Eq. (1.2)
$\mathbf{p}(\boldsymbol{\omega})$	Eq. (2.4)
Φ	Eq. (4.7)
Ψ	Definition 5.5
Ψ_e	Definition 5.5
ψ	Eq. (5.36)
ψ_e	Definition 6.1
Π_{S}	Eq. (B.19)
$\mathrm{P} \subset \mathbb{R}_+^K$	Eq. (4.6)
P_n	Eq. (4.6)
P_+	Eq. (6.10)
$\mathbf{q}(\mathbf{X})$	Eq. (1.2)
$\mathrm{Q} \subset \mathbb{R}$	Sect. 1.3.1
$\mathbb{R}$	Real numbers
$\mathbb{R}_+ \subset \mathbb{R}$	Sect. A.4
$\mathbb{R}_{++} \subset \mathbb{R}_+$	Sect. A.4
$\mathbb{R}^K$	Sect. A.1
$\mathbb{R}^{K\times K}$	Sect. A.2
$\mathbb{R}_{++}^K(\Omega)$	Sect. 2.1
$\sigma(\mathrm{A})$	Matrix spectrum; Definition A.7
$\rho(\mathbf{X})$	Spectral radius; Definition A.7
S_K	Sect. 1.2
$\mathrm{S}_K(\mathbf{X})$	Eq. (1.3)
$\mathrm{SIR}_k(\mathbf{p})$	Eq. (4.2)
$\mathrm{S} \subset \mathbb{R}^K$	Eq. (6.11)
S	Sect. 4.1
$\mathrm{trace}(\mathbf{X})$	Matrix trace; Sect. A.2
$\boldsymbol{\theta}(\mathbf{p})$	Eq. (6.30)
$\mathbf{u} \leq \mathbf{v}$	Partial ordering; Sect. A.1
$\mathbf{p}, \mathbf{q}, \mathbf{s}, \mathbf{u}, \mathbf{v}, \mathbf{z}, \ldots$	Vectors; Sect. A.1

$\mathbf{V}$	Eq. (4.4)
$\mathrm{W}_K(\mathbf{X})$	Eq. (1.14)
$\mathrm{X}_K \subset \mathrm{N}_K$	Definition A.21
$\mathrm{X}_K(\Omega)$	Definition 1.32
$\mathrm{X}_{K,\mathbf{\Gamma}}(\Omega)$	Eq. (1.49)
$\mathrm{X}^s_{K,\mathbf{\Gamma}}(\Omega)$	Sect. 1.4.1
$\mathrm{X}^p_{K,\mathbf{\Gamma}}(\Omega)$	Sect. 1.4.2
$\mathrm{X}^0_{K,\mathbf{\Gamma}}(\Omega)$	Sect. 1.5
$\mathbf{0}$	Zero vector; Sect. A.1

Part I

Theory

1

On the Perron Root of Irreducible Matrices

This chapter deals with the Perron root of nonnegative irreducible matrices. Applications abound with nonnegative and positive matrices so that it is natural to investigate their properties. In doing so, one of the central problems is to what extent the nonnegativity (positivity) is inherited by the eigenvalues and eigenvectors. The principal tools for the analysis of spectral properties of irreducible matrices are provided by Perron–Frobenius theory. A comprehensive reference on nonnegative matrices is [2]. Some basic results are summarized in Appendix A.4. For more information about the Perron–Frobenius theory, the reader is also referred to [3, 4, 5].

We have divided the chapter into two major parts. The purpose of the first part is to characterize the Perron root of irreducible matrices and present some interesting bounds on it. There exists a vast literature addressing the problem of estimating the Perron root of nonnegative irreducible matrices. Tight bounds on the Perron root have attracted a great deal of attention over several decades. A brief (and by no means extensive) summary of some related results can be found at the end of this chapter. In the second part, we consider the Perron root of matrix-valued functions of some parameter vector. In this case, each matrix entry is a continuous nonnegative function defined on some convex parameter set, with the constraint that the matrix is irreducible for every fixed parameter vector. As a result, the Perron root can be viewed as a positive real-valued function defined on a convex set. Now the objective is to provide conditions under which the Perron root is a convex (or concave) function of the parameter vector. Note that the convexity property is a key ingredient in the development of access control and resource allocation strategies for wireless networks.

1.1 Some Basic Definitions

We use $X_K \subset \mathbb{R}^{K\times K}$ to denote the set of all $K \times K$ nonnegative irreducible matrices. Let $\rho(\mathbf{X})$ be the Perron root of $\mathbf{X} \in X_K$. By the Perron–Frobenius theorem (Theorem A.25in Appendix A.4.1), $\rho(\mathbf{X})$ is a simple eigenvalue of $\mathbf{X}$ and is equal to its spectral radius so that

$$\rho(\mathbf{X}) = \max_{\lambda\in\sigma(\mathbf{X})} |\lambda| \quad \text{and} \quad \rho(\mathbf{X}) \in \sigma(\mathbf{X}) \tag{1.1}$$

where $\sigma(\mathbf{X})$ denotes the spectrum of $\mathbf{X}$ (Definition A.7). Due to (1.1), if $\mathbf{X} \in \mathrm{X}_K$, $\rho(\mathbf{X})$ is used to denote both the Perron root of $\mathbf{X}$ and its spectral radius. Moreover, if

$$\begin{aligned} \mathbf{X}\mathbf{p} &= \rho(\mathbf{X})\mathbf{p} \\ \mathbf{X}^T\mathbf{q} &= \rho(\mathbf{X})\mathbf{q} \end{aligned} \tag{1.2}$$

holds for some $\mathbf{q}, \mathbf{p} \in \mathbb{R}^K$, then both $\mathbf{q} := \mathbf{q}(\mathbf{X})$ and $\mathbf{p} := \mathbf{p}(\mathbf{X})$ are positive vectors, and there are no other nonnegative eigenvectors of $\mathbf{X}$ except for positive multiples of $\mathbf{q}$ and $\mathbf{p}$, regardless of the eigenvalue. Unless something else is stated, assume that $\mathbf{q}^T\mathbf{p} = 1$ or, equivalently, $\|\mathbf{w}\|_1 = 1$ where $\mathbf{w} = \mathbf{q} \circ \mathbf{p}$. For readability purposes, let Π_K denote the standard simplex in $\mathbb{R}_+^K$, i.e. we have

$$\Pi_K := \{\mathbf{x} \in \mathbb{R}_+^K : \|\mathbf{x}\|_1 = 1\}\,.$$

Furthermore, we define $\Pi_K^+ := \Pi_K \cap \mathbb{R}_{++}^K$, which contains all positive vectors whose elements sum up to 1. Hence, $\mathbf{w} = \mathbf{q} \circ \mathbf{p} \in \Pi_K^+$ for any $\mathbf{X} \in \mathrm{X}_K$.

Definition 1.1 (Perron Eigenvectors). *Let $\mathbf{q}, \mathbf{p} \in \mathbb{R}_{++}^K$ be left and right positive eigenvectors of $\mathbf{X} \in \mathrm{X}_K$. If additionally $\mathbf{q}, \mathbf{p} \in \Pi_K^+$, then the unique eigenvectors $\mathbf{q}$ and $\mathbf{p}$ are called the left and right Perron eigenvectors of $\mathbf{X} \in \mathrm{X}_K$, respectively (see also Definition A.26).*

Throughout this chapter, we use $\mathbf{q}$ and $\mathbf{p}$ to designate positive left and right eigenvectors of $\mathbf{X} \in \mathrm{X}_K$, respectively. In cases where ambiguity may occur, we write $\mathbf{q}(\mathbf{X})$ and $\mathbf{p}(\mathbf{X})$ to denote these eigenvectors of $\mathbf{X}$.

Caution: In the second part of the book, this notation is not used. In particular, $\mathbf{p}$ will not denote any positive right eigenvector.

1.2 Some Bounds on the Perron Root and Their Applications

This section presents several bounds on the Perron root of irreducible matrices. Some of these results provide a starting point for the development of the theory presented in the subsequent sections of this book, while others establish interesting connections, thereby helping to better understand the complex interrelations in practical systems.

Let $\mathrm{S}_K \subset \mathrm{N}_K$ be the set of all stochastic matrices of size $K \times K$ (Definition A.20). Therefore, each row of $\mathbf{A} \in \mathrm{S}_K$, say row k denoted by $\mathbf{a}^{(k)} \in \mathbb{R}_+^K$, satisfies $\mathbf{a}^{(k)} \in \Pi_K$. For every fixed $\mathbf{X} \in \mathrm{X}_K$, we define an associated subset $\mathrm{S}_K(\mathbf{X})$ of S_K as

$$\mathrm{S}_K(\mathbf{X}) = \{\mathbf{A} \in \mathrm{S}_K : (\mathbf{A})_{k,l} = a_{k,l} = 0 \text{ if and only if } x_{k,l} = 0\}\,. \tag{1.3}$$

Note that since $\mathbf{X}$ is irreducible, every member of $\mathrm{S}_K(\mathbf{X})$ is an irreducible stochastic matrix. Hence, we have $\mathrm{S}_K(\mathbf{X}) \subset \mathrm{X}_K$ for any $\mathbf{X} \in \mathrm{X}_K$.

Although not explicitly stated in this form, the following Perron root characterization can be deduced from [6, Equation 2.6 with 2.8 and 2.9] (see also the bibliographical notes at the end of this chapter).

Theorem 1.2. *Let $\mathbf{X} \in \mathrm{X}_K$ be arbitrary. Then, we have*

$$\sum_{k,l=1}^{K} u_k a_{k,l} \log \frac{x_{k,l}}{a_{k,l}} \le \log \rho(\mathbf{X}) \tag{1.4}$$

for all $\mathbf{A} \in \mathrm{S}_K(\mathbf{X})$ where $\mathbf{u} = (u_1, \dots, u_K) \in \Pi_K^+$ is the left Perron eigenvector of $\mathbf{A}$. Equality holds in (1.4) if and only if

$$(\mathbf{A})_{k,l} = a_{k,l} = \frac{x_{k,l} p_l}{\rho(\mathbf{X}) p_k}, \quad 1 \le k, l \le K \tag{1.5}$$

where $\mathbf{p} \in \mathbb{R}_{++}^K$ is a positive right eigenvector of $\mathbf{X}$.

Proof. Since $\rho(\mathbf{X}/\rho(\mathbf{X})) = \rho(\mathbf{X})/\rho(\mathbf{X}) = 1$, we can assume that $\rho(\mathbf{X}) = 1$. Let $\mathbf{A} \in \mathrm{S}_K(\mathbf{X})$ be fixed and define

$$f(\mathbf{e}) = \sum_{k,l=1}^{K} u_k a_{k,l} \log\Big(\frac{x_{k,l}}{a_{k,l}} \frac{e_l}{e_k}\Big) \tag{1.6}$$

for an arbitrary $\mathbf{e} \in \mathbb{R}_{++}^K$. Note that $f(\mathbf{1})$ is equal to the left hand side of (1.4). Moreover, $f(\mathbf{e})$ is independent of the choice of $\mathbf{e}$ since

$$\begin{aligned}\sum_{k,l=1}^{K} u_k a_{k,l} \log\Big(\frac{e_l}{e_k}\Big) &= \sum_{k,l=1}^{K} u_k a_{k,l} \log e_l - \sum_{k,l=1}^{K} u_k a_{k,l} \log e_k \\ &= \sum_{l=1}^{K} \log e_l \Big(\sum_{k=1}^{K} u_k a_{k,l}\Big) - \sum_{k=1}^{K} u_k \log e_k \Big(\sum_{l=1}^{K} a_{k,l}\Big) \\ &= \sum_{l=1}^{K} u_l \log e_l - \sum_{k=1}^{K} u_k \log e_k = 0\end{aligned}$$

where we used the fact that $\mathbf{A}$ is stochastic and $\mathbf{A}^T\mathbf{u} = \mathbf{u}$. Thus, without loss of generality, we can substitute any positive vector into (1.6). In particular, we can substitute $\mathbf{p}$ (a positive right eigenvector of $\mathbf{X}$) into (1.6) and confine our attention to matrices of the form

$$(\tilde{\mathbf{X}})_{k,l} = \tilde{x}_{k,l} = \frac{x_{k,l}p_l}{\rho(\mathbf{X})p_k} = \frac{x_{k,l}p_l}{p_k}, \quad 1 \le k,l \le K\,.$$

Now as $\tilde{\mathbf{X}}$ is stochastic and $\log x \le x - 1$ for all $x > 0$ with equality if and only if $x = 1$, we obtain

$$\sum_{l=1}^{K} a_{k,l} \log \frac{\tilde{x}_{k,l}}{a_{k,l}} \le \sum_{l=1}^{K} a_{k,l}\Big(\frac{\tilde{x}_{k,l}}{a_{k,l}} - 1\Big) = \sum_{l=1}^{K} \tilde{x}_{k,l} - \sum_{l=1}^{K} a_{k,l} = 1 - 1 = 0$$

for each $1 \le k \le K$, with equality if and only if $\mathbf{A} = \tilde{\mathbf{X}}$. So

$$\sum_{k=1}^{K} u_k \sum_{l=1}^{K} a_{k,l} \log \frac{x_{k,l}}{a_{k,l}} \le 0 = \log \rho(\mathbf{X})$$

with equality attained if and only if $\mathbf{A}$ is given by (1.5).

The following corollary is immediate.

Corollary 1.3. *Let $\mathbf{X} \in \mathrm{X}_K$ be arbitrary and fixed. Then,*

$$\log \rho(\mathbf{X}) = \sup_{\mathbf{A} \in \mathrm{S}_K(\mathbf{X})} \Big(\sum_{k,l=1}^{K} u_k a_{k,l} \log \frac{x_{k,l}}{a_{k,l}} \Big) \tag{1.7}$$

where $\mathbf{u} = (u_1, \ldots, u_K) \in \Pi_K^+$ is the left Perron eigenvector of $\mathbf{A}$.

The Perron root characterization in (1.7) turns out to be of great value in proving the central result of the second part of this chapter, namely a sufficient condition on convexity of the Perron root. Moreover, this characterization provides some interesting insights into the properties of the Perron root. So it is worth dwelling on this for a moment. An interesting problem is the exact relationship between irreducible matrices that have the same maximizers in (1.7). To be precise, let $\mathbf{X} \in \mathrm{X}_K$ be arbitrary and suppose that the supremum in (1.7) is attained at $\mathbf{A}(\mathbf{X}) \in \mathrm{S}_K(\mathbf{X})$. We define

$$\mathrm{A}_K(\mathbf{X}) := \Big\{ \mathbf{Y} \in \mathrm{X}_K : \mathbf{A}(\mathbf{X}) = \arg\sup_{\mathbf{A} \in \mathrm{S}_K(\mathbf{Y})} \Big(\sum_{k,l=1}^{K} u_k a_{k,l} \log \frac{y_{k,l}}{a_{k,l}} \Big) \Big\}. \tag{1.8}$$

Hence, by Theorem 1.2, $\mathrm{A}_K(\mathbf{X})$ contains all irreducible matrices for which the supremum in (1.7) is attained at $\mathbf{A}(\mathbf{X})$. The following observation characterizes this set.

Observation 1.4. *Given an arbitrary $\mathbf{X} \in \mathrm{X}_K$, we have $\mathbf{Y} \in \mathrm{A}_K(\mathbf{X})$ if and only if there exists a diagonal matrix $\mathbf{D}$ with positive diagonal entries such that*

$$\mathbf{Y} = \frac{\rho(\mathbf{Y})}{\rho(\mathbf{X})} \mathbf{D}\mathbf{X}\mathbf{D}^{-1}. \tag{1.9}$$

Proof. Let $\mathbf{A}(\mathbf{Y}) \in \mathrm{S}_K(\mathbf{Y})$ be any matrix such that

$$\mathbf{A}(\mathbf{Y}) = \arg\sup_{\mathbf{A}\in \mathrm{S}_K(\mathbf{Y})} \sum_{k,l=1}^{K} u_k a_{k,l} \log \frac{y_{k,l}}{a_{k,l}} .$$

It follows from (1.9) that $\mathbf{p}(\mathbf{Y}) = \mathbf{D}\mathbf{p}(\mathbf{X})$. Furthermore, considering Theorem 1.2 and (1.9), we obtain

$$\begin{aligned}(\mathbf{A}(\mathbf{Y}))_{k,l} &= \frac{y_{k,l}(\mathbf{D}\mathbf{p})_l}{\rho(\mathbf{Y})(\mathbf{D}\mathbf{p})_k} = \frac{1}{\rho(\mathbf{Y})d_k p_k}\Big(\frac{\rho(\mathbf{Y})}{\rho(\mathbf{X})}\frac{d_k x_{k,l}}{d_l}\Big)d_l p_l = \frac{x_{k,l}p_l}{\rho(\mathbf{X})p_k} \\ &= (\mathbf{A}(\mathbf{X}))_{k,l}\end{aligned}$$

where $\mathbf{p} = \mathbf{p}(\mathbf{X})$. Therefore, $\mathrm{A}_K(\mathbf{Y}) = \mathrm{A}_K(\mathbf{X})$, and the observation follows.

Interestingly, Theorem 1.2 gives rise to a well-known bound on the Perron root of the Hadamard product of two irreducible matrices [7]. Note that if $\mathbf{X}\circ\mathbf{Y} \in \mathrm{X}_K$ (entry-wise multiplication), then $\mathbf{X} \in \mathrm{X}_K$ and $\mathbf{Y} \in \mathrm{X}_K$.

Corollary 1.5. *Let $\mathbf{X}\circ\mathbf{Y} \in \mathrm{X}_K$ be arbitrary. Then,*

$$\rho(\mathbf{X}\circ\mathbf{Y}) \le \rho(\mathbf{X})\cdot\rho(\mathbf{Y}). \tag{1.10}$$

Proof. For every $1 \le k,l \le K$, let $\tilde{\mathbf{X}}$ and $\tilde{\mathbf{Y}}$ be given by

$$\tilde{x}_{k,l} = \begin{cases} x_{k,l} & x_{k,l}y_{k,l} > 0 \\ 0 & x_{k,l}y_{k,l} = 0 \end{cases} \qquad \tilde{y}_{k,l} = \begin{cases} y_{k,l} & x_{k,l}y_{k,l} > 0 \\ 0 & x_{k,l}y_{k,l} = 0 \end{cases}$$

and note that $\mathrm{S}_K(\mathbf{X}\circ\mathbf{Y}) = \mathrm{S}_K(\tilde{\mathbf{X}}) = \mathrm{S}_K(\tilde{\mathbf{Y}})$. Hence, by Corollary 1.3,

$$\begin{aligned}
\log\rho(\mathbf{X}\circ\mathbf{Y}) &= \sup_{\mathbf{A}\in\mathrm{S}_K(\mathbf{X}\circ\mathbf{Y})}\Big(\sum_{k,l=1}^{K} u_k a_{k,l}\log\frac{x_{k,l}y_{k,l}}{a_{k,l}}\Big) \\
&\overset{(a)}{\le} \sup_{\mathbf{A}\in\mathrm{S}_K(\mathbf{X}\circ\mathbf{Y})}\Big(\sum_{k,l=1}^{K} u_k a_{k,l}\log\frac{x_{k,l}y_{k,l}}{a_{k,l}a_{k,l}}\Big) \\
&\overset{(b)}{=} \sup_{\mathbf{A}\in\mathrm{S}_K(\mathbf{X}\circ\mathbf{Y})}\Big(\sum_{k,l=1}^{K} u_k a_{k,l}\log\frac{x_{k,l}}{a_{k,l}} + \sum_{k,l=1}^{K} u_k a_{k,l}\log\frac{y_{k,l}}{a_{k,l}}\Big) \\
&\overset{(c)}{\le} \sup_{\mathbf{A}\in\mathrm{S}_K(\mathbf{X}\circ\mathbf{Y})}\Big(\sum_{k,l=1}^{K} u_k a_{k,l}\log\frac{x_{k,l}}{a_{k,l}}\Big) + \sup_{\mathbf{A}\in\mathrm{S}_K(\mathbf{X}\circ\mathbf{Y})}\Big(\sum_{k,l=1}^{K} u_k a_{k,l}\log\frac{y_{k,l}}{a_{k,l}}\Big) \\
&= \sup_{\mathbf{A}\in\mathrm{S}_K(\tilde{\mathbf{X}})}\Big(\sum_{k,l=1}^{K} u_k a_{k,l}\log\frac{x_{k,l}}{a_{k,l}}\Big) + \sup_{\mathbf{A}\in\mathrm{S}_K(\tilde{\mathbf{Y}})}\Big(\sum_{k,l=1}^{K} u_k a_{k,l}\log\frac{y_{k,l}}{a_{k,l}}\Big) \\
&\overset{(d)}{\le} \log\rho(\mathbf{X}) + \log\rho(\mathbf{Y})
\end{aligned}$$

where (a) is due to the fact $a_{k,l}^2 \leq a_{k,l} \leq 1$ for any $\mathbf{A} \in \mathrm{S}_K(\mathbf{X} \circ \mathbf{Y})$, (b) follows from $\log(xy) = \log x + \log y$ for all $x, y > 0$, (c) holds since $\sup(f+g) \leq \sup f + \sup g$ for any functions f, g, and (d) follows from Corollary 1.3 and the fact that $\mathrm{S}_K(\tilde{\mathbf{X}}) \subseteq \mathrm{S}_K(\mathbf{X})$ and $\mathrm{S}_K(\tilde{\mathbf{Y}}) \subseteq \mathrm{S}_K(\mathbf{Y})$.

Remark 1.6. It is interesting to point out that a necessary condition for equality to hold in (1.10) is that there exists a diagonal matrix $\mathbf{D}$ with positive diagonal elements such that $\mathbf{X} = \frac{\rho(\mathbf{X})}{\rho(\mathbf{Y})}\mathbf{D}\mathbf{Y}\mathbf{D}^{-1}$. This is because equality in (c) can hold only if $\mathbf{Y} \in \mathrm{A}_K(\mathbf{X})$. By Observation 1.4, however, this is true if and only if there exists a diagonal matrix $\mathbf{D}$ with positive diagonal elements such that $\mathbf{X} = \frac{\rho(\mathbf{X})}{\rho(\mathbf{Y})}\mathbf{D}\mathbf{Y}\mathbf{D}^{-1}$.

The next result provides an upper bound on the logarithm of $\rho(\mathbf{X})$, thereby giving rise to another type of Perron root characterization. In Sect. 1.2.3, using different techniques, we generalize this result by considering $F(\rho(\mathbf{X}))$ where $F : \mathbb{R}_{++} \to \mathbb{R}$ pertains to some class of continuous functions, of which the logarithmic function is a special case.

Theorem 1.7. *Let* $\mathbf{X} \in \mathrm{X}_K$, *and let* $\mathbf{w} := \mathbf{w}(\mathbf{X}) = (w_1, \ldots, w_K) = \mathbf{p} \circ \mathbf{q} \in \Pi_K^+$, *where* $\mathbf{q}$ *and* $\mathbf{p}$ *are left and right positive eigenvectors of* $\mathbf{X}$, *respectively. Then, for all* $\mathbf{s} \in \mathbb{R}_{++}^K$,

$$\log \rho(\mathbf{X}) \leq \sum_{k=1}^{K} w_k \log \frac{(\mathbf{X}\mathbf{s})_k}{s_k}, \tag{1.11}$$

with equality if $\mathbf{s} = \mathbf{p}$.

Proof. Let $\mathbf{s} \in \mathbb{R}_{++}^K$ be arbitrary and let $\hat{s}_k = s_k/p_k, 1 \leq k \leq K$, which is well-defined since $p_k > 0$ for each $1 \leq k \leq K$. Since $\mathbf{X}$ is irreducible, $\mathbf{X}\mathbf{s} > 0$ for any $\mathbf{s} > 0$ (see Appendix A.4). Hence, the right-hand side of (1.11) is well-defined. We have

$$\begin{aligned}\sum_{k=1}^{K} w_k \log \hat{s}_k &= \sum_{k=1}^{K} p_k q_k \log \hat{s}_k = \sum_{k=1}^{K} p_k \sum_{l=1}^{K} \frac{x_{l,k} q_l}{\rho(\mathbf{X})} \log \hat{s}_k \\ &= \sum_{l=1}^{K} p_l q_l \sum_{k=1}^{K} \frac{x_{l,k} p_k}{\rho(\mathbf{X}) p_l} \log \hat{s}_k .\end{aligned}$$

Since $\sum_{k=1}^{K} \frac{x_{l,k} p_k}{\rho(\mathbf{X}) p_l} = 1, 1 \leq l \leq K$, we can apply Jensen's inequality to obtain

$$\sum_{k=1}^{K} \frac{x_{l,k} p_k}{\rho(\mathbf{X}) p_l} \log \hat{s}_k \leq \log\Big(\frac{1}{\rho(\mathbf{X}) p_l} \sum_{k=1}^{K} x_{l,k} p_k \hat{s}_k\Big) = \log \frac{(\mathbf{X}\mathbf{s})_l}{p_l} - \log \rho(\mathbf{X}) \tag{1.12}$$

for every $1 \leq l \leq K$ with equality if $\mathbf{s} = \mathbf{p}$. Now combining this with the previous equality and the fact that $\|\mathbf{w}\|_1 = 1$ yields

$$\log \rho(\mathbf{X}) \le \sum_{l=1}^{K} w_l \log \frac{(\mathbf{Xs})_l}{p_l} - \sum_{l=1}^{K} w_l \log \hat{s}_l = \sum_{l=1}^{K} w_l \log \frac{(\mathbf{Xs})_l}{s_l}$$

for all $\mathbf{s} > 0$, with equality if $\mathbf{s} = \mathbf{p}$.

Remark 1.8. In general, $\mathbf{s} = \mathbf{p}$ is not necessary for the equality in (1.11) to hold. For instance, if all rows of $\mathbf{X} \in \mathrm{X}_K$ have only one positive entry, then there is actually no sum in (1.12) for each $1 \le l \le K$. Therefore, in such cases, the right-hand side of (1.12) is identically equal to the left-hand side, regardless of the choice of $\mathbf{s}$. An example of a $K \times K$ irreducible matrix with only one positive entry in each row is the circulant matrix with the first row given by $(0, 0, \ldots, 0, \alpha)$ for some constant $\alpha > 0$ (see (1.16)).

As an immediate consequence of Theorem 1.7, one obtains:

Corollary 1.9. *Let* $\mathbf{X} \in \mathrm{X}_K, \mathbf{p} \in \mathbb{R}^K_{++}$ *and* $\mathbf{w} \in \Pi^+_K$ *be as in Theorem 1.7. Then,*

$$\log \rho(\mathbf{X}) = \inf_{\mathbf{s} \in \mathbb{R}^K_{++}} \sum_{k=1}^{K} w_k \log \frac{(\mathbf{Xs})_k}{s_k} \tag{1.13}$$

The infimum is attained if $\mathbf{s} = \mathbf{p}$.

Recall that Theorem 1.2 has been used to prove an upper bound on $\rho(\mathbf{X} \circ \mathbf{Y})$ (Corollary 1.5). Interestingly, if we replace the entrywise product (or the Hadamard product) by normal matrix multiplication, $\rho(\mathbf{XY})$ can be arbitrarily large on X_K. So even if $\mathbf{X} \in \mathrm{X}_K$ and $\mathbf{Y} \in \mathrm{X}_K$ are fixed and known, not much can be said about $\rho(\mathbf{XY})$. However, if instead of Theorem 1.2, we consider Theorem 1.7, it is possible to derive a lower bound for $\rho(\mathbf{XY})$ on the following set of irreducible matrices generated by arbitrary $\mathbf{X} \in \mathrm{X}_K$:

$$\mathrm{W}_K(\mathbf{X}) := \{\mathbf{Y} \in \mathrm{X}_K : \mathbf{q}(\mathbf{Y}) \circ \mathbf{p}(\mathbf{Y}) = \mathbf{q}(\mathbf{X}) \circ \mathbf{p}(\mathbf{X}) \in \Pi^+_K\} \subset \mathrm{X}_K\,. \tag{1.14}$$

In words, given any $\mathbf{X} \in \mathrm{X}_K$, $\mathrm{W}_K(\mathbf{X})$ is a set of those irreducible matrices $\mathbf{Y}$ such that $\mathbf{q}(\mathbf{Y}) \circ \mathbf{p}(\mathbf{Y}) = \mathbf{q}(\mathbf{X}) \circ \mathbf{p}(\mathbf{X}) = \mathbf{w}(\mathbf{X})$. Note that for any $\mathbf{X} \in \mathrm{X}_K$, there holds

$$\mathbf{X}, \mathbf{X}^T \in \mathrm{W}_K(\mathbf{X})\,.$$

So, if $\mathbf{X}$ is not symmetric, the cardinality of $\mathrm{W}_K(\mathbf{X})$ is at least 2.

Corollary 1.10. *Suppose that* $\mathbf{X} \in \mathrm{X}_K$ *is given, and let* $\mathbf{Y} \in \mathrm{W}_K(\mathbf{X})$ *be arbitrary but chosen such that* $\mathbf{XY} \in \mathrm{X}_K$*. Then,*

$$\rho(\mathbf{X})\rho(\mathbf{Y}) \le \rho(\mathbf{XY})\,, \tag{1.15}$$

with equality if $\mathbf{p}(\mathbf{Y}) = \mathbf{p}(\mathbf{X})$.

Proof. Let $\mathbf{w} = \mathbf{q}(\mathbf{X}) \circ \mathbf{p}(\mathbf{X}) = \mathbf{q}(\mathbf{Y}) \circ \mathbf{p}(\mathbf{Y})$. By assumption, $\mathbf{X}, \mathbf{Y}, \mathbf{XY} \in \mathrm{X}_K$, from which we have $\rho(\mathbf{X}), \rho(\mathbf{Y}), \rho(\mathbf{XY}) > 0$. Therefore, (1.15) is true if and only if

$$\log \rho(\mathbf{X}) + \log \rho(\mathbf{Y}) \leq \log \rho(\mathbf{XY}) .$$

Now by Theorem 1.7,

$$\begin{aligned}\sum_{k=1}^{K} w_k \log \frac{(\mathbf{XYs})_k}{s_k} &= \sum_{k=1}^{K} w_k \log\Big(\frac{(\mathbf{XYs})_k}{(\mathbf{Ys})_k}\frac{(\mathbf{Ys})_k}{s_k}\Big) \\ &= \sum_{k=1}^{K} w_k \log \frac{(\mathbf{XYs})_k}{(\mathbf{Ys})_k} + \sum_{k=1}^{K} w_k \log \frac{(\mathbf{Ys})_k}{s_k} \\ &\geq \log \rho(\mathbf{X}) + \log \rho(\mathbf{Y})\end{aligned}$$

for *all* $\mathbf{s} \in \mathbb{R}^K_{++}$. So, by the Collatz-Wielandt formula (Theorem A.27)),

$$\begin{aligned}\log \rho(\mathbf{X}) + \log \rho(\mathbf{Y}) &\leq \min_{\mathbf{s}\in\mathbb{R}^K_{++}} \sum_{k=1}^{K} w_k \log \frac{(\mathbf{XYs})_k}{s_k} \leq \min_{\mathbf{s}\in\mathbb{R}^K_{++}} \max_{1\leq k\leq K} \log \frac{(\mathbf{XYs})_k}{s_k} \\ &= \log\Big(\min_{\mathbf{s}\in\mathbb{R}^K_{++}} \max_{1\leq k\leq K} \frac{(\mathbf{XYs})_k}{s_k}\Big) = \log \rho(\mathbf{XY}) .\end{aligned}$$

This proves the bound. Considering Theorem 1.7, we see that there is equality in (1.15) if $\mathbf{p}(\mathbf{X}) = \mathbf{p}(\mathbf{Y})$, and the corollary follows.

As a consequence of the corollary and the remark before, one has

$$\rho(\mathbf{X})\rho(\mathbf{X}^T) \leq \rho(\mathbf{X}\mathbf{X}^T)$$

with equality if $\mathbf{X}$ is symmetric.

Note that we have made the assumption $\mathbf{XY} \in \mathrm{X}_K$ since the set of irreducible matrices is not closed under matrix multiplication. For instance, the circulant matrix

$$\mathbf{T} := \begin{pmatrix} 0 & \cdots & 0 & 0 & 1 \\ 1 & \cdots & 0 & 0 & 0 \\ \vdots & \vdots & \vdots & \ddots & \vdots \\ 0 & \cdots & 1 & 0 & 0 \\ 0 & \cdots & 0 & 1 & 0 \end{pmatrix} \in \mathbb{R}^{K\times K}_{+} \tag{1.16}$$

and $\mathbf{T}^{K-1}$ are both irreducible. However, their product $\mathbf{T}\,\mathbf{T}^{K-1} = \mathbf{I}$ is not irreducible.

The last result in this section is a simple application of Theorem 1.7. Again, the result gives rise to a Perron root characterization that is a special case of the Perron root characterizations presented in Sect. 1.2.3.

Theorem 1.11. *Let* $\mathbf{X} \in \mathrm{X}_K, \mathbf{p} \in \mathbb{R}^K_{++}$, *and* $\mathbf{w} \in \Pi^+_K$ *be as in Theorem 1.7. Then,*

$$\rho(\mathbf{X}) \leq \sum_{k=1}^{K} w_k \frac{(\mathbf{Xs})_k}{s_k} \tag{1.17}$$

for all $\mathbf{s} \in \mathbb{R}^K_{++}$, *with equality if and only if* $\mathbf{s} = \mathbf{p}$.

Proof. By Theorem 1.7 and $\|\mathbf{w}\|_1 = 1$, we have

$$0 \le \sum_{k=1}^{K} w_k \log\Big(\frac{(\mathbf{X}\mathbf{s})_k}{s_k}\Big) - \log\rho(\mathbf{X})\sum_{k=1}^{K} w_k = \sum_{k=1}^{K} w_k \log\Big(\frac{(\mathbf{X}\mathbf{s})_k}{\rho(\mathbf{X})s_k}\Big).$$

Now since $\log x \le x - 1$ for all $x > 0$ with equality if and only if $x = 1$, this implies that

$$\begin{aligned} 0 \le \sum_{k=1}^{K} w_k \Big[\frac{(\mathbf{X}\mathbf{s})_k}{\rho(\mathbf{X})s_k} - 1\Big] &= \sum_{k=1}^{K} w_k \frac{(\mathbf{X}\mathbf{s})_k}{\rho(\mathbf{X})s_k} - \sum_{k=1}^{K} w_k \\ &= \frac{1}{\rho(\mathbf{X})}\sum_{k=1}^{K} w_k \frac{(\mathbf{X}\mathbf{s})_k}{s_k} - 1 \end{aligned} \tag{1.18}$$

from which the bound in (1.17) follows. Moreover, equality holds if and only if $\mathbf{s} = \mathbf{p}$.

The following corollary is immediate.

Corollary 1.12. *Let $\mathbf{X} \in \mathcal{X}_K, \mathbf{p} \in \mathbb{R}^K_{++}$, and $\mathbf{w} \in \Pi^+_K$ be as in Theorem 1.7. Then,*

$$\rho(\mathbf{X}) = \inf_{\mathbf{s}\in\mathbb{R}^K_{++}} \sum_{k=1}^{K} w_k \frac{(\mathbf{X}\mathbf{s})_k}{s_k}. \tag{1.19}$$

The infimum is attained if and only if $\mathbf{s} = \mathbf{p}$.

We point out that $\mathbf{s} = \mathbf{p}$ is *necessary* and sufficient for the equality in (1.17) to hold for *all* irreducible matrices, including the circulant matrix in (1.16). This is because $\log x \le x - 1$ for $x > 0$ with equality if and only if $x = 1$. Therefore, there is equality in (1.18) if and only if $\mathbf{s} = \mathbf{p}$. This stands in some contrast to Theorem 1.7.

1.2.1 Concavity of the Perron Root on Some Subsets of Irreducible Matrices

In this section, we apply the above results to obtain bounds on the Perron root of

$$\mathbf{X}(\mu) := (1-\mu)\hat{\mathbf{X}} + \mu\check{\mathbf{X}}, \quad \mu \in [0,1]$$

where $\hat{\mathbf{X}} \in \mathcal{X}_K$ and $\check{\mathbf{X}} \in \mathcal{X}_K$ are given. Note that $\mathbf{X}(\mu) \in \mathcal{X}_K$ for all $0 \le \mu \le 1$. In particular, the results show that the Perron root is concave on some subsets of $\mathcal{X}_K$. Given an arbitrary $\mathbf{X} \in \mathcal{X}_K$, we are particularly interested in $\mathcal{W}_K(\mathbf{X}) \subset \mathcal{X}_K$ defined by (1.14). The first theorem is an application of Theorem 1.7 and asserts that the Perron root is log-concave on $\mathcal{W}_K(\mathbf{X})$ (for the definition of log-concavity, see Appendix B.3).

Theorem 1.13. *Suppose that* $\mathbf{X} \in \mathrm{X}_K$ *is given and* $\hat{\mathbf{X}}, \check{\mathbf{X}} \in \mathrm{W}_K(\mathbf{X})$ *are arbitrary. Then,*

$$\rho(\mathbf{X}(\mu)) \geq \rho(\hat{\mathbf{X}})^{1-\mu} \rho(\check{\mathbf{X}})^{\mu} \tag{1.20}$$

for all $\mu \in (0,1)$.

Proof. Assume that $\mu \in (0,1)$ is fixed, and let $\tilde{\mathbf{p}}$ be a positive right eigenvector of $\mathbf{X}(\mu)$. Furthermore, let $\mathbf{w} = \mathbf{q}(\hat{\mathbf{X}}) \circ \mathbf{p}(\hat{\mathbf{X}}) = \mathbf{q}(\check{\mathbf{X}}) \circ \mathbf{p}(\check{\mathbf{X}})$. Since $\|\mathbf{w}\|_1 = 1$ and $\rho(\mathbf{X}(\mu)) = (\mathbf{X}(\mu)\tilde{\mathbf{p}})_k / \tilde{p}_k$ for each $1 \leq k \leq K$, one has

$$\begin{aligned}\log \rho(\mathbf{X}(\mu)) &= \sum_{k=1}^{K} w_k \log \rho(\mathbf{X}(\mu)) = \sum_{k=1}^{K} w_k \log\Big(\frac{(\mathbf{X}(\mu)\tilde{\mathbf{p}})_k}{\tilde{p}_k}\Big) \\ &= \sum_{k=1}^{K} w_k \log\Big(\frac{(1-\mu)(\hat{\mathbf{X}}\tilde{\mathbf{p}})_k + \mu(\check{\mathbf{X}}\tilde{\mathbf{p}})_k}{\tilde{p}_k}\Big), \quad \mu \in (0,1)\,.\end{aligned}$$

Hence, by concavity of the logarithmic function,

$$\begin{aligned}\log \rho(\mathbf{X}(\mu)) &\geq (1-\mu) \sum_{k=1}^{K} w_k \log\Big(\frac{(\hat{\mathbf{X}}\tilde{\mathbf{p}})_k}{\tilde{p}_k}\Big) + \mu \sum_{k=1}^{K} w_k \log\Big(\frac{(\check{\mathbf{X}}\tilde{\mathbf{p}})_k}{\tilde{p}_k}\Big) \\ &\geq (1-\mu) \log \rho(\hat{\mathbf{X}}) + \mu \log \rho(\check{\mathbf{X}})\,,\end{aligned}$$

where the last step follows from Theorem 1.7.

Interestingly, we can obtain a significantly stronger assertion if instead of Theorem 1.7, we consider Theorem 1.11.

Theorem 1.14. *Let* $\hat{\mathbf{X}}$ *and* $\check{\mathbf{X}}$ *be as in Theorem 1.13. Then,*

$$\rho(\mathbf{X}(\mu)) \geq (1-\mu)\rho(\hat{\mathbf{X}}) + \mu\rho(\check{\mathbf{X}}), \quad \mu \in (0,1)\,. \tag{1.21}$$

Moreover, strict inequality holds if there is no $\alpha > 0$ *such that* $\mathbf{p}(\hat{\mathbf{X}}) = \alpha \check{\mathbf{p}}(\check{\mathbf{X}})$.

Proof. Let $\mu \in (0,1)$ be arbitrary, and let $\mathbf{w} = \mathbf{q}(\hat{\mathbf{X}}) \circ \mathbf{p}(\hat{\mathbf{X}}) = \mathbf{q}(\check{\mathbf{X}}) \circ \mathbf{p}(\check{\mathbf{X}})$. Suppose that $\tilde{\mathbf{p}}$ is a right positive eigenvector of $\mathbf{X}(\mu)$. Proceeding essentially as above yields

$$\begin{aligned}\rho(\mathbf{X}(\mu)) &= \sum_{k=1}^{K} w_k \rho(\mathbf{X}(\mu)) = \sum_{k=1}^{K} w_k \frac{(\mathbf{X}(\mu)\tilde{\mathbf{p}})_k}{\tilde{p}_k} \\ &= (1-\mu) \sum_{k=1}^{K} w_k \frac{(\hat{\mathbf{X}}\tilde{\mathbf{p}})_k}{\tilde{p}_k} + \mu \sum_{k=1}^{K} w_k \frac{(\check{\mathbf{X}}\tilde{\mathbf{p}})_k}{\tilde{p}_k}\,.\end{aligned}$$

Now considering Theorem 1.11 proves the bound. To prove strict concavity, note that by Theorem 1.11,

$$\rho(\hat{\mathbf{X}}) = \sum_{k=1}^{K} w_k \frac{(\hat{\mathbf{X}}\tilde{\mathbf{p}})_k}{\tilde{p}_k} \qquad \text{and} \qquad \rho(\check{\mathbf{X}}) = \sum_{k=1}^{K} w_k \frac{(\check{\mathbf{X}}\tilde{\mathbf{p}})_k}{\tilde{p}_k}$$

if and only if $\tilde{\mathbf{p}} = \alpha_1 \mathbf{p}(\hat{\mathbf{X}})$ and $\tilde{\mathbf{p}} = \alpha_2 \mathbf{p}(\check{\mathbf{X}})$ for some $\alpha_1, \alpha_2 > 0$ or, equivalently, if and only if there exists $\alpha > 0$ such that $\mathbf{p}(\hat{\mathbf{X}}) = \alpha \mathbf{p}(\check{\mathbf{X}})$. Hence,

$$(1-\mu)\sum_{k=1}^{K} w_k \frac{(\hat{\mathbf{X}}\tilde{\mathbf{p}})_k}{\tilde{p}_k} + \mu \sum_{k=1}^{K} w_k \frac{(\check{\mathbf{X}}\tilde{\mathbf{p}})_k}{\tilde{p}_k} > (1-\mu)\rho(\hat{\mathbf{X}}) + \mu\rho(\check{\mathbf{X}}), \quad \mu \in (0,1)\,.$$

if there is no $\alpha > 0$ such that $\mathbf{p}(\hat{\mathbf{X}}) = \alpha \mathbf{p}(\check{\mathbf{X}})$.

For any $\mathbf{X} \in \mathrm{X}_K$, the theorem shows that the Perron root is concave on $\mathrm{W}_K(\mathbf{X})$. It is interesting to notice that $\mathrm{W}_2(\mathbf{X}) = \mathrm{X}_2$ for any $\mathbf{X} \in \mathrm{X}_2$ such that trace($\mathbf{X}$). Consequently, Theorem 1.14 implies that if $K = 2$, the Perron root is concave on the set of *traceless* irreducible matrices.

Furthermore, an examination of Observation 1.4 reveals that $\check{\mathbf{X}} \in \mathrm{A}_K(\hat{\mathbf{X}})$ for some given $\hat{\mathbf{X}}$ if and only if

$$\check{\mathbf{X}} = \frac{\rho(\check{\mathbf{X}})}{\rho(\hat{\mathbf{X}})} \mathbf{D}\hat{\mathbf{X}}\mathbf{D}^{-1}\,.$$

Now since $\mathbf{q}(\check{\mathbf{X}}) = \mathbf{D}^{-1}\mathbf{q}(\hat{\mathbf{X}})$ and $\mathbf{p}(\check{\mathbf{X}}) = \mathbf{D}\mathbf{p}(\hat{\mathbf{X}})$, we see that $\check{\mathbf{X}} \in \mathrm{A}_K(\hat{\mathbf{X}})$ implies

$$\mathbf{w}(\check{\mathbf{X}}) = (\mathbf{D}^{-1}\mathbf{q}(\hat{\mathbf{X}})) \circ \mathbf{D}\mathbf{p}(\hat{\mathbf{X}}) = \mathbf{w}(\hat{\mathbf{X}})\,.$$

Hence, we can conclude that

$$\mathrm{A}_K(\mathbf{X}) \subseteq \mathrm{W}_K(\mathbf{X}) \tag{1.22}$$

for any $\mathbf{X} \in \mathrm{X}_K$. This gives rise to the following corollary.

Corollary 1.15. *Suppose that* $\mathbf{X} \in \mathrm{X}_K$ *is given, and let* $\hat{\mathbf{X}}, \check{\mathbf{X}} \in \mathrm{A}_K(\mathbf{X})$. *Then,*

$$\rho(\mathbf{X}(\mu)) \geq \rho(\hat{\mathbf{X}})^{1-\mu}\rho(\check{\mathbf{X}})^{\mu} \qquad \text{and} \qquad \rho(\mathbf{X}(\mu)) \geq (1-\mu)\rho(\hat{\mathbf{X}}) + \mu\rho(\check{\mathbf{X}})$$

for all $\mu \in (0,1)$.

Proof. Combine (1.22) with Theorems 1.13 and 1.14.

Another consequence of Observation 1.4, the relationship (1.22) and Theorem 1.14 is the following.

Corollary 1.16. *Let* $\mathbf{X} \in \mathrm{X}_K$, *and let* $\mathbf{D} \neq \mathbf{I}$ *be an arbitrary diagonal matrix with positive diagonal entries. Then,*

$$\rho((1-\mu)\mathbf{X} + \mu\mathbf{D}\mathbf{X}\mathbf{D}^{-1}) \geq (1-\mu)\rho(\mathbf{X}) + \mu\rho(\mathbf{D}\mathbf{X}\mathbf{D}^{-1}) = \rho(\mathbf{X}) \tag{1.23}$$

for all $\mu \in (0,1)$.

Finally, as $\mathbf{X}^T \in \mathrm{W}_K(\mathbf{X})$ for any $\mathbf{X} \in \mathrm{X}_K$, it follows from Theorem 1.14 that $\rho((1-\mu)\mathbf{X} + \mu\mathbf{X}^T)$ is a concave function of $\mu \in [0,1]$. If, in addition, $\mathbf{X} \neq \mathbf{X}^T$, we have $\mathbf{p}(\mathbf{X}) \neq \alpha\mathbf{p}(\mathbf{X}^T)$ for any constant α. This leads to the following corollary.

Corollary 1.17. *Let $\mathbf{X}$ with $\mathbf{X} \neq \mathbf{X}^T$ be given. Then, $\rho((1-\mu)\mathbf{X}+\mu\mathbf{X}^T)$ is a strictly concave function of $\mu \in [0,1]$.*

1.2.2 Kullback–Leibler Divergence Characterization

The Kullback-Leibler divergence (KLD) between two probability mass functions is one of the fundamental concepts in information theory and also in other fields like statistics and physics [8, 9, 10]. In fact, the mutual information of two random variables is equal to the KLD between their joint distributions and product distributions [9]. In this section, we characterize the Perron root of irreducible matrices in terms of the KLD generalized to any positive discrete measure. Let us start with the precise definition of the KLD.

Definition 1.18. *Suppose that x is a realization of a discrete random variable with the set of possible values* X. *Then, the KLD between two probability mass functions $f(x), x \in \mathrm{X}$, and $g(x), x \in \mathrm{X}$, is defined as [9]*

$$D(f\|g) := \sum_{x\in\mathrm{X}} f(x) \log \frac{f(x)}{g(x)}, \tag{1.24}$$

where we used the convention that $0 \log \frac{0}{g} = 0$ and $f \log \frac{f}{0} = \infty$.

Note that the KLD is not a distance since it is not symmetric $D(f\|g) \neq D(g\|f)$ and does not satisfy the triangle inequality. Nevertheless, the definition makes sense since $D(f\|g)$ is always nonnegative with $D(f\|g) = 0$ if and only if $f = g$. This immediately follows from the fact that $\log x \leq x - 1$ for every $x > 0$ with equality if and only if $x = 1$. Hence,

$$\begin{aligned} D(f\|g) &= \sum_{x\in\mathrm{X}} f(x) \log \frac{f(x)}{g(x)} = -\sum_{x\in\mathrm{X}} f(x) \log \frac{g(x)}{f(x)} \\ &\geq \sum_{x\in\mathrm{X}} f(x) - \sum_{x\in\mathrm{X}} g(x) = 1 - 1 = 0 \end{aligned} \tag{1.25}$$

with equality if and only if $f = g$.

The definition of the KLD can be generalized in a natural manner to any positive discrete measure as follows.

Definition 1.19. *Let $\mathbf{X} \in \mathrm{X}_K$, and let $\mathbf{A} \in \mathrm{S}_K$, where S_K is the set of stochastic matrices. Then,*

$$D(\mathbf{a}^{(k)}\|\mathbf{x}^{(k)}) = \sum_{l=1}^{K} a_{k,l} \log \frac{a_{k,l}}{x_{k,l}}, \quad 1 \le k \le K \tag{1.26}$$

is the (generalized) KLD between the kth row of $\mathbf{A}$ *and the kth row of* $\mathbf{X}$ *denoted by* $\mathbf{a}^{(k)}$ *and* $\mathbf{x}^{(k)}$*, respectively.*

Note that since $x \log \frac{x}{0} = \infty$ for all $x > 0$ and $0 \log \frac{0}{x} = 0$ for all $x \ge 0$, $\mathbf{A}$ in the definition above is an arbitrary stochastic matrix. Comparing (1.26) with (1.4) reveals that the left-hand side of (1.4) multiplied by -1 is equal to $\sum_{k=1}^{K} u_k D(\mathbf{a}^{(k)}\|\mathbf{x}^{(k)})$. Combining this with Corollary 1.3 gives rise to a relationship between the KLD and the Perron root of nonnegative irreducible matrices.

Observation 1.20. *Let* $\mathbf{X} \in \mathrm{X}_K$ *be arbitrary, and let* $\mathrm{S}_K(\mathbf{X})$ *be a set of irreducible stochastic matrices defined by (1.3). Then,*

$$\log \frac{1}{\rho(\mathbf{X})} = \inf_{\mathbf{A} \in \mathrm{S}_K(\mathbf{X})} \Big(\sum_{k=1}^{K} u_k D(\mathbf{a}^{(k)}\|\mathbf{x}^{(k)}) \Big) \tag{1.27}$$

where $\mathbf{u} \in \Pi_K^+$ *is the left Perron eigenvector of* $\mathbf{A}$*. The infimum in (1.27) is attained if and only if*

$$(\mathbf{A})_{k,l} = a_{k,l} = \frac{x_{k,l} p_l}{\rho(\mathbf{X}) p_k}, \quad 1 \le k, l \le K \tag{1.28}$$

where $\mathbf{p}$ *is a positive right eigenvector of* $\mathbf{X}$*.*

On the other hand, when we consider Theorem 1.7, one obtains the following characterization.

Observation 1.21. *Given* $\mathbf{X} \in \mathrm{X}_K$*, let* $\mathbf{w} = \mathbf{q} \circ \mathbf{p} \in \Pi_K^+$*, where* $\mathbf{q}$ *and* $\mathbf{p}$ *are positive left and right eigenvectors of* $\mathbf{X}$*, respectively. Then,*

$$\log \rho(\mathbf{X}) = \inf_{\mathbf{s} \in \mathbb{R}_{++}^K} \Big(D(\mathbf{w}\|\mathbf{s}) - D(\mathbf{w}\|\mathbf{X}\mathbf{s}) \Big). \tag{1.29}$$

The infimum is attained if $\mathbf{s} = \mathbf{p}$*.*

1.2.3 Some Extended Perron Root Characterizations

As already mentioned, this section generalizes some of the previous results to a certain class of functions that depends on a nonnegative irreducible matrix $\mathbf{X}$. Although the following definition of this function class may appear a bit artificial, the reader will be convinced of its importance in the second part of the book. In particular, it will become clear that for functions from outside of this class, the network problems may be hardly tractable or even not solvable efficiently under real world conditions.

Definition 1.22 (Function Class $\mathcal{G}(\mathbf{X})$). *Let $\mathbf{X} \in X_K$ be given. Say that a continuous function $F : \mathbb{R}_{++} \to \mathbb{R}$ pertains to a function class $\mathcal{G}(\mathbf{X})$ (written as $F \in \mathcal{G}(\mathbf{X})$) if*

(i) F is continuously differentiable and strictly increasing,
(ii) for any fixed $\mathbf{z} \in \Pi_K^+$, the function $H : \mathbb{R}_{++}^K \to \mathbb{R}$ given by

$$H(\mathbf{s}) := \sum_{k=1}^{K} z_k F\Big(\frac{(\mathbf{X}\mathbf{s})_k}{s_k}\Big) \tag{1.30}$$

attains its infimum on $\mathbb{R}_{++}^K$. Moreover, all local minima are global and

$$\nabla H(\mathbf{s}^*) = 0 \tag{1.31}$$

is necessary and sufficient for $\mathbf{s}^* \in \mathbb{R}_{++}^K$ to be a global minimizer.

Remark 1.23. Note that members of $\mathcal{G}(\mathbf{X})$ are not necessarily bijections from $\mathbb{R}_{++}$ onto $\mathbb{R}$. In fact, they are injections (one-to-one maps) from $\mathbb{R}_{++}$ into $\mathbb{R}$ so that their ranges are in general subsets of $\mathbb{R}$. The definition could be also modified to include the case of a function defined on some arbitrary open subset of $\mathbb{R}_{++}$. The second requirement ensures that the set $\{H(\mathbf{s}) : \mathbf{s} \in \mathbb{R}_{++}^K\}$ is bounded below for any fixed vector $\mathbf{z} > 0$ and the greatest lower bound, which is the infimum of H over all positive vectors, is attained for some $\mathbf{s}^* \in \mathbb{R}_{++}^K$ if and only if (1.31) holds. However, this is not equivalent to saying that $H : \mathbb{R}_{++}^K \to \mathbb{R}$ is a convex function. Finally, note that (1.31) is a necessary optimality condition as $\mathbb{R}_{++}^K$ is an open subset of $\mathbb{R}^K$.

Two prominent examples of functions belonging to $\mathcal{G}(\mathbf{X})$ for some $\mathbf{X} \in X_K$ have already been considered in the foregoing sections. These are $F(x) = \log x, x > 0$, and $F(x) = x, x > 0$. It is obvious that in these two special cases, the first requirement in the definition above is satisfied. When F is the linear function, it can be seen that the function H has a global minimum on $\mathbb{R}_{++}^K$ for any choice of $\mathbf{X} \in X_K$ and $\mathbf{z} \in \Pi_K^+$. In contrast, this does not need to be true if F is the logarithmic function. For instance, when $\mathbf{X} \in X_K$ is equal to the circulant matrix given by (1.16), the set $\{H(\mathbf{s}) : \mathbf{s} \in \mathbb{R}_{++}^K\}$ may fail to be bounded below.[1] So, the logarithmic function pertains to $\mathcal{G}(\mathbf{X})$ if $\mathbf{X}$ is confined to some subset of X_K for which H has a global minimum on $\mathbb{R}_{++}^K$ for any $\mathbf{z} \in \Pi_K^+$. Note that the set is not empty since it contains all positive matrices.

If the infimum is attained, the requirement that every local minimum is a global one is satisfied in these two cases as well. This is because, with these choices of F, the problem of minimizing $H(\mathbf{s})$ over $\mathbb{R}_{++}^K$ can be transformed into an equivalent convex problem using the substitution

[1] Note that if $\mathbf{w} = \mathbf{q}(\mathbf{X}) \circ \mathbf{p}(\mathbf{X})$, then the set is bounded below for any $\mathbf{X} \in X_K$ since then, by Theorem 1.7, $H(\mathbf{s}) \geq \log \rho(\mathbf{X})$ for all $\mathbf{s} \in \mathbb{R}_{++}^K$, with $\rho(\mathbf{X}) > 0$ whenever $\mathbf{X} \in X_K$.

$\mathbf{x} = \log \mathbf{s}, \mathbf{s} \in \mathbb{R}^K_{++}$. More precisely, using the results of Chapter 6, the function $H_e(\mathbf{x}) = H(e^{\mathbf{x}}), \mathbf{x} \in \mathbb{R}^K$, can be shown to be convex when either $F(x) = x, x > 0$, or $F(x) = \log x, x > 0$. Therefore, every stationary point $\mathbf{x}^*$ of H_e satisfying $\nabla H_e(\mathbf{x}^*) = 0$ is a global minimizer of $H_e(\mathbf{x})$ over $\mathbb{R}^K$. At the same time, we have $\nabla H_e(\mathbf{x}^*) = 0$ if and only if $\nabla H(\mathbf{s}^*) = 0$ with $\mathbf{x}^* = \log \mathbf{s}^*$, from which we conclude that every stationary point $\mathbf{s}^*$ satisfying (1.31) is a global minimum of $H(\mathbf{s})$ over $\mathbb{R}^K_{++}$. For formal proofs, the reader is referred to Chapter 6. For the purpose of this section, it is sufficient to assume that $F \in \mathcal{G}(\mathbf{X})$.

The first result gives rise to a characterization of $F(\rho(\mathbf{X}))$ in terms of the minima of $H(\mathbf{s})$ over $\mathbb{R}^K_{++}$. Recall that by assumption, $\mathbf{q}^T\mathbf{p} = 1$ implying that $\mathbf{p} \circ \mathbf{q} \in \Pi^+_K$.

Theorem 1.24. *Let $\mathbf{X} \in \mathrm{X}_K$ be given, and let $F \in \mathcal{G}(\mathbf{X})$ be arbitrary. Suppose that $\mathbf{w} = \mathbf{p} \circ \mathbf{q} \in \Pi^+_K$, which is a unique vector. Then,*

$$\sum_{k=1}^K w_k F\left(\frac{(\mathbf{X}\mathbf{s})_k}{s_k}\right) \geq F(\rho(\mathbf{X})) \tag{1.32}$$

for all $\mathbf{s} \in \mathbb{R}^K_{++}$. Equality holds if $\mathbf{s} = \mathbf{p} > 0$ (unique up to positive multiples). Moreover, $\mathbf{p} = \arg\min_{\mathbf{s} \in \mathbb{R}^K_{++}} H(\mathbf{s})$ with $H(\mathbf{s})$ defined by (1.30) if and only if $\mathbf{z} = \mathbf{w} > 0$.

Remark 1.25. In fact, we have $H(\mathbf{s}) = \sum_k z_k(\mathbf{X}\mathbf{s})_k/s_k \geq F(\rho(\mathbf{X}))$ for all $\mathbf{s} \in \mathbb{R}^K_{++}$ if and only if $\mathbf{z} = \mathbf{w}$, which is more than what the theorem asserts. This immediately follows from Lemma 1.30 stated later in this section.

Proof. Consider the following minimization problem:

$$\min_{\mathbf{s} \in \mathbb{R}^K_{++}} H(\mathbf{s}) = \min_{\mathbf{s} \in \mathbb{R}^K_{++}} \sum_{k=1}^K z_k F\left(\frac{(\mathbf{X}\mathbf{s})_k}{s_k}\right)$$

for some given $\mathbf{z} \in \Pi^+_K$. By assumption, the minimum exists and (1.31) is a necessary and sufficient condition for characterizing all global minimizers. Hence, $\mathbf{s}^* > 0$ minimizes H over $\mathbb{R}^K_{++}$ if and only if

$$\sum_{j=1}^K z_j F'\left(\frac{(\mathbf{X}\mathbf{s}^*)_j}{s^*_j}\right)\frac{X_{jk}}{s^*_j} = z_k F'\left(\frac{(\mathbf{X}\mathbf{s}^*)_k}{s^*_k}\right)\frac{(\mathbf{X}\mathbf{s}^*)_k}{(s^*_k)^2}, \quad 1 \leq k \leq K\,. \tag{1.33}$$

Using

$$\mathbf{u}(\mathbf{z}, \mathbf{s}) := \left(\frac{z_1}{s_1}, \frac{z_2}{s_2}, \ldots, \frac{z_K}{s_K}\right) \tag{1.34}$$

we can write (1.33) in matrix form to obtain

$$(\mathbf{F}'(\mathbf{s}^*)\mathbf{X})^T \mathbf{u}(\mathbf{z}, \mathbf{s}^*) = \mathbf{F}'(\mathbf{s}^*)(\mathbf{\Gamma}(\mathbf{s}^*))^{-1}\mathbf{u}(\mathbf{z}, \mathbf{s}^*) \tag{1.35}$$

with $\mathbf{\Gamma}(\mathbf{s}) := \mathrm{diag}(\frac{s_1}{(\mathbf{Xs})_1}, \dots, \frac{s_K}{(\mathbf{Xs})_K})$ and

$$\mathbf{F}'(\mathbf{s}) := \mathrm{diag}\left(F'\left(\frac{(\mathbf{Xs})_1}{s_1}\right), \dots, F'\left(\frac{(\mathbf{Xs})_K}{s_K}\right)\right). \tag{1.36}$$

Now since $(\mathbf{Xp})_k/p_k = \rho(\mathbf{X}), 1 \leq k \leq K$, it follows that $\mathbf{F}'(\mathbf{p}) = F'(\rho(\mathbf{X}))\mathbf{I}$ with $F'(\rho(\mathbf{X})) > 0$. Thus, if $\mathbf{s}^* = \mathbf{p}$, the necessary and sufficient optimality condition (1.35) becomes

$$\mathbf{X}^T\mathbf{u}(\mathbf{z}, \mathbf{p}) = \rho(\mathbf{X})\mathbf{u}(\mathbf{z}, \mathbf{p}).$$

So, $\mathbf{p}$ is a global minimizer if and only if $\mathbf{u}(\mathbf{z}, \mathbf{p}) = \mathbf{q}$. An examination of (1.34) reveals that this is true if and only if $\mathbf{z} = \mathbf{w} = \mathbf{p} \circ \mathbf{q} \in \Pi_K^+$, which is uniquely defined in Π_K^+ since $\mathbf{p} > 0$ and $\mathbf{q} > 0$ are unique eigenvectors of $\mathbf{X} \in \mathrm{X}_K$ up to a scaling factor. This proves the second part of the theorem. However, the lower bound (1.32) immediately follows as $\mathbf{p}$ is a global minimizer if $\mathbf{z} = \mathbf{w}$, and therefore, due to $\|\mathbf{w}\|_1 = 1$, we have

$$\min_{\mathbf{s} \in \mathbb{R}_{++}^K} \sum_{k=1}^K w_k F\left(\frac{(\mathbf{Xs})_k}{s_k}\right) = \sum_{k=1}^K w_k F\left(\frac{(\mathbf{Xp})_k}{p_k}\right) = \sum_{k=1}^K w_k F(\rho(\mathbf{X})) = F(\rho(\mathbf{X}))$$

where $\mathbf{p} > 0$ is unique up to positive multiples.

It should be emphasized that due to positivity of $\mathbf{q}$ and $\mathbf{p}$ for any $\mathbf{X} \in \mathrm{X}_K$, the weight vector $\mathbf{w} = \mathbf{p} \circ \mathbf{q}$ is automatically positive. Hence, K addends appear in (1.32).

1.2.4 Collatz–Wielandt-Type Characterization of the Perron Root

Based upon Theorem 1.24, in this section, we prove a saddle point characterization of the Perron root. Because of similarity to the Collatz–Wielandt formula [3], the characterization is referred to as Collatz–Wielandt-type characterization of the Perron root. Before starting with the analysis, let us precisely define a saddle point of a continuous function.

Definition 1.26 (Saddle Point). *Suppose that $G : \mathrm{D}_1 \times \mathrm{D}_2 \to \mathbb{R}$ is a continuous function where $\mathrm{D}_1, \mathrm{D}_2 \subseteq \mathbb{R}^K$ are some open sets.*[2] *We say that $(\mathbf{s}^*, \mathbf{z}^*) \in \mathrm{D}_1 \times \mathrm{D}_2$ is a saddle point of G if, for all $\mathbf{s} \in \mathrm{D}_1, \mathbf{z} \in \mathrm{D}_2$, there holds*

$$G(\mathbf{s}^*, \mathbf{z}) \leq G(\mathbf{s}^*, \mathbf{z}^*) \leq G(\mathbf{s}, \mathbf{z}^*).$$

The following standard result provides a necessary and sufficient condition for a pair of vectors to be a saddle point.

[2] Again, note that the range of G may be a subset of $\mathbb{R}$.

Theorem 1.27. *Under the assumption that the following maxima and minima exist, a vector pair* $(\mathbf{s}^*, \mathbf{z}^*) \in \mathrm{D}_1 \times \mathrm{D}_2$ *is a saddle point of a continuous function* $G : \mathrm{D}_1 \times \mathrm{D}_2 \to \mathbb{R}$ *if and only if*

$$\min_{\mathbf{s}\in \mathrm{D}_1} \max_{\mathbf{z}\in \mathrm{D}_2} G(\mathbf{s}, \mathbf{z}) = \max_{\mathbf{z}\in \mathrm{D}_2} \min_{\mathbf{s}\in \mathrm{D}_1} G(\mathbf{s}, \mathbf{z})$$

and

$$\begin{aligned} \max_{\mathbf{z}\in \mathrm{D}_2} G(\mathbf{s}^*, \mathbf{z}) &= \min_{\mathbf{s}\in \mathrm{D}_1} \max_{\mathbf{z}\in \mathrm{D}_2} G(\mathbf{s}, \mathbf{z}) \\ \min_{\mathbf{s}\in \mathrm{D}_1} G(\mathbf{s}, \mathbf{z}^*) &= \max_{\mathbf{z}\in \mathrm{D}_2} \min_{\mathbf{s}\in \mathrm{D}_1} G(\mathbf{s}, \mathbf{z}) \,. \end{aligned} \tag{1.37}$$

Now consider the following simple lemma.

Lemma 1.28. *For any* $F \in \mathcal{G}(\mathbf{X})$ *and* $\mathbf{X} \in \mathrm{X}_K$*, there holds*

$$\min_{\mathbf{s}\in\mathbb{R}^K_{++}} \max_{\mathbf{z}\in\Pi^+_K} \sum_{k=1}^{K} z_k F\left(\frac{(\mathbf{X}\mathbf{s})_k}{s_k}\right) = F(\rho(\mathbf{X})) \,. \tag{1.38}$$

Moreover, the minimum is attained if and only if $\mathbf{s} = \mathbf{p}$.

Proof. For any $\mathbf{z} \in \Pi^+_K$ and $\mathbf{s} \in \mathbb{R}^K_{++}$, we have

$$\sum_{k=1}^{K} z_k F\left(\frac{(\mathbf{X}\mathbf{s})_k}{s_k}\right) \le \max_{1\le k\le K} F\left(\frac{(\mathbf{X}\mathbf{s})_k}{s_k}\right) .$$

As $\mathbf{z}$ is positive, equality holds if and only if $F((\mathbf{X}\mathbf{s})_k/s_k) = c, k = 1 \dots K$, for some $c \in \mathbb{R}$. Thus, since F is strictly monotonic and $\mathbf{X}$ is irreducible, we see that the equality holds if and only if $\mathbf{s} = \mathbf{p} > 0$, in which case $F((\mathbf{X}\mathbf{p})_k/p_k) = F(\rho(\mathbf{X}))$ for each $1 \le k \le K$. Moreover, for any $\mathbf{z} \in \Pi^+_K$, one has

$$\begin{aligned} \min_{\mathbf{s}\in\mathbb{R}^K_{++}} \sum_{k=1}^{K} z_k F\left(\frac{(\mathbf{X}\mathbf{s})_k}{s_k}\right) &\le \min_{\mathbf{s}\in\mathbb{R}^K_{++}} \max_{1\le k\le K} F\left(\frac{(\mathbf{X}\mathbf{s})_k}{s_k}\right) \\ &= F\left(\min_{\mathbf{s}\in\mathbb{R}^K_{++}} \max_{1\le k\le K} \frac{(\mathbf{X}\mathbf{s})_k}{s_k}\right) = F\big(\rho(\mathbf{X})\big) \end{aligned} \tag{1.39}$$

where, due to the assumption, the first minimum exists and the last equality follows from the Collatz–Wielandt formula for irreducible matrices (Theorem A.27). Equality holds if and only if $\mathbf{s} = \mathbf{p}$ where $\mathbf{p} > 0$ is unique (up to positive multiples) for any $\mathbf{X} \in \mathrm{X}_K$. Now considering Theorem 1.24 proves the lemma.

For the max-min part of the saddle point characterization, we need the following lemma.

Lemma 1.29. *Let* $\mathbf{X} \in \mathrm{X}_K$ *and* $F \in \mathcal{G}(\mathbf{X})$ *be given. Then, the continuous function* $E : \Pi_K^+ \to \mathbb{R}$ *defined by*

$$E(\mathbf{z}) := \min_{\mathbf{s}\in\mathbb{R}_{++}^K} H(\mathbf{s}) = \min_{\mathbf{s}\in\mathbb{R}_{++}^K} \sum_{k=1}^K z_k F\Big(\frac{(\mathbf{X}\mathbf{s})_k}{s_k}\Big) \tag{1.40}$$

is strictly concave.

Proof. Since $H(\alpha\mathbf{s}) = H(\mathbf{s})$ for any $\alpha > 0$, we can assume that $\|\mathbf{s}\|_1 = 1$. Concavity of $E(\mathbf{z})$ is clear from the properties of the minimum operator. So we only need to show strict concavity. To this end, assume that $E : \Pi_K^+ \to \mathbb{R}$ is not strictly concave. Then, there must exist $\hat{\mathbf{z}} \in \Pi_K^+$ and $\check{\mathbf{z}} \in \Pi_K^+$ with $\hat{\mathbf{z}} \neq \check{\mathbf{z}}$ such that

$$\begin{aligned}E(\mathbf{z}(\mu)) &= (1-\mu)E(\hat{\mathbf{z}}) + \mu E(\check{\mathbf{z}}) \\ &= (1-\mu)\min_{\mathbf{s}\in\mathbb{R}_{++}^K} \underbrace{\sum_{k=1}^K \hat{z}_k F\Big(\frac{(\mathbf{X}\mathbf{s})_k}{s_k}\Big)}_{\hat{H}(\mathbf{s})} + \mu \min_{\mathbf{s}\in\mathbb{R}_{++}^K} \underbrace{\sum_{k=1}^K \check{z}_k F\Big(\frac{(\mathbf{X}\mathbf{s})_k}{s_k}\Big)}_{\check{H}(\mathbf{s})}\end{aligned}$$

for some $\mu \in (0,1)$ where $\mathbf{z}(\mu) = (1-\mu)\hat{\mathbf{z}} + \mu\check{\mathbf{z}} \in \Pi_K^+$. Clearly, the equality holds if and only if one of the following holds.

(i) there exist $\hat{\mathbf{s}} \in \mathbb{R}_{++}^K$ and $\check{\mathbf{s}} \in \mathbb{R}_{++}^K$ with $\hat{\mathbf{s}} \neq \check{\mathbf{s}}$ such that $\hat{H}(\hat{\mathbf{s}}) = \check{H}(\check{\mathbf{s}}) = E(\hat{\mathbf{z}}) = E(\check{\mathbf{z}})$.
(ii) there exists $\mathbf{s}^* \in \mathbb{R}_{++}^K$ such that $\hat{H}(\mathbf{s}^*) = \check{H}(\mathbf{s}^*) = E(\hat{\mathbf{z}}) = E(\check{\mathbf{z}})$.

First, we consider (i). Let $\mu \in (0,1)$ be arbitrary and define $\tilde{\mathbf{s}}(\mu) \in \mathbb{R}_{++}^K$ as $H(\tilde{\mathbf{s}}(\mu)) = E(\mathbf{z}(\mu))$. In words, $\tilde{\mathbf{s}}(\mu)$ minimizes the function H with the weight vector being equal to $\mathbf{z}(\mu)$. Then, we have

$$(1-\mu)\hat{H}(\hat{\mathbf{s}}) + \mu\check{H}(\check{\mathbf{s}}) = E(\mathbf{z}(\mu)) = H(\tilde{\mathbf{s}}(\mu)) = (1-\mu)\hat{H}(\tilde{\mathbf{s}}(\mu)) + \mu\check{H}(\tilde{\mathbf{s}}(\mu))\,.$$

This however contradicts $\hat{\mathbf{s}} \neq \check{\mathbf{s}}$, and hence disproves (i).

Now let us turn our attention to (ii). By assumption, $\mathbf{s}^*$ minimizes H over $\mathbb{R}_{++}^K$ if and only if (1.31) is satisfied. Thus, proceeding essentially as in the proof of Theorem 1.24 shows that H attains its minimum at $\mathbf{s}^* \in \mathbb{R}_{++}^K$ for both $\hat{\mathbf{z}}$ and $\check{\mathbf{z}}$ if and only if

$$(\mathbf{F}'(\mathbf{s}^*)\mathbf{X})^T\mathbf{u}(\tilde{\mathbf{z}},\mathbf{s}^*) = \mathbf{F}'(\mathbf{s}^*)(\boldsymbol{\Gamma}(\mathbf{s}^*))^{-1}\mathbf{u}(\tilde{\mathbf{z}},\mathbf{s}^*),\ \tilde{\mathbf{z}} = \hat{\mathbf{z}} \text{ and } \tilde{\mathbf{z}} = \check{\mathbf{z}}$$

where $\mathbf{F}'(\mathbf{s})$ is defined by (1.36), $\boldsymbol{\Gamma}(\mathbf{s}) = \mathrm{diag}(s_1/(\mathbf{X}\mathbf{s})_1, \dots, s_K/(\mathbf{X}\mathbf{s})_K)$ is positive definite, and $\mathbf{u}(\tilde{\mathbf{z}},\mathbf{s}) = (\tilde{z}_1/s_1, \dots, \tilde{z}_K/s_K)$. Due to strict monotonicity of F, the diagonal elements of $\mathbf{F}'(\mathbf{s})$ are positive for all $\mathbf{s} \in \mathbb{R}_{++}^K$. Therefore, $\mathbf{F}'(\mathbf{s}^*)$ is invertible and

$$\Big(\boldsymbol{\Gamma}(\mathbf{s}^*)\mathbf{F}'(\mathbf{s}^*)^{-1}\mathbf{X}^T\mathbf{F}'(\mathbf{s}^*)\Big)\mathbf{u}(\tilde{\mathbf{z}},\mathbf{s}^*) = \mathbf{u}(\tilde{\mathbf{z}},\mathbf{s}^*)\,.$$

Since $\mathbf{X}$ is irreducible, so also is $\mathbf{A}(\mathbf{s}^*) := \mathbf{\Gamma}(\mathbf{s}^*)\mathbf{F}'(\mathbf{s}^*)^{-1}\mathbf{X}^T\mathbf{F}'(\mathbf{s}^*)$. This in turn implies that $\mathbf{A}(\mathbf{s}^*)$ has unique (up to positive multiples) left and right positive eigenvectors. Hence, since $\hat{\mathbf{z}}, \check{\mathbf{z}} \in \Pi_K^+$, we must have $\mathbf{u}(\hat{\mathbf{z}}, \mathbf{s}^*) = \mathbf{u}(\check{\mathbf{z}}, \mathbf{s}^*)$ or, equivalently, $\hat{\mathbf{z}} = \check{\mathbf{z}}$. But this contradicts $\hat{\mathbf{z}} \neq \check{\mathbf{z}}$, and therefore completes the proof.

Now we are in a position to prove the max-min part of the saddle point characterization.

Lemma 1.30. *Let $\mathbf{X} \in \mathrm{X}_K$ be arbitrary. Then, for any $F \in \mathcal{G}(\mathbf{X})$,*

$$\max_{\mathbf{z}\in\Pi_K^+} \min_{\mathbf{s}\in\mathbb{R}_{++}^K} \sum_{k=1}^K z_k F\left(\frac{(\mathbf{X}\mathbf{s})_k}{s_k}\right) = F(\rho(\mathbf{X})) . \tag{1.41}$$

Moreover,

$$\mathbf{w} = \arg\max_{\mathbf{z}\in\Pi_K^+} \min_{\mathbf{s}\in\mathbb{R}_{++}^K} \sum_{k=1}^K z_k F\left(\frac{(\mathbf{X}\mathbf{s})_k}{s_k}\right) \tag{1.42}$$

if and only if $\mathbf{w} = \mathbf{q} \circ \mathbf{p}$, which is a unique vector in Π_K^+.

Proof. Proceeding as in the proof of Lemma 1.28 yields

$$U(\mathbf{z}) := \min_{\mathbf{s}\in\mathbb{R}_{++}^K} \sum_{k=1}^K z_k F\left(\frac{(\mathbf{X}\mathbf{s})_k}{s_k}\right) \le \min_{\mathbf{s}\in\mathbb{R}_{++}^K} \max_{1\le k\le K} F\left(\frac{(\mathbf{X}\mathbf{s})_k}{s_k}\right) = F(\rho(\mathbf{X}))$$

for any $\mathbf{z} \in \Pi_K^+$. On the other hand, by Theorem 1.24,

$$\min_{\mathbf{s}\in\mathbb{R}_{++}^K} \sum_{k=1}^K w_k F\left(\frac{(\mathbf{X}\mathbf{s})_k}{s_k}\right) = \sum_{k=1}^K w_k F\left(\frac{(\mathbf{X}\mathbf{p})_k}{p_k}\right) = F(\rho(\mathbf{X}))$$

and therefore $\mathbf{w} \in \Pi_K^+$ is a maximizer of U. However, by Lemma 1.29, the function U is strictly concave, and hence $\mathbf{w} = \mathbf{p} \circ \mathbf{q}$ is a unique maximizer in Π_K^+.

Now let us combine these results to obtain a saddle point characterization of the Perron root $\rho(\mathbf{X})$.

Theorem 1.31. *Let $\mathbf{X} \in \mathrm{X}_K$, $F \in \mathcal{G}(\mathbf{X})$ be given. Define $G : \mathbb{R}_{++}^K \times \Pi_K^+ \to \mathbb{R}$ as*

$$G(\mathbf{s}, \mathbf{z}) := \sum_{k=1}^K z_k F\left(\frac{(\mathbf{X}\mathbf{s})_k}{s_k}\right) . \tag{1.43}$$

Then,

(a) the pair $(\mathbf{p}, \mathbf{w}) \in \mathbb{R}_{++}^K \times \Pi_K^+$ is a saddle point of G, and

$$F\big(\rho(\mathbf{X})\big) = \min_{\mathbf{s}\in\mathbb{R}_{++}^K} \max_{\mathbf{z}\in\Pi_K^+} G(\mathbf{s}, \mathbf{z}) = \max_{\mathbf{z}\in\Pi_K^+} \min_{\mathbf{s}\in\mathbb{R}_{++}^K} G(\mathbf{s}, \mathbf{z}) , \tag{1.44}$$

(b) $\mathbf{p} \in \mathbb{R}_{++}$ is unique up to positive multiples,
(c) $\mathbf{w} = \mathbf{q} \circ \mathbf{p}$ is a unique vector in Π_K^+.

1.3 Convexity of the Perron Root

So far we have exclusively dealt with the Perron root of a fixed irreducible matrix. Beginning with this section, we shift our attention to a class of matrix-valued functions that maps a given convex parameter set $\Omega \subseteq \mathbb{R}^K$ into a subset of X_K. This gives rise to the definition of a continuous function that maps Ω into the set of positive reals such that the output values of this function are equal to the Perron roots of the corresponding irreducible matrices. This map is of main interest in this section. To make our statements precise, we need to introduce some new definitions.

1.3.1 Some Definitions

Let $Q_k \subseteq \mathbb{R}, k = 1, \dots, K$, be arbitrary nonempty open intervals on the real line, and let the parameter set Ω be the Cartesian product of these intervals:

$$\Omega := Q_1 \times \cdots \times Q_K \subseteq \mathbb{R}^K . \tag{1.45}$$

So Ω is an open *convex set* (Definition B.15). Suppose that $\{x_{k,l}(\boldsymbol{\omega}) : \Omega \to \mathbb{R}_+, 1 \le k, l \le K, \boldsymbol{\omega} \in \Omega\}$ is a collection of bounded continuous functions. We write these functions in matrix form to obtain

$$\mathbf{X}(\boldsymbol{\omega}) = (x_{k,l}(\boldsymbol{\omega}))_{1 \le k,l \le K}$$

which is nothing but a matrix-valued function from Ω into N_K, i.e., we have $\mathbf{X} : \Omega \to N_K$.

Definition 1.32. *We say that $\mathbf{X} : \Omega \to N_K$ ($\mathbf{X} : \Omega \to X_K$) is nonnegative (irreducible) on Ω if $\mathbf{X}(\boldsymbol{\omega}) \in N_K$ ($\mathbf{X}(\boldsymbol{\omega}) \in X_K$) for every fixed $\boldsymbol{\omega} \in \Omega$. The set of all nonnegative (irreducible) matrix-valued functions on Ω is denoted by $N_K(\Omega)$ ($X_K(\Omega) \subset N_K(\Omega)$).*

Unless otherwise stated, it is assumed that $\mathbf{X} \in X_K(\Omega)$. Obviously, for every fixed $\boldsymbol{\omega} \in \Omega$, $\rho(\mathbf{X}(\boldsymbol{\omega}))$ is the Perron root of $\mathbf{X}(\boldsymbol{\omega}) \in X_K$. Hence, since the spectral radius of any matrix varies continuously with the matrix entries (Theorem A.6), $\rho(\mathbf{X}) : \Omega \to \mathbb{R}_{++}$ is a continuous function. To avoid cumbersome notation, alongside $\rho(\mathbf{X}(\boldsymbol{\omega}))$, we define

$$\lambda_p(\boldsymbol{\omega}) := \rho(\mathbf{X}(\boldsymbol{\omega})), \quad \boldsymbol{\omega} \in \Omega .$$

Moreover, for every fixed $\boldsymbol{\omega} \in \Omega$, $\lambda_p(\boldsymbol{\omega})$ is referred to as the Perron root of $\mathbf{X}(\boldsymbol{\omega})$ (or simply the Perron root). Throughout this chapter, $\hat{\boldsymbol{\omega}} \in \Omega$ and $\check{\boldsymbol{\omega}} \in \Omega$ are two arbitrary fixed parameter vectors, and

$$\boldsymbol{\omega}(\mu) := (1-\mu)\hat{\boldsymbol{\omega}} + \mu\check{\boldsymbol{\omega}}, \quad \mu \in [0,1]$$

is their convex combination. Given any $\mu \in [0,1]$, the Perron roots of $\mathbf{X}(\boldsymbol{\omega}(\mu)), \mathbf{X}(\hat{\boldsymbol{\omega}})$ and $\mathbf{X}(\check{\boldsymbol{\omega}})$ are designated by $\lambda_p(\boldsymbol{\omega}(\mu)), \lambda_p(\hat{\boldsymbol{\omega}})$ and $\lambda_p(\check{\boldsymbol{\omega}})$, respectively.

This section is devoted to the problem of convexity of the Perron root $\lambda_p(\boldsymbol{\omega})$. More precisely, we are going to find out under which conditions on the matrix entries the Perron root is a convex function on the parameter set Ω. Interestingly, even if each entry of $\mathbf{X}(\boldsymbol{\omega})$ is convex on Ω, simple examples show that this property is in general *not* inherited by the Perron root. Indeed, it is shown that for the Perron root to be convex for any choice of $\mathbf{X} \in \mathrm{X}_K(\Omega)$, it is necessary and sufficient that each entry of $\mathbf{X}(\boldsymbol{\omega})$ is log-convex on Ω. For the precise definition of log-convexity and some related results, the reader is referred to Appendix B.3.

Remark 1.33. Note that there is a subtle discrepancy between the standard definition of log-convexity and our definition. Indeed, as the logarithmic function is defined for positive reals, any log-convex function is by definition positive. In contrast, $x_{k,l}(\boldsymbol{\omega})$ is nonnegative, and therefore may take zero value on Ω. To avoid this problem, we consider the extended-value logarithm by taking $\log 0 = -\infty$. Using this convention, the zero function $x_{k,l}(\boldsymbol{\omega}) \equiv 0$ is log-convex on Ω. Furthermore, if $x_{k,l}(\boldsymbol{\omega}) = 0$ for some $\boldsymbol{\omega} \in \Omega$ and $x_{k,l}$ is log-convex, then $x_{k,l}(\boldsymbol{\omega}) \equiv 0$ for all $\boldsymbol{\omega} \in \Omega$. The only reason for the extension is that it enables us to refer to the identically zero function as a log-convex function.

Throughout the book, we use the following definition, which is a straightforward extension of the above notion of log-convexity to matrix-valued functions.

Definition 1.34. *We say that* $\mathbf{X} \in \mathrm{N}_K(\Omega)$ *is log-convex (on* Ω*) if and only if for each* $1 \leq k, l \leq K$ *and all* $\hat{\boldsymbol{\omega}}, \check{\boldsymbol{\omega}} \in \Omega$*, we have*

$$x_{k,l}(\boldsymbol{\omega}(\mu)) \leq x_{k,l}(\hat{\boldsymbol{\omega}})^{1-\mu} x_{k,l}(\check{\boldsymbol{\omega}})^{\mu}, \quad \mu \in (0,1). \tag{1.46}$$

The set of all log-convex matrix-valued functions is denoted by $\mathrm{LC}_K(\Omega)$*. In particular, note that* $x_{k,l}(\boldsymbol{\omega}) \equiv 0$ *for all* $\boldsymbol{\omega} \in \Omega$ *satisfies (1.46), and therefore is log-convex on* Ω*.*

If there is strict inequality in (1.46) for all $\hat{\boldsymbol{\omega}}, \check{\boldsymbol{\omega}} \in \Omega$ with $\hat{\boldsymbol{\omega}} \neq \check{\boldsymbol{\omega}}$, then the function $x_{k,l}$ is said to be strictly log-convex. If $\mathbf{X}(\boldsymbol{\omega})$ is confined to be irreducible on Ω, then $\mathrm{LC}_K(\Omega)$ should be considered to be a subset of $\mathrm{X}_K(\Omega)$. Otherwise, we have $\mathrm{LC}_K(\Omega) \subset \mathrm{N}_K(\Omega)$. It is important to notice that log-convexity of $\mathbf{X}(\boldsymbol{\omega})$ is an additional property on top of nonnegativity or irreducibility. For instance, given any collection of nonnegative and nonzero vectors $\{\mathbf{v}^{(k,l)} \in \mathbb{R}^K_+, \neq \mathbf{0}, 1 \leq k, l \leq K\}$, $\mathbf{X} : \Omega \to \mathrm{N}_K$ given by

$$\big(\mathbf{X}(\boldsymbol{\omega})\big)_{k,l} = x_{k,l}(\boldsymbol{\omega}) = \langle \mathbf{v}^{(k,l)}, \boldsymbol{\omega} \rangle$$

is nonnegative on $\Omega = \mathbb{R}^K_+$ ($\mathbf{X} \in \mathrm{N}_K(\mathbb{R}^K_+)$) and irreducible on $\mathbb{R}^K_{++}$ ($\mathbf{X} \in \mathrm{X}_K(\mathbb{R}^K_{++})$). Yet $\mathbf{X} \notin \mathrm{LC}_K(\mathbb{R}^K_+)$ and $\mathbf{X} \notin \mathrm{LC}_K(\mathbb{R}^K_{++})$. In contrast, if

$$\big(\mathbf{X}(\boldsymbol{\omega})\big)_{k,l} = x_{k,l}(\boldsymbol{\omega}) = \langle \mathbf{v}^{(k,l)}, e^{\boldsymbol{\omega}} \rangle$$

where $e^{\boldsymbol{\omega}} = (e^{\omega_1}, \ldots, e^{\omega_K})$, then $\mathbf{X} \in \mathrm{LC}_K(\mathbb{R}^K) \subset \mathrm{X}_K(\mathbb{R}^K)$.

As far as applications in wireless networks are concerned, $\mathbf{X}(\boldsymbol{\omega})$ usually has the following special form

$$\mathbf{X}(\boldsymbol{\omega}) = \boldsymbol{\Gamma}(\boldsymbol{\omega})\mathbf{V} \tag{1.47}$$

where $\mathbf{V} \in \mathrm{X}_K$ and

$$\boldsymbol{\Gamma}(\boldsymbol{\omega}) := \mathrm{diag}\big(\gamma_1(\omega_1), \ldots, \gamma_K(\omega_K)\big) \tag{1.48}$$

with $\gamma_k : \mathrm{Q}_k \to \mathbb{R}_{++}$ being a twice continuously differentiable and bijective (and hence also strictly monotone) function. In fact, except for Sects. 1.3.2 and 1.3.3, we will restrict our attention to this special form.

In case of wireless applications, the matrix $\mathbf{V}$ in (1.47) may change over time $t \in \mathbb{R}$ according to some stochastic process. Therefore, rather than with $\mathbf{X}(\boldsymbol{\omega})$, one has to deal with $\mathbf{X} : \Omega \times \mathbb{R} \to \mathrm{X}_K$ and $\lambda_p : \Omega \times \mathbb{R} \to \mathbb{R}_{++}$. In order to capture the effect of these variations on the network performance, it is often sufficient to assume that $\mathbf{V}$ is piecewise constant on $\mathbb{R}$, that is, given any $k \in \mathbb{Z}$ and $T > 0$, we have $\mathbf{V}(t) = \mathbf{V}^{(k)}$ for all $t \in [kT, (k+1)T)$ where $\mathbf{V}^{(k)} \in \mathrm{X}_K$ is a randomly chosen matrix. The probability distribution of the random matrix on the set of irreducible matrices is usually not known. However, in many cases, it is reasonable to assume that $\mathbf{V} = \mathbf{V}^{(k)}$ can take on any value on X_K. This gives rise to the following definition

$$\mathrm{X}_{K,\boldsymbol{\Gamma}}(\Omega) := \{\boldsymbol{\Gamma}(\boldsymbol{\omega})\mathbf{V}, \boldsymbol{\omega} \in \Omega : \mathbf{V} \in \mathrm{X}_K\} \subset \mathrm{X}_K(\Omega)\,. \tag{1.49}$$

Note that $\mathbf{X} \in \mathrm{LC}_K(\Omega)$ with $\mathbf{X}(\boldsymbol{\omega}) = \boldsymbol{\Gamma}(\boldsymbol{\omega})\mathbf{V}$ and any fixed $\mathbf{V} \in \mathrm{N}_K$ if and only if $\boldsymbol{\Gamma} \in \mathrm{LC}_K(\Omega)$. Consequently, $\mathrm{X}_{K,\boldsymbol{\Gamma}}(\Omega) \subset \mathrm{LC}_K(\Omega)$ if and only if $\boldsymbol{\Gamma} \in \mathrm{LC}_K(\Omega)$.

Remark 1.35. The notation $\boldsymbol{\Gamma} \in \mathrm{LC}_K(\Omega)$ means that $\gamma_k : \mathrm{Q}_k \to \mathbb{R}_{++}$ is log-convex for each $1 \le k \le K$. The off-diagonal entries of $\boldsymbol{\Gamma}(\boldsymbol{\omega})$ are zero, and hence, by definition, log-convex. The notation $\mathbf{X} \in \mathrm{X}_{K,\boldsymbol{\Gamma}}(\Omega)$ should mean that $\mathbf{X}(\boldsymbol{\omega}) = \boldsymbol{\Gamma}(\boldsymbol{\omega})\mathbf{V}$ for some fixed $\gamma_k : \mathrm{Q}_k \to \mathbb{R}_{++}, 1 \le k \le K$, and $\mathbf{V} \in \mathrm{X}_K$.

1.3.2 Sufficient Conditions

In this section, we provide a sufficient condition for the Perron root $\lambda_p(\boldsymbol{\omega})$ to be log-convex on Ω. Subsequently, we consider the issue of strict log-convexity. The following result shows that if $\mathbf{X} \in \mathrm{LC}_K(\Omega)$, then $\lambda_p(\boldsymbol{\omega})$ is log-convex on Ω, and therefore also convex. In particular, this implies that if $\boldsymbol{\Gamma} \in \mathrm{LC}_K(\Omega)$, then the Perron root of $\mathbf{X}(\boldsymbol{\omega}) = \boldsymbol{\Gamma}(\boldsymbol{\omega})\mathbf{V}$ is log-convex on Ω for *all* $\mathbf{V} \in \mathrm{X}_K$. The converse problem is considered in the next section.

Theorem 1.36. *If* $\mathbf{X} \in \mathrm{LC}_K(\Omega) \subset \mathrm{X}_K(\Omega)$, *then*

$$\lambda_p(\boldsymbol{\omega}(\mu)) \leq \lambda_p(\hat{\boldsymbol{\omega}})^{1-\mu}\lambda_p(\check{\boldsymbol{\omega}})^{\mu} \tag{1.50}$$

for all $\mu \in (0,1)$ *and all* $\hat{\boldsymbol{\omega}}, \check{\boldsymbol{\omega}} \in \Omega$.

Proof. Let $\mu \in (0,1)$ be arbitrary and fixed. As $\mathbf{X} \in \mathrm{X}_K(\Omega)$, it follows that $\mathbf{X}(\boldsymbol{\omega}(\mu)) \in \mathrm{X}_K$, regardless of the choice of $\hat{\boldsymbol{\omega}} \in \Omega$ and $\check{\boldsymbol{\omega}} \in \Omega$. Thus, applying Theorem 1.2 to $\lambda_p(\boldsymbol{\omega}(\mu))$ yields

$$\log \lambda_p(\boldsymbol{\omega}(\mu)) = \sup_{\mathbf{A}\in \mathrm{S}_K} \Big(\sum_{k,l=1}^{K} u_k a_{k,l} \log \frac{x_{k,l}(\boldsymbol{\omega}(\mu))}{a_{k,l}} \Big)$$

where $\mathrm{S}_K := \mathrm{S}_K(\mathbf{X}(\boldsymbol{\omega}(\mu)))$ is defined by (1.3). Since the logarithmic function is strictly increasing and $\mathbf{X}(\boldsymbol{\omega})$ is log-convex on Ω, taking (1.46) into account on the right-hand side of the equality above gives

$$\begin{aligned}
\log \lambda_p(\boldsymbol{\omega}(\mu)) &\leq \sup_{\mathbf{A}\in \mathrm{S}_K} \Big(\sum_{k,l=1}^{K} u_k a_{k,l} \log \frac{x_{k,l}(\hat{\boldsymbol{\omega}})^{1-\mu}\, x_{k,l}(\check{\boldsymbol{\omega}})^{\mu}}{a_{k,l}^{1-\mu}\, a_{k,l}^{\mu}} \Big) \\
&= \sup_{\mathbf{A}\in \mathrm{S}_K} \Big((1-\mu) \sum_{k,l=1}^{K} u_k a_{k,l} \log \frac{x_{k,l}(\hat{\boldsymbol{\omega}})}{a_{k,l}} \\
&\quad + \mu \sum_{k,l=1}^{K} u_k a_{k,l} \log \frac{x_{k,l}(\check{\boldsymbol{\omega}})}{a_{k,l}} \Big).
\end{aligned}$$

Now since $\sup(f+g) \leq \sup f + \sup g$ for any functions f and g, one obtains

$$\begin{aligned}
\log \lambda_p(\boldsymbol{\omega}(\mu)) &\leq (1-\mu) \sup_{\mathbf{A}\in \mathrm{S}_K} \Big(\sum_{k,l=1}^{K} u_k a_{k,l} \log \frac{x_{k,l}(\hat{\boldsymbol{\omega}})}{a_{k,l}} \Big) \\
&\quad + \mu \sup_{\mathbf{A}\in \mathrm{S}_K} \Big(\sum_{k,l=1}^{K} u_k a_{k,l} \log \frac{x_{k,l}(\check{\boldsymbol{\omega}})}{a_{k,l}} \Big) \\
&\overset{(a)}{=} (1-\mu) \sup_{\mathbf{A}\in \mathrm{S}_K(\mathbf{X}(\hat{\boldsymbol{\omega}}))} \Big(\sum_{k,l=1}^{K} u_k a_{k,l} \log \frac{x_{k,l}(\hat{\boldsymbol{\omega}})}{a_{k,l}} \Big) \\
&\quad + \mu \sup_{\mathbf{A}\in \mathrm{S}_K(\mathbf{X}(\check{\boldsymbol{\omega}}))} \Big(\sum_{k,l=1}^{K} u_k a_{k,l} \log \frac{x_{k,l}(\check{\boldsymbol{\omega}})}{a_{k,l}} \Big) \\
&= (1-\mu) \log \lambda_p(\hat{\boldsymbol{\omega}}) + \mu \log \lambda_p(\check{\boldsymbol{\omega}})
\end{aligned}$$

where (a) follows from the fact that the suprema are attained on $\mathrm{S}_K(\mathbf{X}(\hat{\boldsymbol{\omega}})) \subseteq \mathrm{S}_K$ and $\mathrm{S}_K(\mathbf{X}(\check{\boldsymbol{\omega}})) \subseteq \mathrm{S}_K$, respectively. The last equation is an application of Corollary 1.3.

The following result asserts that the Perron root $\lambda_p(\boldsymbol{\omega})$ is strictly log-convex if at least one entry of $\mathbf{X}(\boldsymbol{\omega})$ is strictly log-convex on Ω.

Theorem 1.37. *Let* $\mathbf{X} \in \mathrm{LC}_K(\Omega) \subset \mathrm{X}_K(\Omega)$ *and suppose that at least one entry of* $\mathbf{X}(\boldsymbol{\omega})$ *is strictly log-convex function. Then,*

$$\lambda_p(\boldsymbol{\omega}(\mu)) < \lambda_p(\hat{\boldsymbol{\omega}})^{1-\mu}\lambda_p(\check{\boldsymbol{\omega}})^{\mu}, \quad \hat{\boldsymbol{\omega}}, \check{\boldsymbol{\omega}} \in \Omega \tag{1.51}$$

for all $\mu \in (0,1)$.

Proof. Let everything be as in the proof above, and, without loss of generality, assume that

$$x_{k_0,l_0}(\boldsymbol{\omega}(\mu)) < x_{k_0,l_0}(\hat{\boldsymbol{\omega}})^{1-\mu}x_{k_0,l_0}(\check{\boldsymbol{\omega}})^{\mu}, \quad \mu \in (0,1) \tag{1.52}$$

for some $1 \leq k_0, l_0 \leq K$. Then, due to strict monotonicity of $\log(x), x > 0$,

$$\begin{aligned}
&\log\lambda_p(\boldsymbol{\omega}(\mu)) \\
&= \sup_{\mathbf{A}\in \mathrm{S}_K}\Big(\sum_{k,l=1}^{K} u_k a_{k,l}\log\frac{x_{k,l}(\boldsymbol{\omega}(\mu))}{a_{k,l}}\Big) \\
&\leq \sup_{\mathbf{A}\in \mathrm{S}_K}\Big((1-\mu)\sum_{\substack{k,l=1\\ k\neq k_0, l\neq l_0}}^{K} u_k a_{k,l}\log\frac{x_{k,l}(\hat{\boldsymbol{\omega}})}{a_{k,l}} \\
&\quad +\mu\sum_{\substack{k,l=1\\ k\neq k_0, l\neq l_0}}^{K} u_k a_{k,l}\log\frac{x_{k,l}(\check{\boldsymbol{\omega}})}{a_{k,l}} + u_{k_0}a_{k_0,l_0}\log\frac{x_{k_0,l_0}(\boldsymbol{\omega}(\mu))}{a_{k_0,l_0}}\Big) \\
&\overset{(a)}{<} \sup_{\mathbf{A}\in \mathrm{S}_K}\Big((1-\mu)\sum_{k,l=1}^{K} u_k a_{k,l}\log\frac{x_{k,l}(\hat{\boldsymbol{\omega}})}{a_{k,l}} + \mu\sum_{k,l=1}^{K} u_k a_{k,l}\log\frac{x_{k,l}(\check{\boldsymbol{\omega}})}{a_{k,l}}\Big)
\end{aligned}$$

for all $\mu \in (0,1)$ where (a) follows from (1.52). Now proceeding essentially as above completes the proof.

Note that the condition of Theorem 1.37 is never satisfied when $\mathbf{X} \in \mathrm{X}_K(\Omega)$ is of the form $\mathbf{X}(\boldsymbol{\omega}) = \mathbf{\Gamma}(\boldsymbol{\omega})\mathbf{V}$ since then the value of $x_{k,l}(\boldsymbol{\omega})$ is independent of $\omega_j, j \neq k$. Therefore, $x_{k,l}(\boldsymbol{\omega})$ cannot be strictly log-convex on Ω. In this case, instead of demanding that at least one entry of $\mathbf{X}(\boldsymbol{\omega})$ is strictly log-convex on Ω, we could require that for every $\hat{\boldsymbol{\omega}}, \check{\boldsymbol{\omega}} \in \Omega$, there is an entry $x_{k,l}(\boldsymbol{\omega}(\mu))$ of $\mathbf{X}(\boldsymbol{\omega}(\mu))$ that is a strictly log-convex function of $\mu \in (0,1)$. Obviously, this requirement is satisfied if γ_k is strictly log-convex on Q_k for each $1 \leq k \leq K$ (see also Theorem 2.12).

1.3.3 Convexity of the Feasibility Set

Definition 1.38 (Feasibility Set). *For any* $\mathbf{X} \in \mathrm{X}_K(\Omega)$, *there is an associated set* $\mathrm{F} \subset \Omega$, *called the feasibility set, given by*

$$\mathrm{F} := \{\boldsymbol{\omega} \in \Omega : \lambda_p(\boldsymbol{\omega}) \leq 1\}. \tag{1.53}$$

If there is no $\boldsymbol{\omega} \in \Omega$ *such that* $\lambda_p(\boldsymbol{\omega}) \leq 1$, *then* F *is an empty set.*

In all that follows, it is assumed that $\mathrm{F} \neq \emptyset$, which excludes the trivial case of F being an empty set. The importance of the feasibility set for wireless networks will become obvious later in the second part of the book. The reader is also referred to Chapter 2 where we will introduce the notion of a feasibility set under some additional constraints.

Our main concern is the question whether or not the feasibility set is a convex set. As the Perron root is a continuous map from Ω into the set of reals, a sufficient condition for convexity of F immediately follows from Theorem 1.44 if one considers the fact that the geometric mean is bounded above by the arithmetic mean (see Appendix B.3). Indeed, for all $\mu \in (0,1)$, we have

$$\begin{aligned}\lambda_p(\hat{\boldsymbol{\omega}})^{1-\mu}\lambda_p(\check{\boldsymbol{\omega}})^{\mu} &\leq (1-\mu)\lambda_p(\hat{\boldsymbol{\omega}}) + \mu\lambda_p(\check{\boldsymbol{\omega}}) \\ &\leq \max\{\lambda_p(\hat{\boldsymbol{\omega}}), \lambda_p(\check{\boldsymbol{\omega}})\}, \quad \hat{\boldsymbol{\omega}}, \check{\boldsymbol{\omega}} \in \Omega.\end{aligned} \tag{1.54}$$

Thus, by Theorem 1.36, if $\mathbf{X}(\boldsymbol{\omega})$ is log-convex on Ω and $\max\{\lambda_p(\hat{\boldsymbol{\omega}}), \lambda_p(\check{\boldsymbol{\omega}})\} \leq 1$, then $\lambda_p(\boldsymbol{\omega}(\mu)) \leq 1$ for all $\mu \in (0,1)$. In other words, if $\mathbf{X}(\boldsymbol{\omega})$ is log-convex on Ω, then

$$\boldsymbol{\omega}(\mu) \in \mathrm{F}, \quad \hat{\boldsymbol{\omega}}, \check{\boldsymbol{\omega}} \in \mathrm{F}$$

for all $\mu \in (0,1)$. This is summarized in a corollary.

Corollary 1.39. *If* $\mathbf{X} \in \mathrm{LC}_K(\Omega) \subset \mathrm{X}_K(\Omega)$, *then* $\mathrm{F} \subset \Omega$ *is a convex set.*

It is worth pointing out that the converse does not hold in general. To see this, consider the following simple example.

Example 1.40. Let $\Omega = \mathbb{R}^2_{++}$, and let

$$\mathbf{X}(\boldsymbol{\omega}) = \begin{pmatrix} a\,\omega_1 & b\,\omega_1 \\ b\,\omega_2 & a\,\omega_2 \end{pmatrix}, \quad a \geq 0, b > 0.$$

Clearly, $\mathbf{X}(\boldsymbol{\omega})$ is irreducible for every $\boldsymbol{\omega} \in \mathbb{R}^2_{++}$ but not log-convex on $\mathbb{R}^2_{++}$. The Perron root can be easily calculated to yield

$$\lambda_p(\boldsymbol{\omega}) = \frac{1}{2}\Big(a\,(\omega_1+\omega_2) + \big(a^2\,(\omega_1-\omega_2)^2 + 4\,b^2\,\omega_1\,\omega_2\big)^{1/2}\Big).$$

Thus, the set of all $\boldsymbol{\omega} \in \mathbb{R}^2_{++}$ satisfying $\lambda_p(\boldsymbol{\omega}) = 1$ is given by

$$\omega_2 = f(\omega_1) := \frac{1 - a\,\omega_1}{(b^2 - a^2)\omega_1 + a}, \quad \omega_1 \in (0, 1/a).$$

Finally, the second derivative of $f(x), x \in (0, 1/a)$, yields

$$f''(x) = \frac{2\,b^2(b-a)(a+b)}{[(b^2-a^2)x+a]^3}\,.$$

It is easy to see that the denominator is positive for all $x \in (0, 1/a)$. On the other hand, the sign of the numerator is equal to the sign of $b - a$. Hence, the feasibility set is not convex when $b > a$ ($f(x)$ convex on $(0, 1/a)$) and becomes convex when $a > b$ ($f(x)$ concave on $(0, 1/a)$). If $a = b$, $f(x) = 1/a - x, x \in (0, 1/a)$, is a line segment.

Sometimes it is desired to know whether F is a strictly convex set in the following sense (we refer for instance to the discussion in Sect. 5.4).

Definition 1.41 (Strictly Convex Feasibility Set). F *is said to be strictly convex (or s-convex) if* $\boldsymbol{\omega}(\mu) = (1-\mu)\hat{\boldsymbol{\omega}} + \mu\check{\boldsymbol{\omega}}$ *is interior to* F *(relative to* Ω*) for all* $\mu \in (0, 1)$ *and* $\hat{\boldsymbol{\omega}}, \check{\boldsymbol{\omega}} \in \partial\mathrm{F}$ *where*

$$\partial\mathrm{F} = \{\boldsymbol{\omega} \in \Omega : \lambda_p(\boldsymbol{\omega}) = 1\}\,. \tag{1.55}$$

Remark 1.42. For convenience, in what follows, "the boundary of F" always refers to ∂F, even if F has boundary points other than those in ∂F. According to this convention, F is strictly convex if any boundary point of F cannot be written as a convex combination of two other points of F.

Although convexity is sufficient for most applications, strict convexity provides additional information about the feasibility region. In particular, if F is strictly convex, then, by definition, there exists $\tilde{\boldsymbol{\omega}} \in$ F such that $\lambda_p(\boldsymbol{\omega}(\mu)) < \lambda_p(\tilde{\boldsymbol{\omega}})$ for any $\hat{\boldsymbol{\omega}}, \check{\boldsymbol{\omega}} \in$ F. The following corollary is immediate.

Corollary 1.43. *Let* $\mathbf{X} \in \mathrm{LC}_K(\Omega) \subset \mathrm{X}_K(\Omega)$ *with at least one entry being a strictly log-convex function on* Ω*. Then,* F *is strictly convex.*

As in case of Theorem 1.37, the condition of the corollary is never met when $\mathbf{X} \in \mathrm{X}_{K,\boldsymbol{\Gamma}}(\Omega)$. However, the set is strictly convex if γ_k is strictly log-convex on Q_k for each $1 \le k \le K$.

1.3.4 Necessary Conditions

Having proved that $\mathbf{X} \in \mathrm{LC}_K(\Omega)$ is sufficient for $\lambda_p(\boldsymbol{\omega})$ to be both log-convex and convex on Ω, now we turn our attention to a converse problem. More precisely, we are asking whether $\mathbf{X} \in \mathrm{LC}_K(\Omega)$ is necessary for $\lambda_p(\boldsymbol{\omega})$ to be *convex* on Ω, regardless of the choice of $\mathbf{X} \in \mathrm{X}_K(\Omega)$. In doing so, however, we restrict $\mathbf{X}$ to be a member of $\mathrm{X}_{K,\boldsymbol{\Gamma}}(\Omega)$ defined by (1.49). This is equivalent to saying that $\mathbf{X}(\boldsymbol{\omega}) = \boldsymbol{\Gamma}(\boldsymbol{\omega})\mathbf{V}$ for some $\mathbf{V} \in \mathrm{X}_K$ (see the remark below (1.49)). As a consequence, the problem reduces to finding $\mathbf{V} \in \mathrm{X}_K$ such that convexity of $\lambda_p(\boldsymbol{\omega})$ implies $\boldsymbol{\Gamma} \in \mathrm{LC}_K(\Omega)$.

It is somewhat surprising that if $\lambda_p(\boldsymbol{\omega})$ is required to be *convex* for *all* $\mathbf{X} \in \mathrm{X}_{K,\boldsymbol{\Gamma}}(\Omega)$ and all $K > 1$, then $\boldsymbol{\Gamma}(\boldsymbol{\omega})$ must be log-convex on Ω.

Theorem 1.44. *Let $\gamma_k : \mathrm{Q}_k \to \mathbb{R}_{++}$ be twice continuously differentiable. Suppose that $\lambda_p(\boldsymbol{\omega})$ is convex for all $\mathbf{X} \in \mathrm{X}_{K,\boldsymbol{\Gamma}}(\Omega) \subset \mathrm{X}_K(\Omega)$ and all $K > 1$. Then, $\boldsymbol{\Gamma} \in \mathrm{LC}_K(\Omega)$.*

Proof. Let $\boldsymbol{\Gamma}(\boldsymbol{\omega})$ with $\gamma_k : \mathrm{Q}_k \to \mathbb{R}_{++}$ be arbitrary, and let $\omega_k \in \mathrm{Q}_k$ for $k = 2, \dots, K$ be *fixed*. We choose $\mathbf{V} \in \mathrm{X}_K$ to be

$$\mathbf{V} = \begin{pmatrix} 0 & 0 & \cdots & 0 & 1 \\ \frac{1}{\gamma_2(\omega_2)} & 0 & \cdots & 0 & 0 \\ 0 & \frac{1}{\gamma_3(\omega_3)} & \cdots & 0 & 0 \\ \vdots & \vdots & \ddots & \vdots & \vdots \\ 0 & 0 & \cdots & \frac{1}{\gamma_K(\omega_K)} & 0 \end{pmatrix}$$

so that $\mathbf{X} \in \mathrm{X}_{K,\boldsymbol{\Gamma}}(\Omega)$ takes the form

$$\mathbf{X}(\boldsymbol{\omega}) = \begin{pmatrix} 0 & 0 & \cdots & 0 & \gamma(\omega_1) \\ 1 & 0 & \cdots & 0 & 0 \\ 0 & 1 & \cdots & 0 & 0 \\ \vdots & \vdots & \ddots & \vdots & \vdots \\ 0 & 0 & \cdots & 1 & 0 \end{pmatrix} \tag{1.56}$$

where $\gamma \equiv \gamma_1 : \mathrm{Q}_1 \to \mathbb{R}_{++}$ is twice continuously differentiable. We see that the Perron root of (1.56) is equal to

$$f(\omega_1) = \left(\gamma(\omega_1)\right)^{1/K} . \tag{1.57}$$

So its second derivative yields

$$f''(\omega_1) = \frac{(\frac{1}{K} - 1)\,\gamma'(\omega_1)^2 + \gamma(\omega_1)\,\gamma''(\omega_1)}{K\,\gamma(\omega_1)^{2-\frac{1}{K}}}$$

which is nonnegative for all $\omega_1 \in \mathrm{Q}_1$ if and only if $0 \le (\frac{1}{K} - 1)\gamma'(\omega_1)^2 + \gamma(\omega_1)\,\gamma''(\omega_1)$ for all $\omega_1 \in \mathrm{Q}_1$. This, in turn, is true for all $K > 1$ if and only if $\gamma'(x)^2 \le \gamma(x)\,\gamma''(x)$ for all $x \in \mathrm{Q}_1$ or, equivalently, if and only if γ is a log-convex function. Thus, $\boldsymbol{\Gamma} \in \mathrm{LC}_K(\Omega)$ must hold.

Remark 1.45. Since $\log \gamma(x)^{1/K} = 1/K \log \gamma(x), x \in \mathrm{Q}$, we see that the Perron root (1.57) of (1.56) is *log-convex* if and only if $\gamma(x)\gamma''(x) - (\gamma'(x))^2 \ge 0, x \in \mathrm{Q}$. Therefore, given any fixed $K > 1$, the Perron root $\lambda_p(\boldsymbol{\omega})$ is *log-convex* for *all* $\mathbf{X} \in \mathrm{X}_{K,\boldsymbol{\Gamma}}(\Omega) \subset \mathrm{X}_K(\Omega)$ if and only if $\boldsymbol{\Gamma} \in \mathrm{LC}_K(\Omega)$.

The theorem above asserts that $\boldsymbol{\Gamma} \in \mathrm{LC}_K(\Omega)$ must hold if $\lambda_p(\boldsymbol{\omega})$ is required to be convex for *all* $\mathbf{X} \in \mathrm{X}_{K,\boldsymbol{\Gamma}}(\Omega)$ and all $K > 1$. Thus, the theorem does not say anything when $\mathbf{X}(\boldsymbol{\omega}) = \boldsymbol{\Gamma}(\boldsymbol{\omega})\mathbf{V}$ is either fixed or confined to belong to some subset of $\mathrm{X}_{K,\boldsymbol{\Gamma}}(\Omega)$. In fact, it is shown in Sect. 1.4 that a less stringent property of $\mathbf{\Gamma}(\boldsymbol{\omega})$ is sufficient for the Perron root to be convex on Ω if $\mathbf{V} \in \mathrm{X}_K$ is limited to satisfy either $\mathbf{V} = \mathbf{V}^T$ (symmetry) or $\forall_{\mathbf{x}\in\mathbb{R}^K} \mathbf{x}^T\mathbf{V}\mathbf{x} \geq 0$ (positive semidefinitness).

It is also important to notice that Theorem 1.44 holds even if each function in $\mathbf{X}(\boldsymbol{\omega})$ is positive for all $\boldsymbol{\omega} \in \Omega$, i.e., if $\mathbf{X} \in \mathrm{X}_{K,\mathbf{\Gamma}}(\Omega) \subset \mathrm{P}_K(\Omega)$. To see this, define a positive matrix $\mathbf{X}_\epsilon(\boldsymbol{\omega}), \boldsymbol{\omega} \in \Omega$, as follows

$$\mathbf{X}_\epsilon(\boldsymbol{\omega}) = \mathbf{X}(\boldsymbol{\omega}) + \epsilon\mathbf{1}\mathbf{1}^T, \quad \epsilon > 0$$

where $\mathbf{X}(\boldsymbol{\omega})$ is given by (1.56). Furthermore, let

$$\lambda_p(\epsilon, \boldsymbol{\omega}) = \rho(\mathbf{X}_\epsilon(\boldsymbol{\omega}))$$

and suppose that $\gamma : \mathbb{R} \to \mathbb{R}_{++}$ in (1.56) is chosen such that the Perron root is convex for any positive matrix. Thus, in a special case of $\lambda_p(\epsilon, \boldsymbol{\omega})$ with $\epsilon > 0$, we have

$$\lambda_p(\epsilon, \boldsymbol{\omega}(\mu)) \leq (1-\mu)\lambda_p(\epsilon, \hat{\boldsymbol{\omega}}) + \mu\lambda_p(\epsilon, \check{\boldsymbol{\omega}}), \quad \hat{\boldsymbol{\omega}}, \check{\boldsymbol{\omega}} \in \Omega$$

for all $\mu \in (0,1)$ and $\epsilon > 0$. So, by continuity of $\lambda_p(\epsilon, \boldsymbol{\omega})$ with respect to $\epsilon > 0$ (Theorem A.6), one obtains

$$\begin{aligned}\lambda_p(\boldsymbol{\omega}(\mu)) &= \lim_{\epsilon\to 0} \lambda_p(\epsilon, \boldsymbol{\omega}(\mu)) \\ &\leq (1-\mu)\lim_{\epsilon\to 0} \lambda_p(\epsilon, \hat{\boldsymbol{\omega}}) + \mu \lim_{\epsilon\to 0} \lambda_p(\epsilon, \check{\boldsymbol{\omega}}) = (1-\mu)\lambda_p(\hat{\boldsymbol{\omega}}) + \mu\lambda_p(\check{\boldsymbol{\omega}})\end{aligned}$$

for all $\mu \in (0,)$ where $\lambda_p(\boldsymbol{\omega})$ is given by (1.57). Consequently, since $\lambda_p(\epsilon, \boldsymbol{\omega})$ is convex for all $\epsilon > 0$ (by assumption), so also is $\lambda_p(\boldsymbol{\omega})$. However, by the proof of Theorem 1.44, $\lambda_p(\boldsymbol{\omega})$ given by (1.57) is convex for all $K > 1$ if and only if γ is log-convex. We summarize this in an observation.

Observation 1.46. *If $\lambda_p(\boldsymbol{\omega})$ is convex on Ω for all $\mathbf{X} \in \mathrm{X}_{K,\mathbf{\Gamma}}(\Omega) \subset \mathrm{P}_K(\Omega)$ and all $K > 1$, then $\mathbf{\Gamma} \in \mathrm{LC}_K(\Omega)$.*

A remarkable fact about these results is that although $\lambda_p(\boldsymbol{\omega})$ is required to be convex, we arrive at log-convexity of $\mathbf{\Gamma}(\boldsymbol{\omega})$, which is significantly stronger than convexity. Combining Theorem 1.36 and Theorem 1.44 shows that the following statements are equivalent (if γ is twice continuously differentiable):

(i) $\lambda_p(\boldsymbol{\omega})$ is convex for all $K > 1$ and all $\mathbf{X} \in \mathrm{X}_{K,\mathbf{\Gamma}}(\Omega)$.
(ii) $\lambda_p(\boldsymbol{\omega})$ is log-convex for all $K > 1$ and all $\mathbf{X} \in \mathrm{X}_{K,\mathbf{\Gamma}}(\Omega)$.

1.4 Special Classes of Matrices

We continue the analysis with $\mathbf{X}(\boldsymbol{\omega})$ of the form $\mathbf{X}(\boldsymbol{\omega}) = \mathbf{\Gamma}(\boldsymbol{\omega})\mathbf{V}$ for some $\mathbf{V} \in \mathrm{X}_K$. For simplicity, it is assumed that

$$\gamma(x) := \gamma_1(x) = \cdots = \gamma_K(x), \quad x \in \mathrm{Q}$$

where $\gamma : \mathrm{Q} \to \mathbb{R}_{++}$ is a twice continuously differentiable and bijective function. Hence, throughout this section, $\Omega = \mathrm{Q}^K$. It is emphasized, however, that this assumption does not impact the generality of the analysis. As before, we use F and $\lambda_p(\boldsymbol{\omega})$ to denote the feasibility set and the Perron root of $\mathbf{\Gamma}(\boldsymbol{\omega})\mathbf{V}$ for some $\boldsymbol{\omega} \in \Omega$, respectively.

Obviously, if $\gamma : \mathrm{Q} \to \mathbb{R}_{++}$ is log-convex, then $\mathbf{X} \in \mathrm{LC}_K(\Omega)$. Consequently, by Theorem 1.36, if $\gamma(x)$ is log-convex, the Perron root $\lambda_p(\boldsymbol{\omega})$ is a log-convex function of the parameter vector $\boldsymbol{\omega}$. By Sect. 1.44, it can be inferred that log-convexity of γ is necessary when the Perron root is required to be convex on $\Omega = \mathrm{Q}^K$ for all $\mathbf{V} \in \mathrm{X}_K$ and all $K > 1$. In this section, we put some restrictions on $\mathbf{V} \in \mathrm{X}_K$. In particular, it is shown that the log-convexity requirement can be relaxed to a less stringent requirement when the matrix $\mathbf{V} \in \mathrm{X}_K$ is confined to be either symmetric or positive semidefinite.

1.4.1 Symmetric Matrices

Recall that a square matrix $\mathbf{A} \in \mathbb{R}^{K \times K}$ is said to be symmetric if $\mathbf{A} = \mathbf{A}^T$ (see also Appendix A.3.2). The following theorem provides a necessary and sufficient condition for the Perron root to be convex on $\Omega = \mathrm{Q}^K$ for all $K > 1$ and all $\mathbf{X} \in \mathrm{X}^s_{K,\mathbf{\Gamma}}(\Omega)$ where

$$\mathrm{X}^s_{K,\mathbf{\Gamma}}(\Omega) := \{\mathbf{\Gamma}(\boldsymbol{\omega})\mathbf{V}, \boldsymbol{\omega} \in \Omega : \mathbf{V} \in \mathrm{X}_K, \mathbf{V} = \mathbf{V}^T\}.$$

Theorem 1.47. *Let* $f_\gamma : \mathrm{Q}^2 \to \mathbb{R}_{++}$ *be given by*

$$f_\gamma(x, y) = \sqrt{\gamma(x)\gamma(y)}.$$

Then, the Perron root $\lambda_p(\boldsymbol{\omega})$ *is convex on* $\Omega = \mathrm{Q}^K$ *for all* $\mathbf{X} \in \mathrm{X}^s_{K,\mathbf{\Gamma}}(\Omega)$ *and all* $K > 1$ *if and only if* f_γ *is convex on* Q^2.

Proof. The necessity is easily verified by considering $K = 2$ and $\mathbf{X}(\boldsymbol{\omega}) = \mathbf{\Gamma}(\boldsymbol{\omega})\mathbf{V}$ with $\mathbf{V} = \left(\begin{smallmatrix} 0 & \varrho \\ \varrho & 0 \end{smallmatrix}\right) \in \mathrm{X}_K$. In this case, we have

$$\lambda_p(\boldsymbol{\omega}) = \varrho\sqrt{\gamma(\omega_1)\gamma(\omega_2)}$$

from which the necessary condition immediately follows. To prove the converse, let $\mathbf{X} \in \mathrm{X}^s_{K,\mathbf{\Gamma}}(\Omega)$ be arbitrary and note that due to the symmetry,

$$\lambda_p(\boldsymbol{\omega}(\mu)) = \rho(\mathbf{W}(\mu)), \quad \hat{\boldsymbol{\omega}}, \check{\boldsymbol{\omega}} \in \mathrm{Q}^K, \ \mu \in [0, 1]$$

where $\mathbf{W}(\mu) = \big(w_{k,l}(\mu)\big) := \mathbf{\Gamma}^{\frac{1}{2}}(\boldsymbol{\omega}(\mu))\mathbf{V}\mathbf{\Gamma}^{\frac{1}{2}}(\boldsymbol{\omega}(\mu))$. The entries of $\mathbf{W}(\mu)$ are

$$w_{k,l}(\mu) = \sqrt{\gamma(\omega_k(\mu))}v_{k,l}\sqrt{\gamma(\omega_l(\mu))} = f_\gamma\big(\omega_k(\mu), \omega_l(\mu)\big)v_{k,l}.$$

Hence, by the convexity of f_γ,

$$w_{k,l}(\mu) \leq (1-\mu) f_\gamma(\hat{\omega}_k, \hat{\omega}_l) v_{k,l} + \mu f_\gamma(\breve{\omega}_k, \breve{\omega}_l) v_{k,l}$$

for all $\mu \in (0,1)$. Now due the monotonicity of the Perron root (Theorem A.19), one obtains

$$\begin{aligned} \lambda_p(\boldsymbol{\omega}(\mu)) &\leq \rho\big((1-\mu)\boldsymbol{\Gamma}^{\frac{1}{2}}(\hat{\boldsymbol{\omega}})\mathbf{V}\boldsymbol{\Gamma}^{\frac{1}{2}}(\hat{\boldsymbol{\omega}}) + \mu\boldsymbol{\Gamma}^{\frac{1}{2}}(\breve{\boldsymbol{\omega}})\mathbf{V}\boldsymbol{\Gamma}^{\frac{1}{2}}(\breve{\boldsymbol{\omega}})\big) \\ &\leq (1-\mu)\rho(\boldsymbol{\Gamma}^{\frac{1}{2}}(\hat{\boldsymbol{\omega}})\mathbf{V}\boldsymbol{\Gamma}^{\frac{1}{2}}(\hat{\boldsymbol{\omega}})) + \mu\rho(\boldsymbol{\Gamma}^{\frac{1}{2}}(\breve{\boldsymbol{\omega}})\mathbf{V}\boldsymbol{\Gamma}^{\frac{1}{2}}(\breve{\boldsymbol{\omega}})) \\ &= (1-\mu)\lambda_p(\hat{\boldsymbol{\omega}}) + \mu\lambda_p(\breve{\boldsymbol{\omega}}) \end{aligned}$$

where the second inequality follows from the fact that the spectral radius is convex on the set of symmetric matrices (Theorem A.15).

1.4.2 Symmetric Positive Semidefinite Matrices

Now let us assume that $\mathbf{V}$ is confined to be a symmetric positive semidefinite matrix (Definition A.16). Hence, $\mathbf{X} \in \mathrm{X}^p_{K,\boldsymbol{\Gamma}}(\Omega)$ where $\Omega = \mathrm{Q}^K$ and

$$\mathrm{X}^p_{K,\boldsymbol{\Gamma}}(\Omega) := \left\{\boldsymbol{\Gamma}(\boldsymbol{\omega})\mathbf{V}, \boldsymbol{\omega} \in \Omega : \mathbf{V} \in \mathrm{X}_K, \mathbf{V} = \mathbf{V}^T, \forall_{\mathbf{x}\in\mathbb{R}^K} \mathbf{x}^T\mathbf{V}\mathbf{x} \geq 0\right\}.$$

Note that $\mathbf{V} \in \mathrm{X}_K$ is positive semidefinite if, roughly speaking, its diagonal entries are large enough when compared with the off-diagonal entries. It turns out that $\lambda_p(\boldsymbol{\omega})$ is convex on Ω if $\mathbf{X} \in \mathrm{X}^p_{K,\boldsymbol{\Gamma}}(\Omega)$ and $\gamma : \mathrm{Q} \to \mathbb{R}_{++}$ is a *convex* function.

Theorem 1.48. *Suppose that* $\mathbf{X} \in \mathrm{X}^p_{K,\boldsymbol{\Gamma}}(\Omega)$ *such that* $\gamma : \mathrm{Q} \to \mathbb{R}_{++}$ *is any continuous convex function. Then,* $\lambda_p(\boldsymbol{\omega})$ *is convex on* $\Omega = \mathrm{Q}^K$.

Proof. Let $\mathbf{X}(\boldsymbol{\omega}) = \boldsymbol{\Gamma}(\boldsymbol{\omega})\mathbf{V} \in \mathrm{X}^p_{K,\boldsymbol{\Gamma}}(\Omega)$ be arbitrary. Thus, as $\mathbf{V} \in \mathrm{X}_K$ is symmetric positive semidefinite, we can write $\mathbf{V} = \mathbf{A}\mathbf{A}^T$ with $\mathbf{A} = \mathbf{U}\boldsymbol{\Lambda}^{\frac{1}{2}}$ where $\mathbf{U}$ is orthogonal (Definition A.12) and $\boldsymbol{\Lambda}$ is a real diagonal matrix of the eigenvalues of $\mathbf{V}$. Furthermore,

$$\begin{aligned} \lambda_p(\boldsymbol{\omega}) &= \lambda_{\max}(\boldsymbol{\Gamma}(\boldsymbol{\omega})^{\frac{1}{2}}\mathbf{V}\boldsymbol{\Gamma}(\boldsymbol{\omega})^{\frac{1}{2}}) = \lambda_{\max}(\boldsymbol{\Gamma}(\boldsymbol{\omega})^{\frac{1}{2}}\mathbf{A}\mathbf{A}^T\boldsymbol{\Gamma}(\boldsymbol{\omega})^{\frac{1}{2}}) \\ &= \lambda_{\max}(\mathbf{A}^T\boldsymbol{\Gamma}(\boldsymbol{\omega})^{\frac{1}{2}}\boldsymbol{\Gamma}(\boldsymbol{\omega})^{\frac{1}{2}}\mathbf{A}) \end{aligned} \tag{1.58}$$

where the largest eigenvalue $\lambda_{\max}(\boldsymbol{\Gamma}(\boldsymbol{\omega})^{\frac{1}{2}}\mathbf{V}\boldsymbol{\Gamma}(\boldsymbol{\omega})^{\frac{1}{2}})$ of $\boldsymbol{\Gamma}(\boldsymbol{\omega})^{\frac{1}{2}}\mathbf{V}\boldsymbol{\Gamma}(\boldsymbol{\omega})^{\frac{1}{2}}$ is equal to the induced squared matrix 2-norm of $\boldsymbol{\Gamma}(\boldsymbol{\omega})^{\frac{1}{2}}\mathbf{A}$ (see the definition of induced matrix norms in Appendix A.2). Therefore,

$$\lambda_p(\boldsymbol{\omega}) = \max_{\|\mathbf{v}\|_2=1} \mathbf{v}^T\mathbf{A}^T\boldsymbol{\Gamma}(\boldsymbol{\omega})^{\frac{1}{2}}\boldsymbol{\Gamma}(\boldsymbol{\omega})^{\frac{1}{2}}\mathbf{A}\mathbf{v} = \max_{\|\mathbf{v}\|_2=1} \|\boldsymbol{\Gamma}(\boldsymbol{\omega})^{\frac{1}{2}}\mathbf{A}\mathbf{v}\|_2^2 .$$

The kth element of $\boldsymbol{\Gamma}(\boldsymbol{\omega})^{\frac{1}{2}}\mathbf{A}\mathbf{v}$ is equal to $(\boldsymbol{\Gamma}(\boldsymbol{\omega})^{\frac{1}{2}}\mathbf{A}\mathbf{v})_k = \sqrt{\gamma_k}\sum_l a_{k,l}v_l$. So

$$\|\boldsymbol{\Gamma}(\boldsymbol{\omega})^{\frac{1}{2}}\mathbf{A}\mathbf{v}\|_2^2 = \sum_{k=1}^{K}\gamma(\omega_k)\Big|\sum_{l=1}^{K}a_{k,l}v_l\Big|^2 .$$

Now considering both the convexity of γ and the fact that all sum terms in the equation above are positive yields

$$\begin{aligned}\lambda_p(\boldsymbol{\omega}(\mu)) &= \max_{\|\mathbf{v}\|_2=1}\|\boldsymbol{\Gamma}(\boldsymbol{\omega}(\mu))^{\frac{1}{2}}\mathbf{A}\mathbf{v}\|_2^2 \\ &\le \max_{\|\mathbf{v}\|_2=1}\Big((1-\mu)\|\boldsymbol{\Gamma}(\hat{\boldsymbol{\omega}})^{\frac{1}{2}}\mathbf{A}\mathbf{v}\|_2^2 + \mu\|\boldsymbol{\Gamma}(\check{\boldsymbol{\omega}})^{\frac{1}{2}}\mathbf{A}\mathbf{v}\|_2^2\Big) \\ &\le (1-\mu)\max_{\|\mathbf{v}\|_2=1}\|\boldsymbol{\Gamma}(\hat{\boldsymbol{\omega}})^{\frac{1}{2}}\mathbf{A}\mathbf{v}\|_2^2 + \mu\max_{\|\mathbf{v}\|_2=1}\|\boldsymbol{\Gamma}(\check{\boldsymbol{\omega}})^{\frac{1}{2}}\mathbf{A}\mathbf{v}\|_2^2 \\ &= (1-\mu)\lambda_p(\hat{\boldsymbol{\omega}}) + \mu\lambda_p(\check{\boldsymbol{\omega}}), \quad \hat{\boldsymbol{\omega}},\check{\boldsymbol{\omega}} \in \mathrm{Q}^K\end{aligned}$$

for all $\mu \in (0,1)$, where the following identities

$$\lambda_p(\hat{\boldsymbol{\omega}}) = \max_{\|\mathbf{v}\|_2=1}\|\boldsymbol{\Gamma}(\hat{\boldsymbol{\omega}})^{\frac{1}{2}}\mathbf{A}\mathbf{v}\|_2^2 \qquad \lambda_p(\check{\boldsymbol{\omega}}) = \max_{\|\mathbf{v}\|_2=1}\|\boldsymbol{\Gamma}(\check{\boldsymbol{\omega}})^{\frac{1}{2}}\mathbf{A}\mathbf{v}\|_2^2$$

were used in the last step.

An immediate consequence of the theorem is the following.

Corollary 1.49. *If* $\mathbf{X} \in \mathrm{X}^p_{K,\boldsymbol{\Gamma}}(\Omega)$ *and* $\gamma : \mathrm{Q} \to \mathbb{R}_{++}$ *is any convex function, then* F *is a convex set.*

Interestingly, the feasibility set can be written as the intersection of certain (in general) nonconvex sets. To see this, note that, for any $\mathbf{X} \in \mathrm{X}^p_{K,\boldsymbol{\Gamma}}(\Omega)$ with $\Omega = \mathrm{Q}^K$, one has

$$\mathrm{F} = \{\boldsymbol{\omega} \in \mathrm{Q}^K : \lambda_{\max}\big(\boldsymbol{\Gamma}^{\frac{1}{2}}(\boldsymbol{\omega})\mathbf{V}\boldsymbol{\Gamma}^{\frac{1}{2}}(\boldsymbol{\omega})\big) \le 1\}.$$

Thus, since $\lambda_{\max}\big(\boldsymbol{\Gamma}^{\frac{1}{2}}(\boldsymbol{\omega})\mathbf{V}\boldsymbol{\Gamma}^{\frac{1}{2}}(\boldsymbol{\omega})\big) \le 1$ if and only if $\lambda_{\min}(\boldsymbol{\Gamma}^{-1}(\boldsymbol{\omega}) - \mathbf{V}) \ge 0$ or, equivalently, if and only if

$$0 \le \mathbf{z}^T\big(\boldsymbol{\Gamma}^{-1}(\boldsymbol{\omega}) - \mathbf{V}\big)\mathbf{z} \tag{1.59}$$

for all $\mathbf{z} \in \mathbb{R}^K$, we can write $\mathrm{F} = \bigcap_{\mathbf{z}\in\mathbb{R}^K}\mathrm{M}(\mathbf{z})$ where

$$\mathrm{M}(\mathbf{z}) := \{\boldsymbol{\omega} \in \mathrm{Q}^K : \mathbf{z}^T\big(\boldsymbol{\Gamma}^{-1}(\boldsymbol{\omega}) - \mathbf{V}\big)\mathbf{z} \ge 0\}.$$

Hence, given any symmetric positive semidefinite matrix $\mathbf{V} \in \mathrm{X}_K$, the feasibility set (associated with $\boldsymbol{\Gamma}(\boldsymbol{\omega})\mathbf{V}$) is equal to the intersection of the sets $\mathrm{M}(\mathbf{z})$ with respect to all $\mathbf{z} \in \mathbb{R}^K$. It is interesting to see that although $\mathrm{M}(\mathbf{z})$ is not convex in general, the intersection of these sets is a convex set, provided that γ is a convex function. This is illustrated in Fig. 1.1 for $\gamma(x) = x, x > 0$, in which case the complement of $\mathrm{M}(\mathbf{z})$ in Q^K denoted by $\mathrm{M}^c(\mathbf{z})$ is a convex set. This immediately follows from (1.59) whose right-hand side can be written

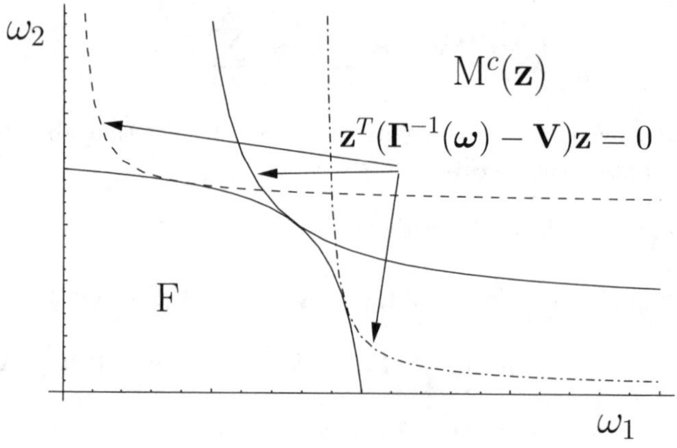

Fig. 1.1. The feasibility set F for some $\mathbf{X} \in \mathrm{X}^p_{K,\mathbf{\Gamma}}(\Omega)$ with $\gamma(x) = x, x > 0, K = 2$ and $\Omega = \mathrm{Q}^2$.

as

$$\mathbf{z}^T\mathbf{\Gamma}^{-1}(\boldsymbol{\omega})\mathbf{z} - \mathbf{z}^T\mathbf{V}\mathbf{z} = \sum_{k=1}^{K} \frac{|z_k|^2}{\gamma(\omega_k)} - \mathbf{z}^T\mathbf{V}\mathbf{z}.$$

Clearly, this function is convex if $\gamma(x) = x, x > 0$. Thus $\mathrm{M}^c(\mathbf{z})$ as the sublevel set of this function with respect to the zero value must be a convex set for any fixed $\mathbf{z} \in \mathbb{R}^K$. The linear case $\gamma(x) = x, x > 0$, is of great practical interest, and hence is separately considered in the next section.

1.5 The Perron Root Under the Linear Mapping

In this section, we further proceed with matrix-valued functions $\mathbf{X}$ of the form $\mathbf{X}(\boldsymbol{\omega}) = \mathbf{\Gamma}(\boldsymbol{\omega})\mathbf{V}$ where $\mathbf{\Gamma}(\boldsymbol{\omega}) = \mathrm{diag}(\gamma(\omega_1), \ldots, \gamma(\omega_K))$. However, in contrast to the previous analysis, it is assumed that $\mathrm{trace}(\mathbf{V}) = 0$. Formally, this is written as

$$\mathbf{X} \in \mathrm{X}^0_{K,\mathbf{\Gamma}}(\Omega) := \{\mathbf{\Gamma}(\boldsymbol{\omega})\mathbf{V}, \boldsymbol{\omega} \in \Omega : \mathbf{V} \in \mathrm{X}_K, \mathrm{trace}(\mathbf{V}) = 0\}.$$

So, in particular, note that $\mathbf{V}$ cannot be positive semidefinite.

Under this assumption, we consider the important special case of the linear function: $\gamma(x) = x, x > 0$. Therefore, in this section, $\Omega = \mathrm{Q}^K = \mathbb{R}^K_{++}$ and

$$\mathbf{\Gamma}(\boldsymbol{\omega}) = \mathrm{diag}(\omega_1, \ldots, \omega_K).$$

Obviously, the linear function is not log-convex so that Theorem 1.36 does not apply in this case. In fact, if $K = 2$ and $\mathbf{V} = \left(\begin{smallmatrix} 0 & \varrho \\ \varrho & 0 \end{smallmatrix}\right)$, the Perron root of $\mathbf{X}(\boldsymbol{\omega}) = \mathrm{diag}(\boldsymbol{\omega})\mathbf{V}$ can be easily found to be $\lambda_p(\boldsymbol{\omega}) = \varrho\sqrt{\omega_1\omega_2}$. Consequently, instead of being convex, $\lambda_p(\boldsymbol{\omega})$ turns out to be concave on $\mathbb{R}^2_{++}$ for all $\mathbf{X} \in \mathrm{X}^0_{2,\boldsymbol{\Gamma}}(\Omega)$. This observation might lead one to think that $\lambda_p(\boldsymbol{\omega})$ is concave in general, that is for all $\mathbf{X} \in \mathrm{X}^0_{K,\boldsymbol{\Gamma}}(\Omega)$ and all $K > 1$. Further results that support this conjecture have been proved in Sect. 1.2.1 where it is shown that the Perron root is concave on some subsets of irreducible matrices.

Observe that if $\lambda_p(\boldsymbol{\omega})$ was concave on $\mathbb{R}^K_{++}$, then not F but its complement in $\mathbb{R}^K_{++}$

$$\mathrm{F}^c = \mathrm{Q}^K \setminus \mathrm{F} = \mathbb{R}^K_{++} \setminus \mathrm{F} \tag{1.60}$$

would be a convex set. If true, this result would have an interesting consequence for optimal scheduling policies for wireless networks (see Sect. 5.4.3). In Sect. 1.5.2, however, we will disprove the conjecture by showing that if $\gamma(x) = x, x > 0$, there exists $K > 3$ and $\mathbf{X} \in \mathrm{X}_{K,\boldsymbol{\Gamma}}(\Omega)$ such that $\lambda_p(\boldsymbol{\omega})$ is not concave. First, though, we will prove two conditions on the feasibility of the parameter vector $\boldsymbol{\omega} \in \mathbb{R}^K_{++}$. These results provide insight into the mutual dependence between distinct entries of $\boldsymbol{\omega} \in \mathrm{F}$.

1.5.1 Some Bounds

We exploit the bound in (1.10) to prove a subset and a superset of F. The following theorem provides a sufficient condition for $\boldsymbol{\omega}$ to be a member of F.

Theorem 1.50. *If*

$$f(\boldsymbol{\omega}) = \sum_{k=1}^{K} \frac{\omega_k \rho(\mathbf{V})}{1 + \omega_k \rho(\mathbf{V})} \le 1 \tag{1.61}$$

then $\boldsymbol{\omega} \in \mathrm{F}$.

Proof. Let $\mathbf{Y}(\boldsymbol{\omega}) = (y_{k,l})$ be given by $y_{k,l} = \omega_k(1 - \delta_{k-l})$, where δ_l denotes the Kronecker delta. Note that $\mathbf{Y}(\boldsymbol{\omega})$ is irreducible for all $\boldsymbol{\omega} \in \mathbb{R}^K_{++}$. Let $\mathbf{p} \in \Pi_K$ be the right Perron eigenvector of $\mathbf{Y}(\boldsymbol{\omega})$. From $\|\mathbf{p}\|_1 = 1$ and

$$\rho(\mathbf{Y}(\boldsymbol{\omega}))p_k = \big(\mathbf{Y}(\boldsymbol{\omega})\mathbf{p}\big)_k = \omega_k \sum_{\substack{l=1 \\ l \ne k}}^{K} p_l = \omega_k \Big(\sum_{l=1}^{K} p_l - p_k\Big) = \omega_k(1 - p_k)$$

we have $p_k = \frac{\omega_k}{\rho(\mathbf{Y}(\boldsymbol{\omega})) + \omega_k}, 1 \le k \le K$. Hence,

$$\sum_{k=1}^{K} p_k = \sum_{k=1}^{K} \frac{\omega_k}{\rho(\mathbf{Y}(\boldsymbol{\omega})) + \omega_k} = 1. \tag{1.62}$$

So, by Corollary 1.5 and $\boldsymbol{\Gamma}(\boldsymbol{\omega})\mathbf{V} = \mathrm{diag}(\boldsymbol{\omega})\mathbf{V} = \mathbf{Y}(\boldsymbol{\omega}) \circ \mathbf{V}$,

$$\lambda_p(\boldsymbol{\omega}) = \rho(\mathbf{D}(\boldsymbol{\omega})\mathbf{V}) = \rho(\mathbf{Y}(\boldsymbol{\omega}) \circ \mathbf{V}) \le \rho\big(\mathbf{Y}(\boldsymbol{\omega})\big)\rho(\mathbf{V}).$$

Combining this with (1.62) yields

$$1 = \sum_{k=1}^{K} \frac{\omega_k}{\rho(\mathbf{Y}(\boldsymbol{\omega})) + \omega_k} \leq \sum_{k=1}^{K} \frac{\omega_k}{\frac{\lambda_p(\boldsymbol{\omega})}{\rho(\mathbf{V})} + \omega_k} = \sum_{k=1}^{K} \frac{\omega_k \rho(\mathbf{V})}{\lambda_p(\boldsymbol{\omega}) + \omega_k \rho(\mathbf{V})}.$$

Thus, if (1.61) holds, then

$$\sum_{k=1}^{K} \frac{\omega_k \rho(\mathbf{V})}{1 + \omega_k \rho(\mathbf{V})} \leq 1 \leq \sum_{k=1}^{K} \frac{\omega_k \rho(\mathbf{V})}{\lambda_p(\boldsymbol{\omega}) + \omega_k \rho(\mathbf{V})}$$

or, equivalently, $\lambda_p(\boldsymbol{\omega}) \leq 1$.

The function $f(\boldsymbol{\omega})$ in (1.61) defines a set $\mathrm{F}_{in} \subseteq \mathrm{F}$ given by

$$\mathrm{F}_{in} := \{\boldsymbol{\omega} \in \mathbb{R}_{++}^K : f(\boldsymbol{\omega}) \leq 1\}.$$

It may be easily verified that for any $\mathbf{V} \in \mathrm{X}_K$, $f(\boldsymbol{\omega})$ is strict concave. Consequently, $\mathrm{F}_{in}^c = \mathbb{R}_{++}^K \setminus \mathrm{F}_{in}$ is a convex set. Now we use (1.10) to prove a necessary condition for the feasibility of $\boldsymbol{\omega} \in \mathrm{Q}^K$.

Theorem 1.51. *If* $\boldsymbol{\omega} \in \mathrm{F}$, *then*

$$1 \leq g(\boldsymbol{\omega}) = \sum_{k=1}^{K} \frac{1}{1 + \rho(\mathbf{V})\omega_k}. \tag{1.63}$$

Proof. Let $\mathbf{Y}(\boldsymbol{\omega})$ be as in the proof above, and let $1/\boldsymbol{\omega}$ be defined as $1/\boldsymbol{\omega} = (1/\omega_1, \dots, 1/\omega_K) > 0$. Since $\mathbf{\Gamma}^{-1}(\boldsymbol{\omega})\mathbf{V} = (\mathrm{diag}(\boldsymbol{\omega}))^{-1}\mathbf{V} = \mathbf{Y}(1/\boldsymbol{\omega}) \circ \mathbf{V}$,

$$\rho(\mathbf{V}) = \rho\big(\mathbf{Y}(1/\boldsymbol{\omega}) \circ \mathbf{V}\,\mathrm{diag}(\boldsymbol{\omega})\big).$$

Thus, by Corollary 1.5 and the fact that $\rho(\mathbf{V}\mathrm{diag}(\boldsymbol{\omega})) = \rho(\mathrm{diag}(\boldsymbol{\omega})\mathbf{V}) = \lambda_p(\boldsymbol{\omega})$, $\log \rho(\mathbf{V}) \leq \log \rho(\mathbf{Y}(1/\boldsymbol{\omega})) + \log \lambda_p(\boldsymbol{\omega})$. Since $\boldsymbol{\omega} \in \mathrm{F}$ or, equivalently, $\lambda_p(\boldsymbol{\omega}) \leq 1$, this implies that

$$\rho(\mathbf{V}) \leq \rho(\mathbf{Y}(1/\boldsymbol{\omega})). \tag{1.64}$$

Now note that $(\mathbf{Y}(1/\boldsymbol{\omega}))_{k,l} = \frac{1}{\omega_k}(1 - \delta_{k-l})$. Hence, proceeding essentially as in the foregoing proof yields

$$\hat{p}_k = \frac{1}{1 + \omega_k \rho(\mathbf{Y}(1/\boldsymbol{\omega}))}$$

where $\hat{\mathbf{p}} \in \Pi_K$ is the right Perron eigenvector of $\mathbf{Y}(1/\boldsymbol{\omega})$. Thus,

$$1 = \sum_{k=1}^{K} \hat{p}_k = \sum_{k=1}^{K} \frac{1}{1 + \omega_k \rho(\mathbf{Y}(1/\boldsymbol{\omega}))}$$

and the theorem follows from (1.64).

The function $g(\boldsymbol{\omega})$ in (1.63) defines a superset F_{out} of F given by

$$\mathrm{F}_{out} := \{\boldsymbol{\omega} \in \mathbb{R}^K_{++} : 1 \leq g(\boldsymbol{\omega})\}.$$

For any fixed $\mathbf{V} \in \mathrm{X}_K$, the function $g(\boldsymbol{\omega})$ in (1.63) is strictly convex so that $\mathrm{F}^c_{out} = \mathbb{R}^K_{++} \setminus \mathrm{F}_{out}$ is a convex set.

Summarizing, we can state that $\mathrm{F}_{in} \subseteq \mathrm{F} \subseteq \mathrm{F}_{out}$ and

$$\mathrm{F}^c_{out} \subseteq \mathrm{F}^c \subseteq \mathrm{F}^c_{in}$$

where both F^c_{in} and F^c_{out} are convex sets. Thus, F^c is embedded into two convex sets. Furthermore, if $K = 2$, the implicit function $g(\boldsymbol{\omega}) = 1$ is given by $\rho(\mathbf{V}) = 1/\sqrt{\omega_1\omega_2}$. Consequently, in the two-dimensional case, $\mathrm{F}^c = \mathrm{F}^c_{out}$ is a convex set.

Finally we use Theorem 1.2 as a starting point to prove a necessary condition on $\boldsymbol{\omega} \in \mathrm{F}$. To this end, let $\hat{\mathbf{A}} \in \mathrm{S}_K(\mathbf{V})$ be a stochastic matrix so that

$$\log \rho(\mathbf{V}) = \sum_{k,l=1}^{K} \hat{u}_k \hat{a}_{k,l} \log\left(\frac{v_{k,l}}{\hat{a}_{k,l}}\right) \tag{1.65}$$

where $\hat{\mathbf{u}} = (\hat{u}_1, \ldots, \hat{u}_K) \in \Pi_K$ is the left Perron eigenvector of $\hat{\mathbf{A}}$. First we consider the following lemma.

Lemma 1.52. *Suppose that* $\mathbf{V}$ *is irreducible. Let* $\hat{\mathbf{A}}$ *and* $\hat{\mathbf{u}}$ *be as above. Then, we have* $\hat{u}_k = y_k \cdot x_k$, $1 \leq k \leq K$, *where* $\mathbf{y}$ *and* $\mathbf{x}$ *with* $\mathbf{y}^T\mathbf{x} = 1$ *are left and right positive eigenvectors of* $\mathbf{V}$, *respectively.*

Proof. By Theorem 1.2, $\hat{\mathbf{A}}$ is

$$(\hat{\mathbf{A}})_{k,l} = \hat{a}_{k,l} = \frac{v_{k,l}x_l}{\rho(\mathbf{V})\, x_k}.$$

By definition, we have $\hat{\mathbf{A}}^T \hat{\mathbf{u}} = \hat{\mathbf{u}}$ and $\mathbf{x} > 0$. Combining this with the equation above yields

$$\hat{u}_k = \sum_{l=1}^{K} \hat{a}_{l,k}\hat{u}_l = \frac{1}{\rho(\mathbf{V})} \sum_{l=1}^{K} \frac{v_{l,k}x_k}{x_l}\hat{u}_l, \quad 1 \leq k \leq K$$

or, equivalently,

$$\rho(\mathbf{V}) \cdot \frac{\hat{u}_k}{x_k} = \sum_{k=1}^{K} v_{l,k}\frac{\hat{u}_l}{x_l}, \quad 1 \leq k \leq K.$$

Hence, the left eigenvectors are of the form $y_k = \frac{\hat{u}_k}{x_k}$. Since $\sum_k \hat{u}_k = 1$, we have $\mathbf{y}^T\mathbf{x} = 1$.

Now we are in a position to prove the announced necessary condition. To keep the result as general as possible, we allow $\gamma : \mathrm{Q} \to \mathbb{R}^K_{++}$ to be any continuous function.

Theorem 1.53. *Let $\mathbf{V} \in \mathrm{X}_K$. If $\boldsymbol{\omega} \in \mathrm{F}$, then we must have*

$$\prod_{k=1}^{K} \big(\gamma(\omega_k)\big)^{x_k y_k} \leq \frac{1}{\rho(\mathbf{V})} \tag{1.66}$$

where, as in Lemma 1.52, $\mathbf{y}$ and $\mathbf{x}$ are left and right positive eigenvectors of $\mathbf{V}$, respectively.

Proof. Let $\hat{\mathbf{A}}$ and $\hat{\mathbf{u}}$ be defined by (1.65). Substituting $\hat{\mathbf{u}}$ and $\hat{\mathbf{A}}$ into (1.4) yields

$$\log \lambda_p(\boldsymbol{\omega}) = \log \rho(\boldsymbol{\Gamma}(\boldsymbol{\omega})\mathbf{V}) \geq \sum_{k,l=1}^{K} \hat{u}_k \hat{a}_{k,l} \log \frac{v_{k,l}}{\hat{a}_{k,l}} + \sum_{k,l=1}^{K} \hat{u}_k \hat{a}_{k,l} \log \gamma(\omega_k)\,.$$

By (1.65), the first term on the right-hand side is equal to $\log \rho(\mathbf{V})$ so that

$$\begin{aligned} \log \lambda_p(\boldsymbol{\omega}) \geq \log \rho(\mathbf{V}) + \sum_{k,l=1}^{K} \hat{u}_k \hat{a}_{k,l} \log \gamma(\omega_k) &= \log \rho(\mathbf{V}) + \sum_{k=1}^{K} \hat{u}_k \log \gamma(\omega_k) \\ &= \log \rho(\mathbf{V}) + \log \prod_{k=1}^{K} \big(\gamma(\omega_k)\big)^{\hat{u}_k} \end{aligned}$$

where we used the fact that $\hat{\mathbf{A}}$ is (row) stochastic. Hence, by Lemma 1.52,

$$\prod_{k=1}^{K} (\gamma(\omega_k))^{x_k y_k} \leq \frac{\lambda_p(\boldsymbol{\omega})}{\rho(\mathbf{V})}.$$

But, if $\boldsymbol{\omega} \in \mathrm{F}$, then $\lambda_p(\boldsymbol{\omega}) \leq 1$, and the theorem follows.

Obviously, if $\gamma(x) = x, x > 0$, the bound reduces to $\prod_{k=1}^{K} (\omega_k)^{x_k y_k} \leq \frac{1}{\rho(\mathbf{V})}$.

1.5.2 Disproof of the Conjecture

Now we disprove the conjecture stated at the beginning of this section. More precisely, it is shown that there exists $\mathbf{X} \in \mathrm{X}^0_{K,\boldsymbol{\Gamma}}(\Omega)$ with $\Omega = \mathbb{R}^K_{++}, K > 1$, and $\gamma(x) = x, x > 0$, such that $\boldsymbol{\omega}(\mu) = (1-\mu)\hat{\boldsymbol{\omega}} + \mu\check{\boldsymbol{\omega}} \notin \mathrm{F}^c$ for some $\mu \in (0,1)$ and $\hat{\boldsymbol{\omega}}, \check{\boldsymbol{\omega}} \in \mathrm{F}^c$ with $\hat{\boldsymbol{\omega}} \neq \check{\boldsymbol{\omega}}$.

First suppose that the conjecture is true, that is, F^c_γ is a convex set. This is equivalent to saying that

$$\lambda_p(\boldsymbol{\omega}(\mu)) \geq 1 \tag{1.67}$$

for all $\mu \in (0,1)$ and all $\hat{\boldsymbol{\omega}}, \check{\boldsymbol{\omega}} \in \mathbb{R}^K_{++}$ with

$$\lambda_p(\hat{\boldsymbol{\omega}}) = \lambda_p(\check{\boldsymbol{\omega}}) = 1\,. \tag{1.68}$$

In words, if both $\hat{\boldsymbol{\omega}}$ and $\check{\boldsymbol{\omega}}$ lie on the boundary $\partial\mathrm{F}$ of F (Definition 1.41 and the remark below), then the entire straight line connecting them must be either outside of the feasibility set or must entirely lie on $\partial\mathrm{F}$.

First we provide a necessary and sufficient condition for (1.67) with (1.68) to be satisfied. Note that in the following lemma, $\mathbf{X}$ is not necessarily a member of $\mathrm{X}^0_{K,\mathbf{r}}(\Omega)$.

Lemma 1.54. *Let $\gamma(x) = x, x > 0$, and let $\mathbf{X} \in \mathrm{X}_{K,\mathbf{r}}(\Omega)$ be arbitrary. Then, we have (1.67) with (1.68) if and only if $\lambda_p(\boldsymbol{\omega})$ is concave on $\Omega = \mathbb{R}^K_{++}$, i.e., if and only if*

$$\lambda_p(\boldsymbol{\omega}(\mu)) \geq (1-\mu)\lambda_p(\hat{\boldsymbol{\omega}}) + \mu\lambda_p(\check{\boldsymbol{\omega}})$$

for all $\mu \in (0,1)$ and $\hat{\boldsymbol{\omega}}, \check{\boldsymbol{\omega}} \in \mathbb{R}^K_{++}$.

Proof. If the spectral radius is concave, then (1.67) with (1.68) immediately follows. So, we only need to prove the converse. To this end, let $\hat{\boldsymbol{\omega}}, \check{\boldsymbol{\omega}} \in \mathbb{R}^K_{++}$ be arbitrary, and let $\hat{\mathbf{s}}, \check{\mathbf{s}} \in \mathbb{R}^K_{++}$ be defined as

$$\mathrm{diag}(\hat{\mathbf{s}})\mathbf{V} = \frac{1}{\lambda_p(\hat{\boldsymbol{\omega}})}\mathrm{diag}(\hat{\boldsymbol{\omega}})\mathbf{V} \qquad \mathrm{diag}(\check{\mathbf{s}})\mathbf{V} = \frac{1}{\lambda_p(\check{\boldsymbol{\omega}})}\mathrm{diag}(\check{\boldsymbol{\omega}})\mathbf{V}\,.$$

Consequently, both $\lambda_p(\hat{\mathbf{s}}) = \rho(\mathrm{diag}(\hat{\mathbf{s}})\mathbf{V})$ and $\lambda_p(\check{\mathbf{s}}) = \rho(\mathrm{diag}(\check{\mathbf{s}})\mathbf{V})$ satisfy (1.68). Let $a, b > 0$ be chosen such that $0 < 1-\mu = \frac{a}{a+b} < 1$ and $0 < \mu = \frac{b}{a+b} < 1$ for $\mu \in (0,1)$. Substituting this into the left-hand side of (1.67) yields

$$\lambda_p(a\hat{\mathbf{s}} + b\check{\mathbf{s}}) \geq a + b \tag{1.69}$$

where we used the fact that $\mathrm{diag}(a\hat{\mathbf{s}} + b\check{\mathbf{s}})\mathbf{V} = a\,\mathrm{diag}(\hat{\mathbf{s}})\mathbf{V} + b\,\mathrm{diag}(\check{\mathbf{s}})\mathbf{V}$ and $\lambda_p(c\mathbf{x}) = c\lambda_p(\mathbf{x})$ for any $c > 0$ and $\mathbf{x} \in \mathbb{R}^K_{++}$. The inequality above holds for *all* $a, b > 0$. Now define $\tilde{a}, \tilde{b} > 0$ as $a = \tilde{a}\lambda_p(\hat{\boldsymbol{\omega}})$ and $b = \tilde{b}\lambda_p(\check{\boldsymbol{\omega}})$. Combining this with (1.69) and $\lambda_p(\boldsymbol{\omega}) = \rho(\mathrm{diag}(\boldsymbol{\omega})\mathbf{V})$ gives

$$\begin{aligned}\lambda_p\big(\tilde{a}\lambda_p(\hat{\boldsymbol{\omega}})\hat{\mathbf{s}} + \tilde{b}\lambda_p(\check{\boldsymbol{\omega}})\check{\mathbf{s}}\big) &= \rho\Big(\tilde{a}\lambda_p(\hat{\boldsymbol{\omega}})\mathrm{diag}(\hat{\mathbf{s}})\mathbf{V} + \tilde{b}\lambda_p(\check{\boldsymbol{\omega}})\mathrm{diag}(\check{\mathbf{s}})\mathbf{V}\Big) \\ &= \rho\Big(\tilde{a}\,\mathrm{diag}(\hat{\boldsymbol{\omega}})\mathbf{V} + \tilde{b}\,\mathrm{diag}(\check{\boldsymbol{\omega}})\mathbf{V}\Big) \\ &= \lambda_p\big(\tilde{a}\,\hat{\boldsymbol{\omega}} + \tilde{b}\,\check{\boldsymbol{\omega}}\big) \geq \tilde{a}\lambda_p\big(\hat{\boldsymbol{\omega}}\big) + \tilde{b}\lambda_p\big(\check{\boldsymbol{\omega}}\big)\end{aligned}$$

for all $\tilde{a}, \tilde{b} > 0$. In particular, this must hold for $\tilde{a} = 1 - \alpha$ and $\tilde{b} = \alpha$ with $\alpha \in (0,1)$, in which case concavity of the spectral radius follows.

It is worth pointing out that if

$$\lambda_p(\boldsymbol{\omega}(\mu)) \geq \min\{\lambda_p(\hat{\boldsymbol{\omega}}), \lambda_p(\check{\boldsymbol{\omega}})\}$$

for all $\mu \in (0,1)$ and all $\hat{\boldsymbol{\omega}}, \check{\boldsymbol{\omega}} \in \mathbb{R}^K_{++}$, then we have (1.67) with (1.68). Thus, by Lemma 1.54 and the fact that every concave function is quasiconcave (for the definition of quasiconcave functions, the reader is referred to [11]), we obtain the following corollary.

Corollary 1.55. *Let $\gamma(x) = x, x > 0$. Then, the following statements are equivalent.*

(a) $\mathrm{F}^c = \mathbb{R}^K_{++} \setminus \mathrm{F}$ is a convex set for all $\mathbf{X} \in \mathcal{X}_{K,\mathbf{\Gamma}}(\Omega)$ and all $K > 1$.
(b) The Perron root is concave on $\mathbb{R}^K_{++}$.
(c) The Perron root is quasiconcave on $\mathbb{R}^K_{++}$.

Now we prove that if $\gamma(x) = x, x > 0$, then the Perron root is not concave *in general*, thereby disproving the conjecture.

Lemma 1.56. *Let $\gamma(x) = x, x > 0$. Then, there exist $K > 1, \mathbf{X} \in \mathcal{X}^0_{K,\mathbf{\Gamma}}(\Omega)$ and $\hat{\boldsymbol{\omega}}, \check{\boldsymbol{\omega}} \in \Omega = \mathbb{R}^K_{++}$ such that*

$$\lambda_p(\boldsymbol{\omega}(\mu)) < (1-\mu)\lambda_p(\hat{\boldsymbol{\omega}}) + \mu\lambda_p(\check{\boldsymbol{\omega}}) \tag{1.70}$$

for some $\mu \in (0,1)$.

Proof. For an arbitrary $c > 0$, let $\mathbf{V}_c = \left(\begin{smallmatrix} 0 & c \\ c & 0 \end{smallmatrix}\right) \geq 0$. Furthermore, define

$$\mathbf{V} = \begin{pmatrix} \mathbf{V}_c & \mathbf{0} \\ \tilde{\mathbf{V}} & \mathbf{V}_c \end{pmatrix} \in \mathbb{R}^{4\times 4}_+$$

where $\tilde{\mathbf{V}} \in \mathbb{R}^{2\times 2}_+$ is an arbitrary nonnegative matrix and $\mathbf{0} \in \mathbb{R}^{2\times 2}_+$ denotes the zero matrix. Furthermore let $\hat{\boldsymbol{\omega}} = (2,2,1,1)$ and $\check{\boldsymbol{\omega}} = (1,1,2,2)$ from which we get

$$\begin{aligned} \lambda_p(\hat{\boldsymbol{\omega}}) &= \rho(\mathrm{diag}(\hat{\boldsymbol{\omega}})\mathbf{V}) = \max\{2\rho(\mathbf{V}_c), \rho(\mathbf{V}_c)\} = 2\,c \\ \lambda_p(\check{\boldsymbol{\omega}}) &= \rho(\mathrm{diag}(\check{\boldsymbol{\omega}})\mathbf{V}) = \max\{\rho(\mathbf{V}_c), 2\rho(\mathbf{V}_c)\} = 2\,c\,. \end{aligned}$$

Thus, $\lambda_p(\boldsymbol{\omega}(\mu)) = \rho(\mathrm{diag}(\boldsymbol{\omega}(\mu))\mathbf{V})$ yields

$$\begin{aligned} \lambda_p(\boldsymbol{\omega}(\mu)) &= \max\{(2(1-\mu)+\mu)\rho(\mathbf{V}_c), ((1-\mu)+2\mu)\rho(\mathbf{V}_c)\} \\ &= c\max\{2-\mu, 1+\mu\} \\ &< 2\,c = (1-\mu)\lambda_p(\hat{\boldsymbol{\omega}}) + \mu\lambda_p(\check{\boldsymbol{\omega}}), \quad \mu \in (0,1)\,. \end{aligned} \tag{1.71}$$

Note that since $\mathbf{V}$ defined above is reducible, so also is $\mathrm{diag}(\boldsymbol{\omega})\mathbf{V}$. Therefore, it remains to show that there exists $\mathbf{X} \in \mathcal{X}^0_{4,\mathbf{\Gamma}}(\mathbb{R}^4_{++})$ for which (1.70) holds. To this end, suppose that $\mathbf{\Delta} \in \mathbb{R}^{4\times 4}_+$ is given by $(\mathbf{\Delta})_{k,l} = 1 - \delta_{k-l}$, where δ_i is the Kronecker delta. Let $\mathbf{X}_\epsilon(\boldsymbol{\omega}) = \mathrm{diag}(\boldsymbol{\omega})\mathbf{V}_\epsilon$ where $\mathbf{V}_\epsilon = \mathbf{V} + \epsilon\mathbf{\Delta}, \epsilon > 0$. Obviously, $\mathbf{X}_\epsilon(\boldsymbol{\omega})$ with $\mathrm{trace}(\mathbf{X}_\epsilon(\boldsymbol{\omega})) = 0$ is irreducible for all $\boldsymbol{\omega} > 0$ and $\epsilon > 0$. Let $\lambda_p(\epsilon, \boldsymbol{\omega})$ be the Perron root of $\mathrm{diag}(\boldsymbol{\omega})\mathbf{V}_\epsilon$ and note that

$$\lim_{\epsilon\to 0} \lambda_p(\epsilon, \boldsymbol{\omega}(\mu)) = \lambda_p(\boldsymbol{\omega}(\mu))$$

for any fixed $\mu \in (0,1)$. Thus, since the Perron root is continuous in $\epsilon > 0$, it follows from (1.71) that there exist $\epsilon > 0$ (and hence an irreducible matrix $\mathbf{V}_\epsilon$) and $\mu \in (0,1)$ such that

$$\lambda_p(\epsilon, \boldsymbol{\omega}(\mu)) < (1-\mu)\lambda_p(\epsilon, \hat{\boldsymbol{\omega}}) + \mu\lambda_p(\epsilon, \check{\boldsymbol{\omega}}) .$$

This completes the proof.

Note that the construction of a counterexample in the proof of Lemma 1.56 requires two traceless irreducible matrices of order at least 2. Thus the proof does not work for $K < 4$. Also note that since $\rho(\mathbf{X}) = \lim_{n\to+\infty} \|\mathbf{X}^n\|^{1/n}$ for any matrix norm (Theorem A.10) and $(\mathbf{X}+\mathbf{Y})^n \geq \mathbf{X}^n + \mathbf{Y}^n$ with $n \geq 1$ for any $\mathbf{X}, \mathbf{Y} \in \mathrm{N}_K$, we actually have $\rho(\mathbf{X}+\mathbf{Y}) \geq \max\{\rho(\mathbf{X}), \rho(\mathbf{Y})\}$. In a special case when $\mathbf{X} = (1-\mu)\mathbf{A}$ and $\mathbf{Y} = \mu\mathbf{B}$ for some $\mathbf{A}, \mathbf{B} \in \mathrm{X}_K$, one obtains

$$\rho((1-\mu)\mathbf{A} + \mu\mathbf{B}) \geq \max\{(1-\mu)\rho(\mathbf{A}), \mu\rho(\mathbf{B})\} .$$

The counterexample in the proof is constructed in such a way that the lower bound is attained for all $\mu \in [0,1]$.

We complete this section by summarizing Lemma 1.54 and Lemma 1.56 in a theorem.

Theorem 1.57. *Suppose that $\gamma(x) = x, x > 0$. Then, F^c is not a convex set in general, i.e., there exist $K > 1, \mathbf{X} \in \mathrm{X}^0_{K,\mathbf{\Gamma}}(\Omega)$ and $\hat{\boldsymbol{\omega}}, \check{\boldsymbol{\omega}} \in \mathrm{F}^c$ such that*

$$\boldsymbol{\omega}(\mu) = (1-\mu)\hat{\boldsymbol{\omega}} + \mu\check{\boldsymbol{\omega}} \notin \mathrm{F}^c$$

for some $\mu \in (0,1)$.

1.6 Some Remarks on Arbitrary Nonnegative Matrices

We finish this chapter by making some remarks on reducible matrices. The weak form of the Perron–Frobenius theorem (Theorem A.30) ensures that the spectral radius of any nonnegative matrix is an eigenvalue of the matrix and that associated eigenvectors are nonnegative. So, in contrast to irreducible matrices, there is no assertion regarding the uniqueness and positivity properties. For this reason, it is not clear which of the presented results carry over to nonnegative matrices. On the other hand, the proof of Theorem A.30 is based on the recognition that any nonnegative matrix can always be written as a limit of a sequence of positive matrices (and hence irreducible ones). Therefore, due to continuity of the spectral radius as a function of the matrix entries, it is justified to conjecture that some of the results remain valid (maybe in a milder form) in case of reducible matrices.

First we show that the main results of Sects. 1.3–1.5 hold for general nonnegative matrices. In fact, this extension is straightforward. In contrast, the problem of extending the results of Sects. 1.2.3 and 1.2.4 to nonnegative matrices is somewhat tricky. This problem is considered in Sect. 1.6.2.

1.6.1 Log-Convexity of the Spectral Radius

Suppose that $\mathbf{X} \geq 0$ is reducible. Then, by Definition A.21 and the fact that interchanging columns and rows of any square matrix does not affect the spectral radius of the matrix, we can assume that $\mathbf{X}$ is of the following form [4] (see also Appendix A.4.3)

$$\mathbf{X} = \begin{pmatrix} \mathbf{X}^{(1)} & \mathbf{0} & \cdots & \mathbf{0} \\ \mathbf{X}^{(2,1)} & \mathbf{X}^{(2)} & \dots & \mathbf{0} \\ \dots & \dots & \dots & \dots \\ \mathbf{X}^{(s,1)} & \mathbf{X}^{(s,2)} & \dots & \mathbf{X}^{(s)} \end{pmatrix} \tag{1.72}$$

where $\mathbf{X}^{(1)}, \dots, \mathbf{X}^{(s)} \geq 0$ are either nonnegative irreducible or zero square matrices. Without loss of generality, let us assume that all diagonal blocks are irreducible. Obviously, if $\mathbf{X}$ is irreducible, then $s = 1$ and $\mathbf{X} = \mathbf{X}^{(1)}$. We refer to (1.72) as the normal form of the matrix $\mathbf{X} \in \mathrm{N}_K$.

An examination of (1.72) reveals that the spectral radius of $\mathbf{X}$ can be written as [4]

$$\rho(\mathbf{X}) = \max\{\rho(\mathbf{X}^{(n)}) : 1 \leq n \leq s\} \tag{1.73}$$

where $\rho(\mathbf{X}^{(n)})$ is used to denote the Perron root of the nth block $\mathbf{X}^{(n)}$. By Theorem A.30, $\rho(\mathbf{X})$ is an eigenvalue of $\mathbf{X}$ but not necessarily a simple one (Definition A.5), which immediately follows from (1.73). Obviously, $\rho(\mathbf{X}^{(n)})$ depends only on the entries of $\mathbf{X}^{(n)}$, which in turn implies that, for any matrix-valued function $\mathbf{X} : \Omega \to \mathrm{N}_K$, $\rho(\mathbf{X}^{(n)}(\boldsymbol{\omega}))$ depends on the parameter vector $\boldsymbol{\omega}$ only through the entries of the nth diagonal block. As a consequence, we can apply Theorem 1.36 to deduce that the Perron root of each diagonal block is log-convex on Ω if $\mathbf{X} \in \mathrm{LC}_K(\Omega) \subset \mathrm{N}_K(\Omega)$. Now since log-convexity is closed under pointwise maximum [11] (see also Sect. 2.3.1), we can make the following two observations.

Observation 1.58. *Let* $\mathbf{X} \in \mathrm{LC}_K(\Omega) \subset \mathrm{N}_K(\Omega)$ *be arbitrary. Then,* $\rho(\mathbf{X}(\boldsymbol{\omega}))$ *is log-convex on* Ω*.*

Observation 1.59. *Let* $\mathbf{X} \in \mathrm{LC}_K(\Omega) \subset \mathrm{N}_K(\Omega)$ *be arbitrary. Then,*

$$\mathrm{F} = \{\boldsymbol{\omega} \in \Omega : \rho(\mathbf{X}(\boldsymbol{\omega})) \leq 1\}$$

is a convex set.

We also see that the irreducibility property is not necessary for the results of Sect. 1.4 to hold. The converse results of Sect. 1.3.4 extend to arbitrary nonnegative matrices as well. The same holds for the results of Sect. 1.5.2.

1.6.2 Characterization of the Spectral Radius

This section aims at extending the results of Sects. 1.2.3 and 1.2.4 to reducible matrices. We will prove that the characterizations of the Perron Root, and in particular the Collatz–Wielandt-type saddle point characterization, remain valid for some subclass of nonnegative matrices that is larger than X_K. However, it will also be apparent that such characterizations are not applicable to general nonnegative matrices.

It was mentioned above that in the case of nonnegative reducible matrices, $\rho(\mathbf{X}) \in \sigma(\mathbf{X})$ does not need to be a simple eigenvalue. Moreover, associated left and right eigenvectors are not necessarily unique up to positive multiples. This prompts us to introduce the notion of the eigenmanifold E_K of $\mathbf{X}$ defined to be

$$\mathrm{E}_K(\mathbf{X}) := \{(\mathbf{q}, \mathbf{p}) \in \mathbb{R}_+^K \times \mathbb{R}_+^K : \mathbf{X}\mathbf{p} = \rho(\mathbf{X})\mathbf{p}, \mathbf{X}^T\mathbf{q} = \rho(\mathbf{X})\mathbf{q}, \mathbf{q}^T\mathbf{p} = 1\}.$$

Furthermore, we define $\mathrm{E}_K^+(\mathbf{X}) := \mathrm{E}_K(\mathbf{X}) \cap (\mathbb{R}_{++}^K \times \mathbb{R}_{++}^K)$, which of course may be an empty set if $\mathbf{X}$ is reducible. Note that the notation $(\mathbf{q}, \mathbf{p}) \in \mathrm{E}_K(\mathbf{X})$ means that $\mathbf{q}$ and $\mathbf{p}$ are nonnegative left and right eigenvectors of $\mathbf{X}$ associated with the same eigenvalue, which is equal to $\rho(\mathbf{X})$.

In what follows, given $\mathbf{X} \in \mathrm{N}_K$, the function $F : \mathbb{R}_{++} \to \mathbb{R}$ is assumed to belong to the function class $\mathcal{G}(\mathbf{X})$ specified in Definition 1.22. For the function $H : \mathbb{R}_{++}^K \to \mathbb{R}$ defined by (1.30) to be well defined, the matrix $\mathbf{X}$ is assumed to be confined to the set N_K^+ defined to be

$$\mathrm{N}_K^+ := \{\mathbf{X} \in \mathrm{N}_K : \exists_{\mathbf{s} \in \mathbb{R}_{++}^K} \ \mathbf{X}\mathbf{s} > 0\}.$$

Since $\mathbf{X}$ is nonnegative, we see that if $\mathbf{X}\mathbf{s} > 0$ holds for some arbitrary $\mathbf{s} \in \mathbb{R}_{++}^K$, then we must have $\mathbf{X}\mathbf{s} > 0$ for all $\mathbf{s} \in \mathbb{R}_{++}^K$. Obviously, for any $K > 1$, there holds $\mathrm{X}_K \subseteq \mathrm{N}_K^+ \subseteq \mathrm{N}_K$. The set N_K^+ is however a proper subset of N_K for any $K > 1$ ($\mathrm{N}_K^+ \subset \mathrm{N}_K$) since any nonnegative matrix with only one positive entry is not a member of N_K^+. On the other hand, for all $K > 1$, N_K^+ is a proper superset of X_K ($\mathrm{X}_K \subset \mathrm{N}_K^+$). This is because for all $K > 1$, there exists a nonnegative (row) stochastic matrix having one column with all entries being equal to zero. Clearly, such matrices are not irreducible but they belong to N_K^+. Finally, we point out that since $\mathbf{X} \in \mathrm{N}_K^+$ is not necessarily irreducible, the second condition of Definition 1.22 must be modified to read as follows: For any $\mathbf{z} \in \Pi_K^+$, the function $H : \mathbb{R}_{++}^K \to \mathbb{R}$ defined by (1.30) has a finite infimum on $\mathbb{R}_{++}^K$. If the infimum is attained for some $\mathbf{s}^* \in \mathbb{R}_{++}^K$, then every local minimum is global, and (1.31) provides a necessary and sufficient condition for characterizing $\mathbf{s}^*$. So the modification takes into account the possibility that H may have no minimum on $\mathbb{R}_{++}^K$.

First we are going to extend Theorem 1.24. As there is no guarantee that $\mathbf{X}$ has a *positive* eigenvector, it is clear that the theorem cannot hold in this form for general nonnegative matrices. The problem is that the equality in (1.32) cannot hold with $\mathbf{s} = \mathbf{p}$ if $\mathbf{p}$ has at least one zero element. On the other hand, however, we see from the proof of Theorem A.30 that $\mathbf{X}$ can be written as a limit of positive matrices, each of which has positive left and right eigenvectors. So one may expect that the bound in (1.32) holds and can be approached arbitrarily closely. Below this intuitive approach is made precise and rigorous.

Theorem 1.60. *Let $\mathbf{X} \in \mathrm{N}_K^+$ and $F \in \mathcal{G}(\mathbf{X})$ be arbitrary, and let $\mathbf{w} = \mathbf{p} \circ \mathbf{q} \in \Pi_K$ for some $(\mathbf{q}, \mathbf{p}) \in \mathrm{E}_K(\mathbf{X})$. Then,*

$$\inf_{\mathbf{s} \in \mathbb{R}_{++}^K} \sum_{k=1}^{K} w_k F\Big(\frac{(\mathbf{X}\mathbf{s})_k}{s_k}\Big) = F(\rho(\mathbf{X})). \tag{1.74}$$

Moreover, if $\mathbf{X}$ has a positive right eigenvector $\mathbf{p}$, the infimum is attained for $\mathbf{s} = \mathbf{p} > 0$.

Proof. Let $\mathbf{X} \in \mathrm{N}_K^+$ be arbitrary, and let $\mathbf{X}_\epsilon = \mathbf{X} + \epsilon \mathbf{1}\mathbf{1}^T$ for all $\epsilon \geq 0$. Hence, $\mathbf{X}_\epsilon \in \mathrm{X}_K$ for all $\epsilon > 0$, and

$$\frac{(\mathbf{X}_\epsilon \mathbf{s})_k}{s_k} = \frac{(\mathbf{X}\mathbf{s})_k}{s_k} + \epsilon \frac{\|\mathbf{s}\|_1}{s_k}, \quad \mathbf{s} \in \mathbb{R}_{++}^K, \quad 1 \leq k \leq K\,. \tag{1.75}$$

Now let $\{\mathbf{w}(\epsilon_n)\}_{n \in \mathbb{N}}$ be any sequence in Π_K^+ with $\lim_{n\to\infty} \epsilon_n = 0$ such that

$$\lim_{n\to\infty} \|\mathbf{w}(\epsilon_n) - \mathbf{w}\|_2 = 0\,. \tag{1.76}$$

By strict monotonicity and continuity of F as well as by continuity of the spectral radius as a function of matrix elements (Theorem A.6), it follows from Theorem 1.24 that

$$\begin{aligned} F(\rho(\mathbf{X})) &= \lim_{n\to\infty} F(\rho(\mathbf{X}_{\epsilon_n})) = \lim_{n\to\infty} \min_{\mathbf{s} \in \mathbb{R}_{++}^K} \sum_{k=1}^{K} w_k(\epsilon_n) F\Big(\frac{(\mathbf{X}_{\epsilon_n}\mathbf{s})_k}{s_k}\Big) \\ &\geq \lim_{n\to\infty} \inf_{\mathbf{s} \in \mathbb{R}_{++}^K} \sum_{k=1}^{K} w_k(\epsilon_n) F\Big(\frac{(\mathbf{X}\mathbf{s})_k}{s_k}\Big) \\ &= \inf_{\mathbf{s} \in \mathbb{R}_{++}^K} \sum_{k=1}^{K} w_k F\Big(\frac{(\mathbf{X}\mathbf{s})_k}{s_k}\Big)\,. \end{aligned} \tag{1.77}$$

On the other hand, however,

$$
\begin{aligned}
F(\rho(\mathbf{X})) &= \lim_{n\to\infty} \min_{\mathbf{s}\in\mathbb{R}^K_{++}} \sum_{k=1}^K w_k(\epsilon_n) F\Big(\frac{(\mathbf{X}_{\epsilon_n}\mathbf{s})_k}{s_k}\Big) \\
&= \lim_{n\to\infty} \inf_{\mathbf{s}\in\mathbb{R}^K_{++}} \Big[\sum_{k=1}^K (w_k(\epsilon_n) - w_k) F\Big(\frac{(\mathbf{X}_{\epsilon_n}\mathbf{s})_k}{s_k}\Big) \\
&\quad + \sum_{k=1}^K w_k \Big(F\Big(\frac{(\mathbf{X}_{\epsilon_n}\mathbf{s})_k}{s_k}\Big) - F\Big(\frac{(\mathbf{X}\mathbf{s})_k}{s_k}\Big)\Big) + \sum_{k=1}^K w_k F\Big(\frac{(\mathbf{X}\mathbf{s})_k}{s_k}\Big)\Big] \\
&\le \lim_{n\to\infty} \inf_{\mathbf{s}\in\mathbb{R}^K_{++}} \Big[\underbrace{\|\mathbf{w}(\epsilon_n) - \mathbf{w}\|_2}_{a(n)} \Big\| F\Big(\frac{(\mathbf{X}_{\epsilon_n}\mathbf{s})_k}{s_k}\Big)\Big\|_2 \\
&\quad + \underbrace{\|\mathbf{w}\|_2 \Big\| F\Big(\frac{(\mathbf{X}_{\epsilon_n}\mathbf{s})_k}{s_k}\Big) - F\Big(\frac{(\mathbf{X}\mathbf{s})_k}{s_k}\Big)\Big\|_2}_{b(n)} + \sum_{k=1}^K w_k F\Big(\frac{(\mathbf{X}\mathbf{s})_k}{s_k}\Big)\Big] \\
&= \inf_{\mathbf{s}\in\mathbb{R}^K_{++}} \sum_{k=1}^K w_k F\Big(\frac{(\mathbf{X}\mathbf{s})_k}{s_k}\Big)
\end{aligned}
$$

where the last step follows since, by (1.75) and (1.76), the nonnegative sequences $\{a(n)\}$ and $\{b(n)\}$ tends to zero as $n \to \infty$. So combining the inequality above with (1.77) yields (1.74). By Theorem 1.24, if there exists a positive eigenvector $\mathbf{p}$ associated with $\rho(\mathbf{X})$, the infimum is attained for $\mathbf{s} = \mathbf{p} > 0$.

As stated in the theorem, the infimum in (1.74) is attained if a positive right eigenvector of $\mathbf{X} \in \mathrm{N}^+_K$ associated with $\rho(\mathbf{X})$ can be found. An elegant characterization of the set of such matrices is provided, for instance, by [4]. In Appendix A.4.3, we have summarized some of these results. In particular, Theorem A.32 characterizes the set of nonnegative matrices with positive right eigenvectors in terms of isolated and maximal diagonal blocks of the matrix (1.72) (Definition A.31). The conclusion of the theorem is that $\mathbf{X}$ has a positive right eigenvector associated with $\rho(\mathbf{X})$ if and only if the corresponding normal form (1.72) satisfies the following two conditions.

(i) $\rho(\mathbf{X})$ is an eigenvalue of each isolated diagonal block, and
(ii) $\rho(\mathbf{X})$ is not an eigenvalue of the remaining diagonal blocks.

Using Definition A.31, the conditions can be expressed in an equivalent way as follows: Each isolated diagonal block is maximal and there are no other maximal diagonal blocks. Let B_K denote the set of all nonnegative matrices whose normal forms satisfy the above two conditions.

It is important to emphasize that $\mathbf{X} \in \mathrm{B}_K$ does not need to have a positive left eigenvector associated with $\rho(\mathbf{X})$. Therefore, although the infimum in (1.74) is attained when $\mathbf{X} \in \mathrm{B}_K$, the weight vector $\mathbf{w}$ may have zero components. From practical point of view, it is interesting to know whether

$\mathbf{w} = \mathbf{q} \circ \mathbf{p}$ is positive or not (see also the discussion in Sect. 5.5). This is equivalent to asking whether $\mathrm{E}_K^+(\mathbf{X})$ is an empty set or not. Based on Theorem A.32, it is relatively easy to show (Theorem A.34 in Appendix A.4.3) that a necessary and sufficient condition for $\mathrm{E}_K^+(\mathbf{X})$ to be nonempty set is that the normal form of $\mathbf{X}$ is block-irreducible matrix (Definition A.33 in Appendix A.4.3) and each diagonal block is maximal in the sense of Definition A.31. We use $\overline{\mathrm{B}}_K$ to denote the set of such block-irreducible matrices. Therefore, we have $\mathrm{E}_K^+(\mathbf{X}) \neq \emptyset$ if and only if $\mathbf{X} \in \overline{\mathrm{B}}_K$.

Combining these observations, we can strengthen Theorem 1.60 as follows: For any $\mathbf{X} \in \mathrm{B}_K \subset \mathrm{N}_K^+$, there holds

$$\min_{\mathbf{s}\in\mathbb{R}_{++}^K} \sum_{k=1}^{K} w_k F\left(\frac{(\mathbf{Xs})_k}{s_k}\right) = F(\rho(\mathbf{X}))\,. \tag{1.78}$$

The minimum is attained if $\mathbf{s} = \mathbf{p}$ where $\mathbf{p} > 0$ is any positive right eigenvector of $\mathbf{X}$. Moreover, if $\mathbf{X} \in \overline{\mathrm{B}}_K$, we have $\mathbf{w} = \mathbf{q} \circ \mathbf{p} > 0$ for any $(\mathbf{q}, \mathbf{p}) \in \mathrm{E}_K^+(\mathbf{X}) \neq \emptyset$.

1.6.3 Collatz–Wielandt-Type Characterization of the Spectral Radius

For any $\mathbf{X} \in \mathrm{N}_K$, we have

$$\sup_{\mathbf{s}\in\mathbb{R}_{++}^K} \min_{1\le k\le K} \frac{(\mathbf{Xs})_k}{s_k} \le \inf_{\mathbf{s}\in\mathbb{R}_{++}^K} \max_{1\le k\le K} \frac{(\mathbf{Xs})_k}{s_k} = \rho(\mathbf{X}) \tag{1.79}$$

where strict inequality can be shown to hold for some nonnegative reducible matrices. For instance, consider a 2×2 diagonal matrix $\mathbf{X} = \mathrm{diag}(x_1, x_2)$ with $0 < x_1 \le x_2$. Then,

$$\sup_{\mathbf{s}\in\mathbb{R}_{++}^2} \min_{1\le k\le 2} \frac{(\mathbf{Xs})_k}{s_k} = x_1 \qquad \inf_{\mathbf{s}\in\mathbb{R}_{++}^2} \max_{1\le k\le 2} \frac{(\mathbf{Xs})_k}{s_k} = x_2\,,$$

and hence if $x_1 < x_2$, we have strict inequality in (1.79).

Remark 1.61. It is essential that the supremum and the infimum in (1.79) are taken over $\mathbb{R}_{++}^K$. Otherwise, we could obtain $0/0$ expressions. In order to avoid this problem, one usually considers the functions $\underline{f}_{\mathbf{X}}(\mathbf{s}) = \min_{1\le k\le K, s_k\neq 0}(\mathbf{Xs})_k/s_k$ and $\overline{f}_{\mathbf{X}}(\mathbf{s}) = \max_{1\le k\le K, s_k\neq 0}(\mathbf{Xs})_k/s_k$, which are both well defined on $\mathbb{R}_+^K$. In this case, it is well-known that [12, 3]

$$\sup_{\mathbf{s}\in\mathbb{R}_+^K} \underline{f}_{\mathbf{X}}(\mathbf{s}) = \inf_{\mathbf{s}\in\mathbb{R}_+^K} \overline{f}_{\mathbf{X}}(\mathbf{s}) = \rho(\mathbf{X})\,.$$

The supremum and the minimum are attained for $\mathbf{s} = \mathbf{p} \ge 0$.

From (1.79), it follows that the Collatz-Wielandt type characterization of Sect. 1.2.4 cannot be extended to general nonnegative matrices. However, the inf-max part of (1.79) can be utilized to extend Lemma 1.28 to the set N_K^+. Indeed, proceeding essentially as in the proof of the lemma shows that for any $\mathbf{X} \in \mathrm{N}_K^+$ and $F \in \mathcal{G}(\mathbf{X})$, we have

$$\inf_{\mathbf{s}\in\mathbb{R}_{++}^K} \sup_{\mathbf{z}\in\Pi_K^+} \sum_{k=1}^K z_k F\left(\frac{(\mathbf{X}\mathbf{s})_k}{s_k}\right) = F(\rho(\mathbf{X}))\,. \tag{1.80}$$

Moreover, by the discussion in Sect. 1.6.2, when $\mathbf{X} \in \overline{\mathrm{B}}_K$, the infimum and the supremum are attained with $\mathbf{s} = \mathbf{p} > 0$ and $\mathbf{z} = \mathbf{w} = \mathbf{q} \circ \mathbf{p} > 0$, respectively.

On the other hand, due to (1.79), Lemma 1.30 cannot be extended to general nonnegative matrices. However, by the previous section, we know that there exists a positive right eigenvector associated with $\rho(\mathbf{X})$ if and only if $\mathbf{X} \in \mathrm{B}_K$. Therefore, with Theorem 1.60 in hand, we can proceed essentially as in the proof of Lemma 1.30 to show that for any $\mathbf{X} \in \mathrm{B}_K$ and $F \in \mathcal{G}(\mathbf{X})$,

$$\sup_{\mathbf{z}\in\Pi_K^+} \min_{\mathbf{s}\in\mathbb{R}_{++}^K} \sum_{k=1}^K z_k F\left(\frac{(\mathbf{X}\mathbf{s})_k}{s_k}\right) = F(\rho(\mathbf{X}))\,. \tag{1.81}$$

Again, if $\mathbf{X} \in \overline{\mathrm{B}}_K$, the supremum is attained with $\mathbf{z} = \mathbf{w} = \mathbf{q} \circ \mathbf{p} > 0$.

Summarizing we can say that the saddle point characterization of Theorem 1.31 holds (with some minor modifications) for any $\mathbf{X} \in \overline{\mathrm{B}}_K$.

Theorem 1.62. *Let $\mathbf{X} \in \overline{\mathrm{B}}_K$ and $F \in \mathcal{G}(\mathbf{X})$ be given. Define $G : \mathbb{R}_{++}^K \times \Pi_K^+ \to \mathbb{R}$ as*

$$G(\mathbf{s},\mathbf{z}) := \sum_{k=1}^K z_k F\left(\frac{(\mathbf{X}\mathbf{s})_k}{s_k}\right). \tag{1.82}$$

Then,

(i) the pair $(\mathbf{p},\mathbf{w}) \in \mathbb{R}_{++}^K \times \Pi_K^+$ is a saddle point of G and

$$F\big(\rho(\mathbf{X})\big) = \min_{\mathbf{s}\in\mathbb{R}_{++}^K} \max_{\mathbf{z}\in\Pi_K^+} G(\mathbf{s},\mathbf{z}) = \max_{\mathbf{z}\in\Pi_K^+} \min_{\mathbf{s}\in\mathbb{R}_{++}^K} G(\mathbf{s},\mathbf{z})\,, \tag{1.83}$$

(ii) $\mathbf{w} = \mathbf{q} \circ \mathbf{p} \in \Pi_K^+$ for any $(\mathbf{q},\mathbf{p}) \in \mathrm{E}_K^+(\mathbf{X}) \neq \emptyset$.

1.7 Bibliograpical Notes

All the results presented in this chapter were obtained by the authors in the course of working on problems in wireless networks [13, 14, 15, 16, 17, 18, 19]. To the best of our knowledge, some of these results were novel at the time of publishing, but some others turned out to have been known in the mathematics community for a while. From the mathematical point of view, however, they are still of some interest due to the different approach as well as the different line of arguments. Moreover, we feel that some proofs are more elementary, simpler, and shorter. Below the reader will find a short list of references where we found alternative proofs of the results presented in this chapter or some closely related results.

The Perron root characterization in Theorem 1.2 is an adapted form of the variational principle for pressure expressed in terms of nonnegative matrices. As aforementioned, this characterization can be deduced from [6, Equation 2.6 with 2.8 and 2.9]. In this form (but without proof), the theorem can be found in [23, Theorem 3.1]. Roughly speaking, both papers deal with the first and second partial derivatives of the Perron root with respect to the entries of nonnegative matrices. The proof of Theorem 1.2 is elementary and seems to be novel.

The assertion of Theorem 1.7 appears in [24, Equation 3.3] for positive matrices, but the result of [24] extends to arbitrary irreducible matrices. In contrast to [24], however, the proof presented here is elementary and therefore may be of interest in its own right. Theorem 4.1 in [24] is closely related to Theorem 1.11. However, the first one seems to apply only to matrices of the form $\mathbf{X} = \mathbf{AYB}$ where $\mathbf{A}$ and $\mathbf{B}$ are diagonal positive definite and $\mathbf{Y}$ is positive semidefinite. There is an extension of this result [24, Theorem 4.2] showing that the inequalities proved in [24, Theorem 4.1] hold for any nonnegative irreducible matrix $\mathbf{X}$ such that $\mathbf{X}^{-1}$ is an M-matrix (Definition A.38). In this book, Theorem 1.11 immediately follows from Theorem 1.7 by considering the fact that $\log x \leq x-1$ for all $x > 0$ with equality if and only if $x = 1$. Therefore, both results apply to an arbitrary nonnegative irreducible matrix. We point out that [24] provides a bunch of interesting results about the spectral radius of $\mathbf{DX}$ where $\mathbf{D}$ is diagonal positive definite and $\mathbf{X}$ is nonnegative.

The problem of convexity of the Perron root (Sects. 1.3) has also attracted some attention in the literature. In particular, it seems that Theorem 1.36 was first proved by [25] for nonnegative matrices whose entries are continuous functions of a scalar parameter on some interval. However, the extension to a parameter vector defined on some convex set is straightforward. Obviously, [25] used different techniques since Theorem 1.2 was not known at this time. In the engineering community, the result was rediscovered by [26]. Kingman's theorem was used by [27] to prove inequalities of the form $\phi(e^{\mathbf{A}+\mathbf{B}}) \leq \phi(e^{\mathbf{A}}e^{\mathbf{B}})$ where $\mathbf{A}$ and $\mathbf{B}$ are complex matrices and ϕ is a real-valued continuous function of the eigenvalues of its matrix argument. In a special case, ϕ is a spectral radius, $\mathbf{A}$ is nonnegative, and $\mathbf{B}$ is diagonal real. There are also some interesting results on log-convexity of spectral functions.

In [28], it was shown that $f(\mathbf{D}) = \log(e^{\mathbf{D}}\mathbf{A})$ where $\mathbf{D}$ is diagonal and $\mathbf{A}$ nonnegative is convex on the set of diagonal matrices. Using different tools, the convexity of $g(\mathbf{D}) = \max\{\operatorname{Re}\lambda : \lambda \in \sigma(\mathbf{A}+\mathbf{D})\}$ is shown in [29, 28, 30, 31].

In [32] (see also [31]), the convexity property of f and g was related to the convexity of certain sets of M-matrices, which in turn are related to the feasibility set defined in this book.

2

On the Positive Solution to a Linear System with Nonnegative Coefficients

This chapter deals with a positive solution $\mathbf{p}$ to the following system of linear equations with nonnegative coefficients:

$$\mathbf{p} = \mathbf{u} + \mathbf{X}\mathbf{p}\,. \tag{2.1}$$

Here and hereafter, $\mathbf{u} \in \mathbb{R}^K_{++}$ is a given *positive* vector, $\mathbf{X} \in \mathbb{R}^{K\times K}_{+}$ is a given nonnegative matrix (not necessarily irreducible), and $\mathbf{p} \in \mathbb{R}^K_{++}$ is a sought vector, provided that it exists.

2.1 Basic Concepts and Definitions

Before starting with the analysis, we need to address the fundamental problem of the existence of a positive solution $\mathbf{p}$ to (2.1). This problem is addressed in Appendix A.4.4. In particular, by Theorem A.35, we know that a necessary and sufficient condition for $\mathbf{p} \geq 0, \mathbf{p} \neq \mathbf{0}$, to exist is that $\rho(\mathbf{X}) < 1$ where $\rho(\mathbf{X})$ is the spectral radius of $\mathbf{X}$. Moreover, as $\mathbf{u}$ is positive, there is a unique solution $\mathbf{p}$, which is strictly positive and given by

$$\mathbf{p} = (\mathbf{I} - \mathbf{X})^{-1}\mathbf{u}\,.$$

Theorem A.30 asserts that $\lambda_p := \rho(\mathbf{X})$ is an eigenvalue of $\mathbf{X}$, that is to say $\lambda_p \in \sigma(\mathbf{X})$ where $\sigma(\mathbf{X})$ is used to denote the spectrum of $\mathbf{X}$ (Definition A.7).

Remark 2.1. Note that except for the nonnegativity, there are no additional constraints on $\mathbf{X}$. In particular, $\mathbf{X}$ does not need to be irreducible. However, it is worth pointing out that if $\mathbf{X}$ is irreducible and its Perron root $\lambda_p = \rho(\mathbf{X}) > 0$ satisfies $\lambda_p < 1$, then $\mathbf{u} \neq \mathbf{0}$ does not need to be positive for (2.1) to have a unique positive solution $\mathbf{p}$. This is one part of the assertion of Theorem A.36.

Analogous to the previous chapter, we allow the entries of $\mathbf{X}$ to continuously depend on some parameter vector $\boldsymbol{\omega} \in \Omega$ where the parameter set Ω is defined by (1.45) and is an open convex subset of $\mathbb{R}^K$. The only difference is that here the matrix is not required to be irreducible for all parameter vectors. In fact, $\mathbf{X}(\boldsymbol{\omega})$ can even be identically the zero matrix, in which case, however, the problems addressed in this chapter are trivial. To be precise, let

$$\mathbf{X}(\boldsymbol{\omega}) := (x_{k,l}(\boldsymbol{\omega}))_{1\leq k,l\leq K}$$

be a matrix-valued function whose entries $x_{k,l} : \Omega \to \mathbb{R}_+$ are *continuous* functions defined on Ω. Considering Definition 1.32, this is formally written as $\mathbf{X} \in \mathrm{N}_K(\Omega)$, in which case $\mathbf{X}$ is said to be nonnegative on Ω. To conform with the applications in wireless networks, we let each entry of the vector $\mathbf{u}$ in (2.1) be a continuous positive function of the parameter vector $\boldsymbol{\omega}$ as well. We indicate this by writing $\mathbf{u} \in \mathbb{R}^K_{++}(\Omega)$.

Now it follows from (2.1) and Theorem A.35 that, for any fixed $\boldsymbol{\omega} \in \Omega$, there exists a unique positive vector $\mathbf{p}(\boldsymbol{\omega})$ satisfying[1]

$$\mathbf{p}(\boldsymbol{\omega}) = \mathbf{X}(\boldsymbol{\omega})\mathbf{p}(\boldsymbol{\omega}) + \mathbf{u}(\boldsymbol{\omega}) \tag{2.2}$$

if and only if

$$\lambda_p(\boldsymbol{\omega}) := \rho(\mathbf{X}(\boldsymbol{\omega})) < 1\,. \tag{2.3}$$

Moreover, for any $\boldsymbol{\omega} \in \Omega$ with $\lambda_p(\boldsymbol{\omega}) < 1$,

$$\mathbf{p}(\boldsymbol{\omega}) = \big(\mathbf{I} - \mathbf{X}(\boldsymbol{\omega})\big)^{-1}\mathbf{u}(\boldsymbol{\omega})\,. \tag{2.4}$$

Let F be the set of those parameter vectors $\boldsymbol{\omega} \in \Omega$ for which a positive solution $\mathbf{p}(\boldsymbol{\omega})$ to (2.2) exists. Formally, we have

$$\mathrm{F} := \{\boldsymbol{\omega} \in \Omega : \lambda_p(\boldsymbol{\omega}) < 1\}\,. \tag{2.5}$$

Note that each entry of the vector $\mathbf{p}(\boldsymbol{\omega})$ is a *continuous* map from F into the set of positive reals $\mathbb{R}_{++}$. This is because if $\boldsymbol{\omega} \in \mathrm{F}$, then the Neumann series $\sum_{l=0}^{\infty}(\mathbf{X}(\boldsymbol{\omega}))^l$ converges (Theorem A.11) and $(\mathbf{I} - \mathbf{X}(\boldsymbol{\omega}))^{-1} = \sum_{l=0}^{\infty}(\mathbf{X}(\boldsymbol{\omega}))^l$. Therefore, since a concatenation of continuous maps is continuous, it follows from

$$p_k(\boldsymbol{\omega}) = \mathbf{e}_k^T\big(\mathbf{I} - \mathbf{X}(\boldsymbol{\omega})\big)^{-1}\mathbf{u}(\boldsymbol{\omega}),\ \boldsymbol{\omega} \in \mathrm{F}, 1 \leq k \leq K \tag{2.6}$$

that $p_k : \mathrm{F} \to \mathbb{R}_{++}$ is continuous. In particular, this implies that the l^1-norm of $\mathbf{p}(\boldsymbol{\omega})$ given by

$$\|\mathbf{p}(\boldsymbol{\omega})\|_1 = \sum_{k=1}^{K} p_k(\boldsymbol{\omega}) = \mathbf{1}^T(\mathbf{I} - \mathbf{X}(\boldsymbol{\omega}))^{-1}\mathbf{u}(\boldsymbol{\omega}),\ \boldsymbol{\omega} \in \mathrm{F} \tag{2.7}$$

is a continuous function on F as well.

[1] However, as $x_{k,l} : \Omega \to \mathbb{R}_+$ are not one-to-one maps, there may exist $\hat{\omega}, \check{\omega} \in \Omega, \hat{\omega} \neq \check{\omega}$, such that $\mathbf{p}(\hat{\omega}) = \mathbf{p}(\check{\omega})$.

In this chapter, we analyze both $p_k(\boldsymbol{\omega})$ and $\|\mathbf{p}(\boldsymbol{\omega})\|_1$ as functions of the parameter vector $\boldsymbol{\omega} \in \mathrm{F}$. In doing so, most of our interest is devoted to matrix-valued functions $\mathbf{X}(\boldsymbol{\omega})$ of the form $\mathbf{X}(\boldsymbol{\omega}) = \boldsymbol{\Gamma}(\boldsymbol{\omega})\mathbf{V}$ with $\mathbf{V} \in \mathrm{N}_K$ and

$$\boldsymbol{\Gamma}(\boldsymbol{\omega}) = \mathrm{diag}\big(\gamma_1(\omega_1), \ldots, \gamma_K(\omega_K)\big)\,.$$

Here and hereafter, $\gamma_k : \mathrm{Q}_k \to \mathbb{R}_{++}$ is a continuous strictly monotone and bijective function and $\mathrm{Q}_k \subseteq \mathbb{R}$ is some open interval (see also Sect. 1.3.1). Formally, this is denoted by $\mathbf{X} \in \mathrm{N}_{K,\boldsymbol{\Gamma}}(\Omega)$ where

$$\mathrm{N}_{K,\boldsymbol{\Gamma}}(\Omega) := \{\boldsymbol{\Gamma}(\boldsymbol{\omega})\mathbf{V}, \boldsymbol{\omega} \in \Omega : \mathbf{V} \in \mathrm{N}_K\} \subset \mathrm{N}_K(\Omega) \tag{2.8}$$

is the set of all nonnegative matrix-valued functions $\mathbf{X}(\boldsymbol{\omega})$ of the form $\mathbf{X}(\boldsymbol{\omega}) = \boldsymbol{\Gamma}(\boldsymbol{\omega})\mathbf{V}$ for some given $\gamma_k : \mathrm{Q}_k \to \mathbb{R}_{++}, k = 1, \ldots, K$. In this special case, it will also be assumed that $\mathbf{u}(\boldsymbol{\omega}) = (\gamma_1(\omega_1), \ldots, \gamma_K(\omega_K))$. Exceptions are only Sects. 2.3.1 and 2.3.2, where $\mathbf{X}(\boldsymbol{\omega})$ and $\mathbf{u}(\boldsymbol{\omega})$ are not confined to this special form.

2.2 Feasibility Sets

The set F defined by (2.5) contains all parameter vectors such that a positive solution to our system of linear equations exists. For this reason, if there are no additional constraints on $\mathbf{p}$, F is referred to as the feasibility set. Notice that the definition is analogous to Definition 1.38, except that now the spectral radius must be strictly smaller than 1. Therefore, the parameter vectors satisfying $\lambda_p(\boldsymbol{\omega}) = 1$ are not members of F.[2]

In wireless networks, however, some additional constraints on $\mathbf{p}$ are imposed, which gives rise to the definition of some subset of F as the feasibility set. Constraints on the l^1-norm of $\mathbf{p}(\boldsymbol{\omega})$ are common to applications in wireless communications networks. More precisely, we say that $\mathbf{p}(\boldsymbol{\omega})$ is constrained in the l^1-norm if

$$\|\mathbf{p}(\boldsymbol{\omega})\|_1 \le P_{\mathrm{t}}, \quad \boldsymbol{\omega} \in \Omega$$

must hold for some given constant $P_{\mathrm{t}} > 0$, referred to as a sum (or total) constraint. Consequently, in this case, the parameter vector $\boldsymbol{\omega} \in \Omega$ is feasible if and only if $\boldsymbol{\omega} \in \mathrm{F}(P_{\mathrm{t}})$ where

$$\mathrm{F}(\alpha) = \{\boldsymbol{\omega} \in \mathrm{F} : \|\mathbf{p}(\boldsymbol{\omega})\|_1 \le \alpha, \alpha > 0\} \subseteq \mathrm{F}\,. \tag{2.9}$$

Notice that due to continuity of $\|\mathbf{p}(\boldsymbol{\omega})\|_1$, $\mathrm{F}(\alpha)$ is monotonic in $\alpha > 0$ with respect to set inclusion in the following sense: For any $0 < \alpha \le \beta$, there holds $\mathrm{F}(\alpha) \subseteq \mathrm{F}(\beta)$. Therefore, since $\mathrm{F}(\alpha) \subseteq \mathrm{F}$ for all $\alpha > 0$, we have

[2] In the previous chapter, F is the set of all the parameter vectors for which the *homogenous* system of linear equations $(\mathbf{I} - \mathbf{X}(\boldsymbol{\omega}))\mathbf{p}(\boldsymbol{\omega}) = 0$, with $\mathbf{X}(\boldsymbol{\omega})$ being irreducible for all $\boldsymbol{\omega} \in \Omega$, has a positive solution $\mathbf{p}(\boldsymbol{\omega})$.

$$\mathrm{F} = \bigcup_{\alpha>0} \mathrm{F}(\alpha) \tag{2.10}$$

where the union is taken with respect to all $\alpha > 0$.

Another common situation encountered in wireless networks is that of constraining each element of $\mathbf{p}(\boldsymbol{\omega})$ individually. Therefore, if there are positive constants $P_1, \ldots, P_K$ such that $p_k(\boldsymbol{\omega}) \le P_k$ must hold for each $1 \le k \le K$, we say that $\mathbf{p}(\boldsymbol{\omega})$ is subject to *individual constraints.* Clearly, in this case, the set of all feasible parameter vectors is given by

$$\mathrm{F}(P_1, \ldots, P_K) := \bigcap_{\alpha \in \{P_1, \ldots, P_K\}} \mathrm{F}_k(\alpha) \tag{2.11}$$

where

$$\mathrm{F}_k(\alpha) := \{\boldsymbol{\omega} \in \mathrm{F} : p_k(\boldsymbol{\omega}) \le \alpha\}\,. \tag{2.12}$$

These two types of constraints are often combined by imposing both individual and sum constraints on $\mathbf{p}(\boldsymbol{\omega})$. Therefore, in this case, the feasibility set becomes

$$\mathrm{F}(P_{\mathrm{t}}; P_1, \ldots, P_K) := \mathrm{F}(P_{\mathrm{t}}) \cap \mathrm{F}(P_1, \ldots, P_K)\,. \tag{2.13}$$

Note that $\mathrm{F}(P_{\mathrm{t}}; P_1, \ldots, P_K) = \mathrm{F}(P_{\mathrm{t}})$ if $P_{\mathrm{t}} \le P_k$ for each $1 \le k \le K$, and $\mathrm{F}(P_{\mathrm{t}}; P_1, \ldots, P_K) = \mathrm{F}(P_1, \ldots, P_K)$ if $\sum_k P_k \le P_{\mathrm{t}}$. Thus, both $\mathrm{F}(P_{\mathrm{t}})$ and $\mathrm{F}(P_1, \ldots, P_K)$ can be viewed as special cases of $\mathrm{F}(P_{\mathrm{t}}; P_1, \ldots, P_K)$.

Remark 2.2. In what follows, we exclude the trivial case where the feasibility set is an empty set.

It is important to emphasize that the geometry of the feasibility sets depends on the choice of $\mathbf{X}(\boldsymbol{\omega})$ and $\mathbf{u}(\boldsymbol{\omega}), \boldsymbol{\omega} \in \Omega$. In particular, the feasibility set is not convex in general. To illustrate the definitions, let us consider an elementary example.

Example 2.3. Let $\mathbf{X}(\boldsymbol{\omega}) = 0$ for all $\boldsymbol{\omega} \in \Omega$ and $\mathbf{u}(\boldsymbol{\omega}) = (\gamma(\omega_1), \ldots, \gamma(\omega_K))$ where $\gamma : \mathrm{Q} \to \mathbb{R}_{++}$ is any continuous bijective function. We see that (2.4) reduces to $\mathbf{p}(\boldsymbol{\omega}) = (\gamma(\omega_1), \ldots, \gamma(\omega_K))$, and hence one obtains

$$\begin{aligned} \mathrm{F} &= \Omega = \mathrm{Q}^K \\ \mathrm{F}(P_{\mathrm{t}}) &= \{\boldsymbol{\omega} \in \mathrm{F} : \textstyle\sum_k \gamma(\omega_k) \le P_{\mathrm{t}}\} \\ \mathrm{F}(P_1, \ldots, P_K) &= \{\boldsymbol{\omega} \in \mathrm{F} : \gamma(\omega_k) \le P_k, 1 \le k \le K\}\,. \end{aligned}$$

Clearly, F and $\mathrm{F}(P_1, \ldots, P_K)$ are both convex sets, regardless of the choice of $\gamma(x)$. In contrast, $\mathrm{F}(P_{\mathrm{t}})$ is not convex in general. A sufficient condition for $\mathrm{F}(P_{\mathrm{t}})$ (and also $\mathrm{F}(P_{\mathrm{t}}; P_1, \ldots, P_K)$) to be a convex set is that $\gamma(x)$ is convex.

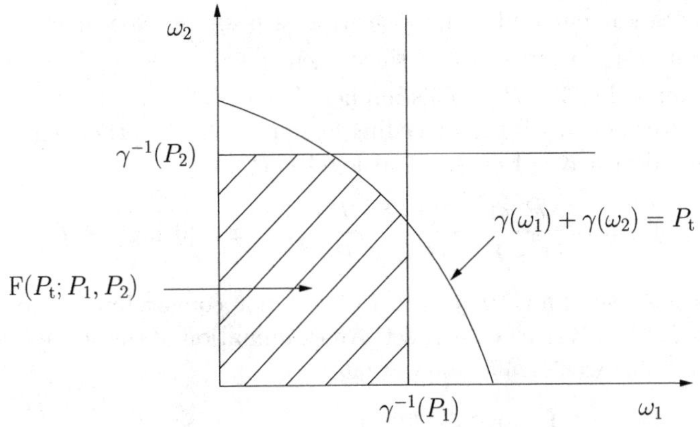

Fig. 2.1. Illustration of Example 2.3: The feasibility set $F(P_t; P_1, P_2)$ with $\mathbf{X}(\boldsymbol{\omega}) \equiv 0$, $\gamma(x) = e^x - 1, x > 0$, and $\mathbf{u}(\boldsymbol{\omega}) = (e^{\omega_1} - 1, e^{\omega_2} - 1)$. The constraints P_1, P_2 and P_t are chosen to satisfy $0 < P_1, P_2 < P_t$ and $P_t < P_1 + P_2$.

An important example of a convex function is $\gamma(x) = e^x - 1, x > 0$. Assuming $\mathbf{X}(\boldsymbol{\omega}) = 0$ for all $\boldsymbol{\omega} \in \Omega = \mathbb{R}^2_{++}$ and $\mathbf{u}(\boldsymbol{\omega}) = (e^{\omega_1} - 1, e^{\omega_2} - 1)$, Fig. 2.1 depicts the feasibility set $F(P_t; P_1, P_2) \subset \mathbb{R}^2_{++}$ defined by (2.13) for some P_1, P_2 and P_t.

Unfortunately, as the example below shows, convexity of $\gamma(x)$ is not sufficient for $F(P_t)$ to be a convex set if $\mathbf{X}(\boldsymbol{\omega}) = \boldsymbol{\Gamma}(\boldsymbol{\omega})\mathbf{V} \neq \mathbf{0}$.

Example 2.4. Suppose that $\mathbf{X}(\boldsymbol{\omega}) = \left(\begin{smallmatrix} 0 & \gamma(\omega_1)\varrho \\ \gamma(\omega_2)\varrho & 0 \end{smallmatrix}\right)$ for some $\varrho \geq 0$. Furthermore, assume that $\mathbf{u}(\boldsymbol{\omega}) = (\gamma(\omega_1), \gamma(\omega_2))$ and $\gamma(x) = e^x - 1, x > 0$. Thus,

$$\Omega = \mathbb{R}^K_{++}$$
$$F = \{\boldsymbol{\omega} \in \Omega : \lambda_p(\boldsymbol{\omega}) = \varrho\sqrt{(e^{\omega_1} - 1)(e^{\omega_2} - 1)} < 1\}.$$

Now we claim that F is not a convex set if $\varrho > 0$. To see this, we write $\lambda_p(\boldsymbol{\omega}) = 1$ with $\varrho > 0$ as a function of $\omega_1 > 0$ to obtain $\omega_2 = f(\omega_1) = \log \frac{1+\varrho^2 e^{\omega_1} - \varrho^2}{\varrho^2(e^{\omega_1}-1)}$, $\varrho > 0$. The function $f(x), x > 0$, is twice differentiable and its second derivative is is strictly positive for all $x > 0$. Consequently, instead of the feasibility set F, its complement in $\mathbb{R}^K_{++}$ ($F^c = \mathbb{R}^K_{++} \setminus F$) is convex.

Now let us consider $F(P_t)$ with $\varrho \geq 0$. Applying (2.7) to our special case yields

$$\|\mathbf{p}(\boldsymbol{\omega})\|_1 = \frac{e^{\omega_1} + e^{\omega_2} - 2 + 2\varrho(e^{\omega_1} - 1)(e^{\omega_2} - 1)}{1 - \varrho^2(e^{\omega_1} - 1)(e^{\omega_2} - 1)}, \quad \boldsymbol{\omega} \in F.$$

Hence, writing $\|\mathbf{p}(\boldsymbol{\omega})\|_1 = P_t$ as a function of $\omega_1 \in [0, \log(1+P_t)]$, one obtains

$$\omega_2 = g(\omega_1) = \log \frac{(1-\varrho)\ (2 + P_t + P_t\,\varrho) + e^{\omega_1}\ (\varrho\ (2 + P_t\,\varrho) - 1)}{1 + \varrho\big(e^{\omega_1} - 1\big)\big(2 + P_t\varrho\big)}$$

where the argument under the logarithm is positive. Now if $\varrho = 0$, $g(x)$ is concave on $x \in [0, \log(1+P_t)]$ since then $g''(x) = -\frac{e^x(2+P_t)}{(2+P_t-e^x)^2}$ is strictly negative on $[0, \log(1+P_t)]$. This implies that $\mathrm{F}(P_t)$ is a convex set, which is in total agreement with the preceding example. On the other hand, if $\varrho = 1$, the second derivative of $g(x), x \in [0, \log(1+P_t)]$, is

$$g''(x) = \frac{e^x\ (1+P_t)\ (2+P_t)}{\left(1+P_t - e^x\ (2+P_t)\right)^2}, \quad , x \in [0, \log(1+P_t)]$$

which is positive. Thus, if $\varrho = 1$, $\mathrm{F}(P_t)$ is not convex but its complement $\mathrm{F}^c(P_t) = \mathbb{R}_+^K \setminus \mathrm{F}(P_t)$ is a convex set. An examination of the second derivative of $g(x), x \in [0, \log(1+P_t)]$, shows that

$$g''(x) \begin{cases} < 0 & \varrho < h(P_t) \\ > 0 & \varrho > h(P_t) \\ = 0 & \varrho = h(P_t) \end{cases} \qquad h(x) = \frac{\sqrt{1+x}-1}{x}.$$

Since $h(x) \to 0$ as $x \to \infty$, we have $g''(x) > 0$ for any $\varrho > 0$, which complies with the discussion above that $f(x)$ is convex for any $\varrho > 0$. On the other hand, if $x \to 0$, then $h(x) \to 1/2$. So, at small values of P_t, convexity of $\mathrm{F}(P_t)$ changes to convexity of $\mathrm{F}^c(P_t)$ around the value $\varrho \approx 1/2$.

The example above demonstrates that the feasibility set may be a nonconvex set even if each entry of $\mathbf{X}(\boldsymbol{\omega})$ is convex on Ω. As a consequence, a stronger property than convexity is necessary to guarantee convexity of F. In the following section, we show that if $\mathbf{X}(\boldsymbol{\omega})$ is log-convex on Ω (see Definition 1.34), then $p_k(\boldsymbol{\omega})$ is a log-convex function of $\boldsymbol{\omega} \in \mathrm{F}$ for each $1 \leq k \leq K$.

2.3 Convexity Results

In this section, we show that if $\mathbf{X} \in \mathrm{N}_K(\Omega)$ and $\mathbf{u} \in \mathbb{R}_{++}^K(\Omega)$ are both log-convex on Ω, then $p_k : \mathrm{F} \to \mathbb{R}_{++}$ given by (2.6) is log-convex for each $1 \leq k \leq K$. This in turn implies that the feasibility set $\mathrm{F}(P_t; P_1, \ldots, P_K)$ is a convex set, regardless of the choice of $P_1, \ldots, P_K > 0$ and $P_t > 0$. Following that, we consider the problem of strict convexity.

Recall that according to Definition 1.34, the notation $\mathbf{X} \in \mathrm{LC}_K(\Omega)$ means that $\mathbf{X} \in \mathrm{N}_K(\Omega)$ is log-convex on Ω. Furthermore, note that by this definition, the identically zero function is a log-convex function (see also the remark in Sect. 1.3). In an analogous manner, we say that $\mathbf{u} \in \mathbb{R}_{++}^K(\Omega)$ is log-convex on Ω if each entry of the vector $\mathbf{u}(\boldsymbol{\omega})$ is a continuous log-convex function defined on Ω. Let us indicate this by writing $\mathbf{u} \in \mathrm{lc}(\Omega) \subset \mathbb{R}_{++}^K(\Omega)$.

2.3.1 Log-Convexity of the Positive Solution

Let $\boldsymbol{\omega}(\mu)$ with $\mu \in [0, 1]$ be a convex combination of two arbitrary vectors $\hat{\boldsymbol{\omega}}, \check{\boldsymbol{\omega}} \in \Omega$:

$$\boldsymbol{\omega}(\mu) = (1-\mu)\hat{\boldsymbol{\omega}} + \mu\check{\boldsymbol{\omega}}, \quad \mu \in [0,1].$$

Unless otherwise stated, assume that $\hat{\boldsymbol{\omega}}, \check{\boldsymbol{\omega}} \in \mathrm{F} \subseteq \Omega$, which implies that both $p_k(\hat{\boldsymbol{\omega}}) > 0$ and $p_k(\check{\boldsymbol{\omega}}) > 0$ exists.

Theorem 2.5. *Let $\mathbf{X} \in \mathrm{LC}_K(\Omega) \subset \mathrm{N}_K(\Omega)$ and $\mathbf{u} \in \mathrm{lc}(\Omega) \subset \mathbb{R}^K_{++}(\Omega)$ be arbitrary. Then, $p_k(\boldsymbol{\omega})$ is log-convex on* F *for each $1 \le k \le K$, i.e., we have*

$$p_k(\boldsymbol{\omega}(\mu)) \le p_k(\hat{\boldsymbol{\omega}})^{1-\mu} p_k(\check{\boldsymbol{\omega}})^{\mu}, \quad 1 \le k \le K \tag{2.14}$$

for all $\mu \in (0,1)$ and $\hat{\boldsymbol{\omega}}, \check{\boldsymbol{\omega}} \in \mathrm{F}$.

Proof. Let $\hat{\boldsymbol{\omega}}, \check{\boldsymbol{\omega}} \in \mathrm{F}$ be arbitrary. Then, by Theorem 1.36 as well as by Sect. 1.6, we know that $\lambda_p(\boldsymbol{\omega}(\mu)) < 1$ for all $\mu \in (0,1)$. Thus, for every $\mu \in (0,1)$, there exists a unique positive $p_k(\boldsymbol{\omega}(\mu))$ given by (see (2.6))

$$p_k(\boldsymbol{\omega}(\mu)) = \mathbf{e}_k^T \left[\mathbf{I} - \mathbf{X}(\boldsymbol{\omega}(\mu))\right]^{-1} \mathbf{u}(\boldsymbol{\omega}(\mu)), \quad 1 \le k \le K.$$

Now let $\mu \in (0,1)$ be arbitrary but fixed. Since $\boldsymbol{\omega}(\mu) \in \mathrm{F}$, we can expand $(\mathbf{I} - \mathbf{X}(\boldsymbol{\omega}(\mu)))^{-1}$ into a Neumann series (see Theorem A.11) to obtain

$$\left[\mathbf{I} - \mathbf{X}(\boldsymbol{\omega}(\mu))\right]^{-1} = \sum_{l=0}^{\infty} \left(\mathbf{X}(\boldsymbol{\omega}(\mu))\right)^l.$$

From this it follows that

$$\begin{aligned} p_k(\boldsymbol{\omega}(\mu)) &= \mathbf{e}_k^T \sum_{l=0}^{\infty} \left(\mathbf{X}(\boldsymbol{\omega}(\mu))\right)^l \mathbf{u}(\boldsymbol{\omega}(\mu)) = \sum_{l=0}^{\infty} \mathbf{e}_k^T \left(\mathbf{X}(\boldsymbol{\omega}(\mu))\right)^l \mathbf{u}(\boldsymbol{\omega}(\mu)) \\ &= \sum_{l=0}^{\infty} g_l\left(\boldsymbol{\omega}(\mu)\right). \end{aligned}$$

By assumption, all the entries of $\mathbf{X}(\boldsymbol{\omega})$ and $\mathbf{u}(\boldsymbol{\omega})$ are log-convex on F. Hence, (2.14) immediately follows from the equation above when one considers the following properties of log-convex functions:

(i) If two positive functions f and g are log-convex, then $f+g$ and $f \cdot g$ are log-convex.
(ii) For any convergent sequence f_n of log-convex functions, the limit $f = \lim_{n\to\infty} f_n$ is log-convex provided that the limit is strictly positive.

Due to (i), $g_l : \mathrm{F} \to \mathbb{R}_{++}$ is log-convex for each $l \ge 0$ and $\sum_{l=0}^{M} g_l(\boldsymbol{\omega})$ is log-convex for any $M > 0$. Furthermore, since $\sum_{l=0}^{M} g_l(\boldsymbol{\omega})$ is increasing in M and g_l is positive, it must converge to a positive limit as $M \to +\infty$. Hence, by (ii), $p_k(\boldsymbol{\omega})$ is log-convex on F and (2.14) must hold.

Remark 2.6. Recall that the spectral radius of $\mathbf{X} \in \mathrm{N}_K(\Omega)$ can be expressed as follows (Theorem A.10)

$$\lambda_p(\omega) = \lim_{m\to+\infty} \|\mathbf{X}(\omega)^m\|^{1/m}.$$

Thus, considering the two properties (i) and (ii) of log-convex functions in the proof of Theorem 2.5 and the fact that if f is log-convex, so also is f^α for every positive α, shows that if the entries of $\mathbf{X}(\omega)$ are log-convex functions on Ω, then $\lambda_p(\omega)$ is log-convex on Ω. This leads to an alternative proof of log-convexity of the spectral root (see for instance [26]).

A trivial but important consequence of the theorem is the following.

Corollary 2.7. *If* $\mathbf{X} \in \mathrm{LC}_K(\Omega) \subset \mathrm{N}_K(\Omega)$ *and* $\mathbf{u} \in \mathrm{lc}(\Omega) \subset \mathbb{R}^K_{++}$, *then* $\|\mathbf{p}(\omega)\|_1 = \sum_{k=1}^K p_k(\omega)$ *is log-convex on* F, *that is to say,*

$$\|\mathbf{p}(\omega(\mu))\|_1 \le \|\mathbf{p}(\hat{\omega})\|_1^{1-\mu}\|\mathbf{p}(\breve{\omega})\|_1^{\mu} \tag{2.15}$$

for all $\mu \in (0,1)$ *and* $\hat{\omega}, \breve{\omega} \in \mathrm{F}$.

Proof. As log-convex functions are closed under addition, it is clear that the log-convexity property carries over to the l^1-norm of $\mathbf{p}(\omega)$.

More generally, we can say that if $\mathbf{X} \in \mathrm{LC}_K(\Omega)$ and $\mathbf{u} \in \mathrm{lc}(\Omega)$, then

$$F(p_1(\omega(\mu)),\dots,p_K(\omega(\mu))) \le F(p_1(\hat{\omega}),\dots,p_K(\hat{\omega}))^{1-\mu}F(p_1(\breve{\omega}),\dots,p_K(\breve{\omega}))^{\mu}$$

for all $\mu \in (0,1)$ and $\hat{\omega}, \breve{\omega} \in \mathrm{F}$ where $F : \mathbb{R}^K_{++} \to \mathbb{R}_{++}$ is any function that preserves log-convexity. Standard examples of such functions are

1. weighted sum: $F(x_1,\dots,x_K) = \sum_{k=1}^K w_k x_k$,
2. weighted pointwise multiplication: $F(x_1,\dots,x_K) = \prod_{k=1}^K w_k x_k$, and
3. pointwise maximum and supremum: $F(x_1,\dots,x_K) = \max_{1\le k\le K} x_k$.

The weighted sum operation and the pointwise multiplication operation preserve log-convexity as log-convex functions are closed under both addition and multiplication. The claim about the pointwise maximum operation follows since

$$\begin{aligned}\max\{p_k(\omega(\mu)) :&1 \le k \le K\}\\ &\le \max\{p_k(\hat{\omega})^{1-\mu}p_k(\breve{\omega})^{\mu} : 1 \le k \le K\}\\ &\le \max\{p_k(\hat{\omega})^{1-\mu} : 1 \le k \le K\}\max\{p_k(\breve{\omega})^{\mu} : 1 \le k \le K\}\\ &= \max\{p_k(\hat{\omega}) : 1 \le k \le K\}^{1-\mu}\max\{p_k(\breve{\omega}) : 1 \le k \le K\}^{\mu}\end{aligned}$$

for all $\mu \in (0,1)$ and $\hat{\omega}, \breve{\omega} \in \mathrm{F}$.

2.3.2 Convexity of the Feasibility Set

Since the geometric mean is bounded above by the arithmetic mean, we have

$$p_k(\hat{\omega})^{1-\mu} p_k(\check{\omega})^{\mu} \leq (1-\mu) p_k(\hat{\omega}) + \mu p_k(\check{\omega}) \leq \max\{p_k(\hat{\omega}), p_k(\check{\omega})\}$$

for all $\hat{\omega}, \check{\omega} \in \mathrm{F}$ and $\mu \in (0,1)$. Thus, if $p_k(\boldsymbol{\omega})$ is log-convex on F, then the inequality above implies that $\mathrm{F}_k(P_k)$ defined by (2.12) is a convex set. By Theorem 2.5, we know that if $\mathbf{X}(\boldsymbol{\omega})$ and $\mathbf{u}(\boldsymbol{\omega})$ are both log-convex on Ω, then $p_k : \mathrm{F} \to \mathbb{R}_{++}$ is log-convex for each $1 \leq k \leq K$. Consequently, since the intersection of convex sets is convex, it follows from (2.11) that $\mathrm{F}(P_1, \ldots, P_K)$ is a convex set if $\mathbf{X} \in \mathrm{LC}_K(\Omega)$ and $\mathbf{u} \in \mathrm{lc}(\Omega)$. By Corollary 2.7 and (2.13), we see that this also true for $\mathrm{F}(P_t)$ and $\mathrm{F}(P_t; P_1, \ldots, P_K)$. We summarize these observations in a corollary.

Corollary 2.8. *Suppose that* $\mathbf{X} \in \mathrm{LC}_K(\Omega) \subset \mathrm{N}_K(\Omega)$ *and* $\mathbf{u} \in \mathrm{lc}(\Omega) \subset \mathbb{R}^K_{++}(\Omega)$. *Then,* $\mathrm{F}(P_1, \ldots, P_K), \mathrm{F}(P_t)$ *and* $\mathrm{F}(P_t; P_1, \ldots, P_K)$ *are convex sets, regardless of the choice of* $P_t, P_1, \ldots, P_K > 0$.

To illustrate the results, let us consider a simple example.

Example 2.9. Let $\mathbf{X}(\boldsymbol{\omega})$ and $\mathbf{u}(\boldsymbol{\omega})$ be defined as in Example 2.4 except that now $\gamma(x) = \exp(x), x \in \mathbb{R}$. Clearly, the exponential function is log-convex on $\mathbb{R}$. Thus, by Theorem 1.36 (note that the matrix $\mathbf{X}(\boldsymbol{\omega})$ is irreducible for all $\boldsymbol{\omega} \in \mathbb{R}^2$), the Perron root is log-convex and, by Corollary 1.39, F is a convex set. In contrast to the previous example, all pairs satisfying $\lambda_p(\boldsymbol{\omega}) = \varrho\sqrt{e^{\omega_1} e^{\omega_2}} = 1$ lie on a line given by $\omega_2 = -\omega_1 - 2\log\varrho$, which, of course, is both convex and concave.

The nonnegative solution (2.4) yields

$$\mathbf{p}(\boldsymbol{\omega}) = \begin{pmatrix} \frac{e^{\omega_1} + \varrho e^{\omega_1+\omega_2}}{1-\varrho^2 e^{\omega_1+\omega_2}} \\ \frac{e^{\omega_2} + \varrho e^{\omega_1+\omega_2}}{1-\varrho^2 e^{\omega_1+\omega_2}} \end{pmatrix}, \quad \varrho^2 e^{\omega_1+\omega_2} < 1\,.$$

By Theorem 2.5, both entries are log-convex on $\mathbb{R}^2$. All pairs (ω_1, ω_2) satisfying $p_1(\boldsymbol{\omega}) = P_1$ and $p_2(\boldsymbol{\omega}) = P_2$ are $\omega_2 = f(\omega_1) = \log[(P_1 - e^{\omega_1})/(\varrho(1 + \varrho P_1))] - \omega_1, \omega_1 < \log P_1$, and $\omega_2 = g(\omega_1) = \log[P_2/(1 + e^{\omega_1}\varrho + e^{\omega_1}\varrho^2 P_2)], \omega_2 < \log P_2$, respectively. It may be verified that $f(x)$ is concave on $(-\infty, \log P_1)$ and $g(x)$ is concave on $\mathbb{R}$ implying that $\mathrm{F}_1(P_1), \mathrm{F}_2(P_2)$ and $\mathrm{F}(P_1, P_2)$ are all convex sets. Similarly, $\|\mathbf{p}(\boldsymbol{\omega})\|_1 = P_t$ can be rewritten to give $\omega_2 = h(\omega_1) = \log[(P_t - e^{\omega_1})/(1 + 2e^{\omega_1}\varrho + e^{\omega_1}\varrho^2 P_2)], \omega_1 < \log P_t$. Again, $h(x)$ can be seen to be concave on $(-\infty, \log P_t)$, from which convexity of $\mathrm{F}(P_t)$ follows.

In the example above, instead of $\gamma(x) = \gamma_1(x) = \gamma_2(x) = e^x, x \in \mathbb{R}$, we could consider any log-convex functions $\gamma_1 : \mathrm{Q}_1 \to \mathbb{R}_{++}$ and $\gamma_2 : \mathrm{Q}_2 \to \mathbb{R}_{++}$. In such a case, the unique positive solution $\mathbf{p}(\boldsymbol{\omega})$ exists if and only if $\boldsymbol{\omega} \in \mathrm{F} = \{\boldsymbol{\omega} \in \Omega : \lambda_p(\boldsymbol{\omega}) = \varrho\sqrt{\gamma_1(\omega_1)\gamma_2(\omega_2)} < 1\}$ and is given by

$$\mathbf{p}(\boldsymbol{\omega}) = \begin{pmatrix} \frac{\gamma_1(\omega_1)+\varrho\gamma_1(\omega_1)\gamma_2(\omega_2)}{1-\varrho^2\gamma_1(\omega_1)\gamma_2(\omega_2)} \\ \frac{\gamma_2(\omega_2)+\varrho\gamma_1(\omega_1)\gamma_2(\omega_2)}{1-\varrho^2\gamma_1(\omega_1)\gamma_2(\omega_2)} \end{pmatrix}, \quad \boldsymbol{\omega} \in \mathrm{F}\,. \tag{2.16}$$

It may be verified that if γ_1 and γ_2 are both log-convex, then each entry of $\mathbf{p}(\boldsymbol{\omega})$ is log-convex on F. This in turn implies that the feasibility set $\mathrm{F}(P_{\mathrm{t}}; P_1, \ldots, P_K)$ is a convex set, regardless of the choice of $P_{\mathrm{t}} > 0$ and $P_1, \ldots, P_K > 0$.

Finally, it is worth pointing out that the results presented in this chapter straightforwardly extends to the case when $p_k(\boldsymbol{\omega})$ is either subject to $\|\mathbf{p}(\boldsymbol{\omega})\|_1 \leq P_{\mathrm{t}}(\boldsymbol{\omega})$ or $p_k(\boldsymbol{\omega}) \leq P_k(\boldsymbol{\omega}), 1 \leq k \leq K$, for all $\boldsymbol{\omega} \in \Omega$ where $P_{\mathrm{t}} : \Omega \to \mathbb{R}_{++}$ and $P_k : \Omega \to \mathbb{R}_{++}$ are given continuous *concave* functions. So if $p_k(\boldsymbol{\omega})$ is convex for each $1 \leq k \leq K$, then $\{\boldsymbol{\omega} \in \Omega : \|\mathbf{p}(\boldsymbol{\omega})\|_1 \leq P_{\mathrm{t}}(\boldsymbol{\omega})\}$ and $\{\boldsymbol{\omega} \in \Omega : p_k(\boldsymbol{\omega}) \leq P_k(\boldsymbol{\omega})\}, 1 \leq k \leq K$, are convex sets.

2.3.3 Strict Log-Convexity

When $\mathbf{X}(\boldsymbol{\omega})$ and $\mathbf{u}(\boldsymbol{\omega})$ are log-convex on Ω, Theorem 2.5 asserts that $p_k(\boldsymbol{\omega})$ is a log-convex function of $\boldsymbol{\omega} \in \mathrm{F}$. In this section, we strengthen this result by proving conditions on strict log-convexity. In the second part of the book, we will exploit these results to prove some interesting properties of the addressed power control problem.

For the analysis in this section, it is assumed that $\mathbf{X} \in \mathrm{N}_K(\Omega)$ and $\mathbf{u} \in \mathbb{R}^K_{++}(\Omega)$ are restricted to be of the following form:

$$\begin{aligned} \mathbf{u}(\boldsymbol{\omega}) &= \boldsymbol{\Gamma}(\boldsymbol{\omega})\mathbf{z} \\ \mathbf{X}(\boldsymbol{\omega}) &= \boldsymbol{\Gamma}(\boldsymbol{\omega})\mathbf{V} \quad \text{with} \quad \mathrm{trace}(\mathbf{V}) = 0 \end{aligned}\,. \tag{2.17}$$

Here and hereafter, $\mathbf{z} = (z_1, \ldots, z_K)$ is any fixed positive vector, $\mathbf{V} \in \mathrm{N}_K$ and $\gamma_k : \mathrm{Q}_k \to \mathbb{R}_{++}, k = 1 \ldots K$ are continuous and strictly monotonic (bijective) functions. Formally, we have $\mathbf{X} \in \mathrm{N}^0_{K,\boldsymbol{\Gamma}}(\Omega)$ which is the subset of $\mathrm{N}_{K,\boldsymbol{\Gamma}}(\Omega)$ defined by (2.8) such that $\mathrm{trace}(\mathbf{V}) = 0$.

Lemma 2.10. *Let* $\mathbf{X} \in \mathrm{N}^0_{K,\boldsymbol{\Gamma}}(\Omega)$ *and* $u(\boldsymbol{\omega}) = \boldsymbol{\Gamma}(\boldsymbol{\omega})\mathbf{z}, \boldsymbol{\omega} \in \Omega$, *be arbitrary. Then,* $\mathbf{p} : \mathrm{F} \to \mathbb{R}^K_{++}$ *defined by (2.4) is a bijection.*

Proof. Due to the bijectivity of $\gamma_k : \mathrm{Q}_k \to \mathbb{R}_{++}$ and the uniqueness of the positive solution in (2.4), it immediately follows from

$$\mathbf{p}(\boldsymbol{\omega}) = (\mathbf{I} - \boldsymbol{\Gamma}(\boldsymbol{\omega})\mathbf{V})^{-1}\boldsymbol{\Gamma}(\boldsymbol{\omega})\mathbf{z} = (\boldsymbol{\Gamma}(\boldsymbol{\omega})^{-1} - \mathbf{V})^{-1}\mathbf{z}$$

that $\mathbf{p}(\boldsymbol{\omega})$ is a bijection from F onto $\mathbb{R}^K_{++}$ for any $\mathbf{z} > 0$.

It is important to emphasize that the positivity of the vector $\mathbf{z}$ is crucial for the results to hold. In contrast, the assumption $\mathrm{trace}(\mathbf{V}) = 0$ is merely motivated by practical applications and could be easily dropped. Note that due to this assumption, $(\mathbf{Vs})_k$ for any $\mathbf{s} \in \mathbb{R}^K$ is independent of s_k for each $1 \leq k \leq K$.

In what follows, we extensively exploit the following special form of Hölder's inequality (Theorem A.2): For any $\mu \in (0,1)$ and $\mathbf{u}, \mathbf{v} \in \mathbb{R}^K_+$, there holds

$$\langle \mathbf{u}, \mathbf{v} \rangle \leq \|\mathbf{u}\|_p \|\mathbf{v}\|_q, \quad p = \frac{1}{1-\mu} \text{ and } q = \frac{1}{\mu}, \tag{2.18}$$

with equality if and only if there exists a constant $c > 0$ such that

$$v_k = c\, u_k^{p-1} = c\, u_k^{\frac{\mu}{1-\mu}}, \quad 1 \leq k \leq K .$$

Finally, recall that $\gamma_k : \mathrm{Q}_k \to \mathbb{R}_{++}, 1 \leq k \leq K$, is said to be strictly log-convex if $\gamma_k(x(\mu)) < \gamma_k(\hat{x})^{1-\mu}\gamma_k(\check{x})^{\mu}$ for all $\mu \in (0,1)$ and $\hat{x}, \check{x} \in \mathrm{Q}_k$ with $\hat{x} \neq \check{x}$ and $x(\mu) = (1-\mu)\hat{x} + \mu\check{x}$. Similarly, we say that $p_k : \mathrm{F} \to \mathbb{R}_{++}$ given by (2.6) is strictly log-convex for some $1 \leq k \leq K$ if $p_k(\boldsymbol{\omega}(\mu)) < p_k(\hat{\boldsymbol{\omega}})^{1-\mu} p_k(\check{\boldsymbol{\omega}})^{\mu}$ for all $\mu \in (0,1)$ and $\hat{\boldsymbol{\omega}}, \check{\boldsymbol{\omega}} \in \mathrm{F}$ with $\hat{\boldsymbol{\omega}} \neq \check{\boldsymbol{\omega}}$. The following result is a straightforward extension of Theorem 2.5 to the case of strictly log-convex functions $\gamma_1, \ldots, \gamma_K$.

Theorem 2.11. *Let $\mathbf{V} \geq 0$ be arbitrary, and let $\gamma_k : \mathrm{Q} \to \mathbb{R}_{++}$ be strictly log-convex for each $1 \leq k \leq K$. Then, for all $\hat{\boldsymbol{\omega}}, \check{\boldsymbol{\omega}} \in \mathrm{F}$ with $\hat{\boldsymbol{\omega}} \neq \check{\boldsymbol{\omega}}$, there exists an index $1 \leq k_0 \leq K$ such that $p_{k_0}(\boldsymbol{\omega}(\mu)) < p_{k_0}(\hat{\boldsymbol{\omega}})^{1-\mu} p_{k_0}(\check{\boldsymbol{\omega}})^{\mu}$ for all $\mu \in (0,1)$.*

Proof. Let $\hat{\boldsymbol{\omega}}, \check{\boldsymbol{\omega}} \in \mathrm{F}$ be arbitrary, and let k_0 be an index such that $\hat{\omega}_{k_0} \neq \check{\omega}_{k_0}$. By Theorem 2.5, we know that $\boldsymbol{\omega}(\mu) = (1-\mu)\hat{\boldsymbol{\omega}} + \mu\check{\boldsymbol{\omega}} \in \mathrm{F}$ for all $\mu \in (0,1)$. Therefore, for any $\mu \in (0,1)$, it follows from (2.2) that

$$p_{k_0}(\boldsymbol{\omega}(\mu)) = \gamma_{k_0}(\omega_{k_0}(\mu))\big(\mathbf{Vp}(\boldsymbol{\omega}(\mu)) + \mathbf{z}\big)_{k_0} .$$

So, by strict log-convexity of γ_{k_0} and positivity of the vector $\mathbf{z}$, we have

$$p_{k_0}(\boldsymbol{\omega}(\mu)) < \gamma_{k_0}(\hat{\omega}_{k_0})^{1-\mu}\gamma_{k_0}(\check{\omega}_{k_0})^{\mu}\big(\mathbf{Vp}(\boldsymbol{\omega}(\mu)) + \mathbf{z}\big)_{k_0} .$$

Considering Theorem 2.5 and Hölder's inequality (2.18) yields

$$\begin{aligned} p_{k_0}(\boldsymbol{\omega}(\mu)) &< \gamma_{k_0}(\hat{\omega}_{k_0})^{1-\mu}\gamma_{k_0}(\check{\omega}_{k_0})^{\mu}\Big[\sum_{l=1}^{K}\big(v_{k_0,l}p_l(\hat{\boldsymbol{\omega}})\big)^{1-\mu}\big(v_{k_0,l}p_l(\check{\boldsymbol{\omega}})\big)^{\mu} + z_{k_0}\Big] \\ &\leq \gamma_{k_0}(\hat{\omega}_{k_0})^{1-\mu}\gamma_{k_0}(\check{\omega}_{k_0})^{\mu}\Big[\big(\mathbf{Vp}(\hat{\boldsymbol{\omega}})\big)_{k_0}^{1-\mu}\big(\mathbf{Vp}(\check{\boldsymbol{\omega}})\big)_{k_0}^{\mu} + z_{k_0}^{1-\mu} z_{k_0}^{\mu}\Big] \\ &= \langle \hat{\mathbf{u}}, \check{\mathbf{u}} \rangle \end{aligned}$$

where[3]

[3] For any vector $\mathbf{u} \in \mathbb{R}^K$ and any constant $c \in \mathbb{R}$, $(\mathbf{u})_k^c = [(\mathbf{u})_k]^c = u_k^c, 1 \leq k \leq K$.

$$\hat{\mathbf{u}} = \begin{pmatrix} (\mathbf{\Gamma}(\hat{\boldsymbol{\omega}})\mathbf{V}\mathbf{p}(\hat{\boldsymbol{\omega}}))_{k_0}^{1-\mu} \\ (\mathbf{\Gamma}(\hat{\boldsymbol{\omega}})\mathbf{z})_{k_0}^{1-\mu} \end{pmatrix} \qquad \check{\mathbf{u}} = \begin{pmatrix} (\mathbf{\Gamma}(\check{\boldsymbol{\omega}})\mathbf{V}\mathbf{p}(\check{\boldsymbol{\omega}}))_{k_0}^{\mu} \\ (\mathbf{\Gamma}(\check{\boldsymbol{\omega}})\mathbf{z})_{k_0}^{\mu} \end{pmatrix}.$$

By repeated application of (2.18), we obtain

$$\begin{aligned} p_{k_0}(\boldsymbol{\omega}(\mu)) &< \|\hat{\mathbf{u}}\|_{\frac{1}{1-\mu}} \|\check{\mathbf{u}}\|_{\frac{1}{\mu}} \\ &= \Big(\mathbf{\Gamma}(\hat{\boldsymbol{\omega}})\mathbf{V}\mathbf{p}(\hat{\boldsymbol{\omega}}) + \mathbf{\Gamma}(\hat{\boldsymbol{\omega}})\mathbf{z}\Big)_{k_0}^{1-\mu} \Big(\mathbf{\Gamma}(\check{\boldsymbol{\omega}})\mathbf{V}\mathbf{p}(\check{\boldsymbol{\omega}}) + \mathbf{\Gamma}(\check{\boldsymbol{\omega}})\mathbf{z}\Big)_{k_0}^{\mu} \\ &= p_{k_0}(\hat{\boldsymbol{\omega}})^{1-\mu} p_{k_0}(\check{\boldsymbol{\omega}})^{\mu}. \end{aligned}$$

This completes the proof.

Remarkably, there are no additional restrictions on $\mathbf{V} \geq 0$. As shown below, we obtain a similar property if we drop the requirement on strict log-convexity of $\gamma_k, 1 \leq k \leq K$, and instead put some constraints on $\mathbf{V}$.

Theorem 2.12. *Let $\gamma_k : \mathrm{Q}_k \to \mathbb{R}_{++}$ be log-convex for each $1 \leq k \leq K$. Suppose that $\mathbf{V} \in \mathbb{R}_+^{K \times K}$ is chosen such that for each $1 \leq l \leq K$, there exists $k \neq l$ with $v_{k,l} > 0$. Then, for any fixed $\hat{\boldsymbol{\omega}}, \check{\boldsymbol{\omega}} \in \mathrm{F}$ with $\hat{\boldsymbol{\omega}} \neq \check{\boldsymbol{\omega}}$, there exists $k_0, 1 \leq k_0 \leq K$, so that $p_{k_0}(\boldsymbol{\omega}(\mu)) < p_{k_0}(\hat{\boldsymbol{\omega}})^{1-\mu} p_{k_0}(\check{\boldsymbol{\omega}})^{\mu}$ for all $\mu \in (0,1)$.*

Proof. Let $\hat{\boldsymbol{\omega}}, \check{\boldsymbol{\omega}} \in \mathrm{F}$ with $\hat{\boldsymbol{\omega}} \neq \check{\boldsymbol{\omega}}$ be arbitrary. Since $\mathbf{p}(\boldsymbol{\omega})$ is a bijection (Lemma 2.10), we have $\mathbf{p}(\hat{\boldsymbol{\omega}}) \neq \mathbf{p}(\check{\boldsymbol{\omega}})$. Choose $l_0, 1 \leq l_0 \leq K$, such that

$$p_{l_0}(\hat{\boldsymbol{\omega}}) \neq p_{l_0}(\check{\boldsymbol{\omega}}) \tag{2.19}$$

and let $k_0 \neq l_0$ be any index with $v_{k_0,l_0} > 0$. Note that by assumption, there exists such an index. Using

$$\hat{\mathbf{u}} = \begin{pmatrix} \sum_{\substack{l=1 \\ l \neq l_0}}^{K} \big(\gamma_{k_0}(\hat{\omega}_{k_0}) v_{k_0,l} p_l(\hat{\boldsymbol{\omega}})\big)^{1-\mu} \\ (\gamma_{k_0}(\hat{\omega}_{k_0}) z_{k_0})^{1-\mu} \\ \big(\gamma_{k_0}(\hat{\omega}_{k_0}) v_{k_0,l_0} p_{l_0}(\hat{\boldsymbol{\omega}})\big)^{1-\mu} \end{pmatrix} \quad \check{\mathbf{u}} = \begin{pmatrix} \sum_{\substack{l=1 \\ l \neq l_0}}^{K} \big(\gamma_{k_0}(\check{\omega}_{k_0}) v_{k_0,l} p_l(\check{\boldsymbol{\omega}})\big)^{\mu} \\ (\gamma_{k_0}(\check{\omega}_{k_0}) z_{k_0})^{\mu} \\ \big(\gamma_{k_0}(\check{\omega}_{k_0}) v_{k_0,l_0} p_{l_0}(\check{\boldsymbol{\omega}})\big)^{\mu} \end{pmatrix}$$

and considering log-convexity of $\gamma_k, 1 \leq k \leq K$, one obtains

$$\begin{aligned} p_{k_0}(\boldsymbol{\omega}(\mu)) &= \Big(\mathbf{\Gamma}(\boldsymbol{\omega}(\mu))\mathbf{V}\mathbf{p}(\boldsymbol{\omega}(\mu)) + \mathbf{\Gamma}(\boldsymbol{\omega}(\mu))\mathbf{z}\Big)_{k_0} \overset{(a)}{\leq} \langle \hat{\mathbf{u}}, \check{\mathbf{u}} \rangle \\ &\overset{(b)}{\leq} \|\hat{\mathbf{u}}\|_{\frac{1}{1-\mu}} \|\check{\mathbf{u}}\|_{\frac{1}{\mu}} = p_{k_0}(\hat{\boldsymbol{\omega}})^{1-\mu} p_{k_0}(\check{\boldsymbol{\omega}})^{\mu} \end{aligned}$$

for any $\mu \in (0,1)$, where (a) follows from Theorem 2.5 and (b) from (2.18). Therefore, since $v_{k_0,l_0} > 0$ and $z_{k_0} > 0$, we can have equality in (b) only if $p_{l_0}(\hat{\boldsymbol{\omega}}) = p_{l_0}(\check{\boldsymbol{\omega}})$ which contradicts (2.19), and hence completes the proof.

It is important to emphasize that Theorems 2.11 and 2.12 do not imply the existence of an index k such that $p_k(\boldsymbol{\omega})$ is strictly log-convex on F. In fact, the theorems only asserts that for any fixed $\hat{\boldsymbol{\omega}}, \check{\boldsymbol{\omega}} \in \mathrm{F}, \hat{\boldsymbol{\omega}} \neq \check{\boldsymbol{\omega}}$, there is an index k such that $p_k(\boldsymbol{\omega}(\mu)) < p_k(\hat{\boldsymbol{\omega}})^{1-\mu} p_k(\check{\boldsymbol{\omega}})^{\mu}$ for all $\mu \in (0,1)$. However, this is sufficient to deduce the following corollary.

Corollary 2.13. *Suppose that at least one of the following holds.*

(i) For each $1 \le k \le K$, $\gamma_k : \mathrm{Q}_k \to \mathbb{R}_{++}$ *is strictly log-convex.*
(ii) Each column of the matrix $\mathbf{V}$ *has at least one positive entry.*

Then $\|\mathbf{p}(\boldsymbol{\omega})\|_1$ *is strictly log-convex on* F.

Proof. Let $\hat{\omega}, \check{\omega} \in \mathrm{F}$ with $\hat{\omega} \neq \check{\omega}$ be arbitrary. For any fixed $\mu \in (0,1)$, we have

$$\begin{aligned}\|\mathbf{p}(\boldsymbol{\omega}(\mu))\|_1 &= \sum_{k=1}^{K} p_k(\boldsymbol{\omega}(\mu)) \overset{(a)}{<} \sum_{k=1}^{K} (p_k(\hat{\boldsymbol{\omega}}))^{1-\mu}(p_k(\check{\boldsymbol{\omega}}))^{\mu} \\ &\overset{(b)}{\le} \Big(\sum_{k=1}^{K} p_k(\hat{\boldsymbol{\omega}})\Big)^{1-\mu}\Big(\sum_{k=1}^{K} p_k(\check{\boldsymbol{\omega}})\Big)^{\mu} = \|\mathbf{p}(\hat{\boldsymbol{\omega}})\|_1^{1-\mu}\|\mathbf{p}(\check{\boldsymbol{\omega}})\|_1^{\mu}\end{aligned}$$

where (a) is either due to Theorem 2.11 or due to Theorem 2.12 depending on whether (i) or (ii) holds, and (b) follows from (2.18).

Obviously, condition (ii) of the corollary, which is equivalent to the condition of Theorem 2.12, is weaker than irreducibility. For instance, the following two reducible matrices satisfy the condition of Theorem 2.12:

$$\mathbf{V} = \begin{pmatrix} 0\,1\,0\,0 \\ 1\,0\,0\,0 \\ 0\,0\,0\,1 \\ 0\,0\,1\,0 \end{pmatrix} \qquad \text{and} \qquad \mathbf{V} = \begin{pmatrix} 0\,1\,1 \\ 1\,0\,0 \\ 0\,0\,0 \end{pmatrix}.$$

With these particular choices of $\mathbf{V}$ and with $\gamma_1(x) = \gamma_2(x) = \gamma_3(x) = e^x, x \in \mathbb{R}$, we have (respectively)

$$\begin{aligned} p_1(\boldsymbol{\omega}) &= e^{\omega_1} p_2(\boldsymbol{\omega}) + e^{\omega_1} z_1 \\ p_2(\boldsymbol{\omega}) &= e^{\omega_2} p_1(\boldsymbol{\omega}) + e^{\omega_2} z_2 \\ p_3(\boldsymbol{\omega}) &= e^{\omega_3} p_4(\boldsymbol{\omega}) + e^{\omega_3} z_3 \\ p_4(\boldsymbol{\omega}) &= e^{\omega_4} p_3(\boldsymbol{\omega}) + e^{\omega_4} z_4 \end{aligned} \qquad \begin{aligned} p_1(\boldsymbol{\omega}) &= e^{\omega_1} p_2(\boldsymbol{\omega}) + e^{\omega_1} p_3(\boldsymbol{\omega}) + e^{\omega_1} z_1 \\ p_2(\boldsymbol{\omega}) &= e^{\omega_2} p_1(\boldsymbol{\omega}) + e^{\omega_2} z_2 \\ p_3(\boldsymbol{\omega}) &= e^{\omega_3} z_3 \,. \end{aligned}$$

In the first case, we see that $p_1(\boldsymbol{\omega})$ and $p_2(\boldsymbol{\omega})$ are strictly log-convex with respect to (ω_1, ω_2) but they are independent of (ω_3, ω_4). For $p_3(\boldsymbol{\omega})$ and $p_4(\boldsymbol{\omega})$, the situation is reversed so that $\|\mathbf{p}(\boldsymbol{\omega})\|_1$ remains strictly log-convex on F. In the second example, we can write $p_1(\boldsymbol{\omega})$ as

$$p_1(\boldsymbol{\omega}) = e^{\omega_1} \frac{e^{\omega_2} z_2 + e^{\omega_3} z_3 + z_1}{1 - e^{\omega_1} e^{\omega_2}}$$

which is strictly log-convex on F. In contrast, the transpose matrix $\mathbf{V} = \left(\begin{smallmatrix} 0\,1\,0 \\ 1\,0\,0 \\ 1\,0\,0 \end{smallmatrix}\right)$ does not satisfy the conditions of Theorem 2.12 since $v_{k,3} = 0$ for each $1 \le k \le K$. In this case, the nonnegative solution $\mathbf{p}(\boldsymbol{\omega}), \boldsymbol{\omega} \in \mathbb{R}^3$, is given by

$$p_1(\boldsymbol{\omega}) = e^{\omega_1}p_2(\boldsymbol{\omega}) + e^{\omega_1}z_1$$
$$p_2(\boldsymbol{\omega}) = e^{\omega_2}p_1(\boldsymbol{\omega}) + e^{\omega_2}z_2$$
$$p_3(\boldsymbol{\omega}) = e^{\omega_3}p_1(\boldsymbol{\omega}) + e^{\omega_3}z_3 \,.$$

We see that whereas $p_1(\boldsymbol{\omega})$ and $p_2(\boldsymbol{\omega})$ are independent of ω_3, $p_3(\boldsymbol{\omega})$ is a log-convex function of ω_3, though not strictly log-convex. Therefore, there is no index k such that $p_k(\boldsymbol{\omega})$ is strictly log-convex along the third coordinate of $\boldsymbol{\omega}$ (with ω_1 and ω_2 being fixed).

Finally we show that if $\mathbf{V}$ is irreducible, $p_k(\boldsymbol{\omega})$ is strictly log-convex on F for each $1 \leq k \leq K$, regardless of whether γ_k is strictly log-convex or only log-convex.

Theorem 2.14. *Let $\gamma_k : \mathrm{Q}_k \to \mathbb{R}_{++}, 1 \leq k \leq K$, be log-convex, and let $\mathbf{V} \in \mathrm{X}_K$. Then, $p_k(\boldsymbol{\omega})$ is strictly log-convex on* F *for each $1 \leq k \leq K$.*

Proof. Let $\hat{\boldsymbol{\omega}}, \check{\boldsymbol{\omega}} \in \mathrm{F}$ with $\hat{\boldsymbol{\omega}} \neq \check{\boldsymbol{\omega}}$ be arbitrary. Suppose that the theorem is false. Then, there exists k_0 and $\mu_0 \in (0,1)$ such that

$$p_{k_0}(\boldsymbol{\omega}(\mu_0)) = p_{k_0}(\hat{\boldsymbol{\omega}})^{1-\mu_0}p_{k_0}(\check{\boldsymbol{\omega}})^{\mu_0} \,.$$

So, by log-convexity of γ_k, Theorem 2.5 and Hölder's inequality,

$$\begin{aligned}
&p_{k_0}(\boldsymbol{\omega}(\mu_0)) \\
&\quad = \gamma_{k_0}(\omega_{k_0}(\mu_0))\Big(\sum_{l=1}^{K} v_{k_0,l}p_l(\boldsymbol{\omega}(\mu_0)) + z_{k_0}\Big) \\
&\quad \overset{(a)}{\leq} \gamma_{k_0}(\omega_{k_0}(\mu_0))\Big(\sum_{l=1}^{K} v_{k_0,l}p_l(\hat{\boldsymbol{\omega}})^{1-\mu_0}p_l(\check{\boldsymbol{\omega}})^{\mu_0} + z_{k_0}\Big) \\
&\quad \leq \gamma_{k_0}(\hat{\omega}_{k_0})^{1-\mu_0}\gamma_{k_0}(\check{\omega}_{k_0})^{\mu_0}\left(\sum_{l=1}^{K}(v_{k_0,l}p_l(\hat{\boldsymbol{\omega}}))^{1-\mu_0}(v_{k_0,l}p_l(\check{\boldsymbol{\omega}}))^{\mu_0} + z_{k_0}\right) \\
&\quad \leq \gamma_{k_0}(\hat{\omega}_{k_0})^{1-\mu_0}\gamma_{k_0}(\check{\omega}_{k_0})^{\mu_0} \\
&\qquad \cdot\left[\Big(\sum_{l=1}^{K} v_{k_0,l}p_l(\hat{\boldsymbol{\omega}})\Big)^{1-\mu_0}\Big(\sum_{l=1}^{K} v_{k_0,l}p_l(\check{\boldsymbol{\omega}})\Big)^{\mu_0} + z_{k_0}^{1-\mu_0}z_{k_0}^{\mu_0}\right] \\
&\quad \overset{(b)}{\leq} \Big(\gamma_{k_0}(\hat{\omega}_{k_0})\Big(\sum_{l=1}^{K} v_{k_0,l}p_l(\hat{\boldsymbol{\omega}}) + z_{k_0}\Big)\Big)^{1-\mu_0}\Big(\gamma_{k_0}(\check{\omega}_{k_0})\Big(\sum_{l=1}^{K} v_{k_0,l}p_l(\check{\boldsymbol{\omega}}) + z_{k_0}\Big)\Big)^{\mu_0} \\
&\quad = p_{k_0}(\hat{\boldsymbol{\omega}})^{1-\mu_0}p_{k_0}(\check{\boldsymbol{\omega}})^{\mu_0} \,.
\end{aligned}$$

So, in each step, we have equality. Now let $\mathsf{N}_1 \subset \{1,\ldots,K\}$ be a set of those indices l for which $v_{k_0,l} > 0$. As $\mathbf{V}$ is irreducible, we have $\mathsf{N}_1 \neq \emptyset$. Hence, since $\mathbf{z}$ is positive, it follows from (2.18) that there can be equality in (b) only if $\forall_{l \in \mathsf{N}_1}\, p_l(\hat{\boldsymbol{\omega}}) = p_l(\check{\boldsymbol{\omega}})$. Now suppose that $\mathsf{N}_2 \subset \{1,\ldots,K\}$ with $\mathsf{N}_2 \neq \mathsf{N}_1$ is a set of all indices l such that there exists $k_1 \in \mathsf{N}_1, k_1 \neq k_0$, with $v_{k_1,l} > 0$. Again, due to irreducibility of $\mathbf{V}$, it holds $\mathsf{N}_2 \neq \emptyset$. Moreover, since there is equality in (a) if only if $p_{k_1}(\boldsymbol{\omega}(\mu_0)) = p_{k_1}(\hat{\boldsymbol{\omega}})^{1-\mu_0} p_{k_1}(\check{\boldsymbol{\omega}})^{\mu_0}$ for each $k_1 \in \mathsf{N}_1$, we can reason along the same lines as above to show that $\forall_{l \in \mathsf{N}_2}\, p_l(\hat{\boldsymbol{\omega}}) = p_l(\check{\boldsymbol{\omega}})$. Now since $\mathbf{V}$ is irreducible, we can proceed in this way until there are no indices left to obtain

$$\forall_{1 \le k \le K}\ p_k(\hat{\boldsymbol{\omega}}) = p_k(\check{\boldsymbol{\omega}}).$$

Clearly, since $\mathbf{z}$ is positive and $\mathbf{p}(\boldsymbol{\omega})$ is a bijection, this implies that $\hat{\boldsymbol{\omega}} = \check{\boldsymbol{\omega}}$, which contradicts $\hat{\boldsymbol{\omega}} \neq \check{\boldsymbol{\omega}}$ and therefore completes the proof.

Figure 2.2 depicts $\|\mathbf{p}(\boldsymbol{\omega}(\mu))\|_1$ as a function of $\mu \in [0,1]$ for three different log-convex functions $\gamma(x) = \gamma_1(x) = \ldots = \gamma_K(x), x > 0$, and a randomly chosen irreducible matrix $\mathbf{V}$. Since $\gamma(x) = e^x/(1-e^x)$ is strictly log-convex on $\mathrm{Q} = (-\infty, 0)$ and $\gamma(x) = 1/x$ is strictly log-convex on $(0, +\infty)$, it follows from Theorem 2.11 that $\|\mathbf{p}(\boldsymbol{\omega})\|_1$ is strictly log-convex on Q^K. In contrast, $\gamma(x) = e^x$ is not strictly log-convex on $\mathbb{R}$. Nevertheless, since $\mathbf{V}$ is irreducible, Theorem 2.14 asserts that the l^1-norm is strictly log-convex.

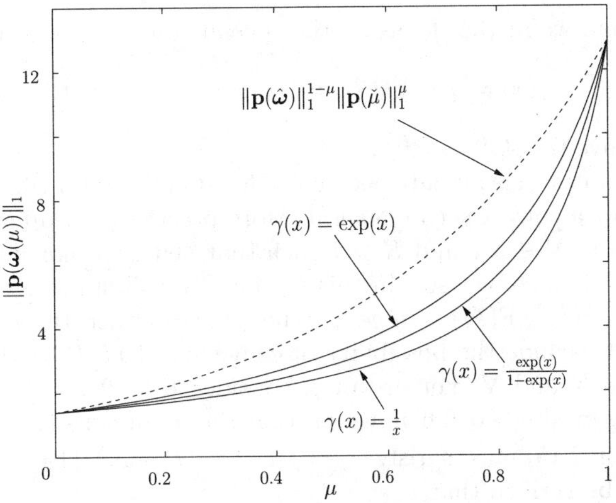

Fig. 2.2. The l^1-norm $\|\mathbf{p}(\boldsymbol{\omega}(\mu))\|_1$ as a function of $\mu \in [0,1]$ for some fixed $\hat{\boldsymbol{\omega}}, \check{\boldsymbol{\omega}} \in \mathrm{Q}^K$ chosen such that $\|\mathbf{p}(\hat{\boldsymbol{\omega}})\|_1$ and $\|\mathbf{p}(\check{\boldsymbol{\omega}})\|_1$ are independent of the choice of γ.

2.3.4 Strict Convexity of the Feasibility Sets

The results in the preceding section may be used to deduce strict convexity of the feasibility set in the following sense (see also Definition 1.41).

Definition 2.15. $\mathrm{F}(P_t)$ *(respectively,* $\mathrm{F}(P_1, \ldots, P_K)$*) is said to be strictly convex (or s-convex) if* $\boldsymbol{\omega}(\mu) = (1-\mu)\hat{\boldsymbol{\omega}} + \mu\check{\boldsymbol{\omega}}$ *is interior to* $\mathrm{F}(P_t)$ *(respectively,*

$\mathrm{F}(P_1,\ldots,P_K)$) *for all* $\mu \in (0,1)$ *and* $\hat{\omega}, \check{\omega} \in \partial\mathrm{F}(P_t)$ *(respectively,* $\hat{\omega}, \check{\omega} \in \partial\mathrm{F}(P_1,\ldots,P_K)$*),* $\hat{\omega} \neq \check{\omega}$*, where*

$$\begin{aligned} \partial\mathrm{F}(P_t) &= \{\boldsymbol{\omega} \in \mathrm{F} : \|\mathbf{p}(\boldsymbol{\omega})\|_1 = P_t\} \\ \partial\mathrm{F}(P_1,\ldots,P_K) &= \{\boldsymbol{\omega} \in \mathrm{F} : \exists_{1\le k\le K}\, p_k(\boldsymbol{\omega}) = P_k\}\,. \end{aligned} \tag{2.20}$$

Under the setup of Corollary 2.13, $\mathrm{F}(P_t)$ is a strictly convex set for all $P_t > 0$ since then $\|\mathbf{p}(\boldsymbol{\omega})\|_1$ is strictly log-convex. These conditions however are not necessary for $\mathrm{F}(P_t)$ to be a strictly convex set (see Example 2.3). As far as $\mathrm{F}(P_1,\ldots,P_K)$ is concerned, the set is strictly convex when $p_k(\boldsymbol{\omega})$ is strictly log-convex for each $1 \le k \le K$. Therefore, we have the following corollary.

Corollary 2.16. *Under the setup of Theorem 2.14,* $F(P_1,\ldots,P_K)$ *is a strictly convex set for any* $P_1,\ldots,P_K > 0$*.*

Of course, if $\mathrm{F}(P_1,\ldots,P_K)$ is strictly convex, so also is $F(P_t; P_1,\ldots,P_K)$.

2.4 The Linear Case

In this section, we further focus on the special case (2.17) except that now

$$\gamma(x) = \gamma_1(x) = \cdots = \gamma_K(x) = x, \quad x > 0\,.$$

Hence, we have $\Omega = \mathrm{Q}^K = \mathbb{R}^K_{++}$.

The linear case has already been considered in Sect. 1.5 where it is shown that F^c is not a convex set in general. More precisely, Theorem 1.57 asserts that there exist $\mathbf{V} \in \mathcal{X}_K$ and $K > 1$ such that neither F nor its complement $\mathrm{F}^c = \mathbb{R}^K_{++} \setminus \mathrm{F}$ is a convex set. In this section, we will use this result to show that $\mathrm{F}^c(P_t) = \mathbb{R}^K_{++} \setminus \mathrm{F}(P_t)$ is *in general* not convex either. However, note that this does not exclude the possibility of convexity of $\mathrm{F}^c(P_t)$ for some special choices of P_t, K and $\mathbf{V}$. For instance, consider $K = 2, \mathbf{z} = (1,1)$ and $\mathbf{V} = \left(\begin{smallmatrix} 0 & \varrho \\ \varrho & 0 \end{smallmatrix}\right)$ for any fixed $\varrho > 0$. Then, we see that the set of pairs $(\omega_1, \omega_2) \in \partial\mathrm{F}(P_t)$ (see Definition 2.15) must satisfy $\omega_2 = f(\omega_1) = (P_t - \omega_1)/(1 + 2\varrho\omega_1 + \varrho^2\omega_1 P_t)$. Now it may be verified that

$$f'(x) = \frac{-(1+\varrho P_t)^2}{(1+\varrho(2+\varrho P_t)x)^2}, \quad x > 0\,.$$

Thus, as the numerator is independent of x and the denominator is increasing in $x > 0$, we must have $f''(x) \ge 0$ for every $x > 0$. From this, it follows that $f(x)$ is not concave but convex on $\mathbb{R}_{++}$. As a consequence of this, $\mathrm{F}^c(P_t) = \mathbb{R}^2_{++} \setminus \mathrm{F}(P_t)$ is a convex set if $K = 2$ and $\gamma_1(x) = \gamma_2(x) = x, x > 0$.

As in Sect. 1.5, this simple example might suggest that $\mathrm{F}^c(P_t)$ is a convex set in general, which in turn would allow us to draw some interesting conclusions with respect to optimal scheduling in wireless networks. Unfortunately, simple reasoning shows that such a general statement is not possible.

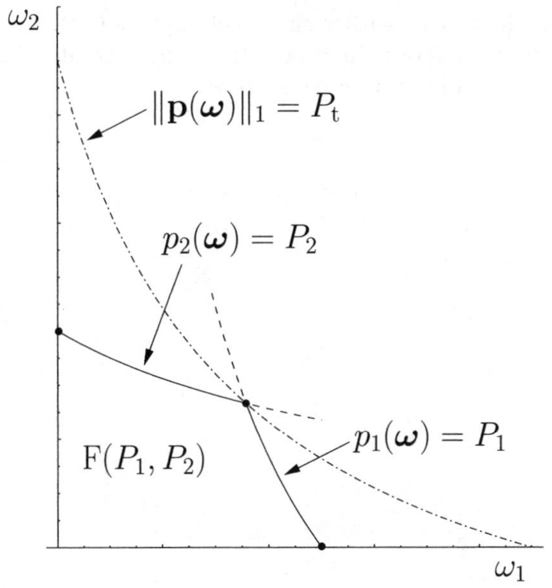

Fig. 2.3. $\mathrm{F}(P_1, P_2)$ is equal to the intersection of $\mathrm{F}_1(P_1)$ and $\mathrm{F}_2(P_2)$. Thus, $\mathrm{F}^c(P_1, P_2)$ is equal to the union of $\mathrm{F}_1^c(P_1)$ and $\mathrm{F}_2^c(P_2)$, each of which is a convex set if $\gamma(x) = x, x > 0$. However, the union of these sets is not convex in general.

Theorem 2.17. *There exist at least one* $P_t > 0$ *and an irreducible matrix* $\mathbf{V} \geq 0$ *for some* $K > 1$ *such that* $\mathrm{F}^c(P_t)$ *is not convex.*

Proof. The proof is by contradiction. So, assume that $\mathrm{F}^c(P_t)$ is convex for all $P_t > 0, K > 1$ and all $\mathbf{V} \in \mathrm{X}_K$. Therefore, as the intersection of convex sets is convex, it follows from (see (2.10))

$$\mathrm{F}^c = \bigcap_{P_t > 0} \mathrm{F}^c(P_t)$$

that F^c is a convex set for all $K > 1$ and all $\mathbf{V} \in \mathrm{X}_K$. However, this contradicts Theorem 1.57, and therefore prove the assertion.

Notice that the theorem only deals with the feasibility set when $\mathbf{p}(\boldsymbol{\omega})$ is constrained in the l^1-norm. When each element of $\mathbf{p}(\boldsymbol{\omega})$ is constrained individually, the complement of the feasibility set defined by (2.11) is not convex even if $K = 2$. Indeed, proceeding essentially as before shows that $p_1(\boldsymbol{\omega}) = P_1$ and $p_2(\boldsymbol{\omega}) = P_2$ are both convex if they are written explicitely as functions of ω_1. However, even though $\mathrm{F}_1^c(P_1)$ and $\mathrm{F}_2^c(P_2)$ are both convex sets, the set

$$\mathrm{F}^c(P_1, P_2) = (\mathrm{F}_1(P_1) \cap \mathrm{F}_2(P_2))^c = \mathrm{F}_1^c(P_1) \cup \mathrm{F}_2^c(P_2)$$

does not need to be convex as the union of convex sets is not convex in general. This is illustrated in Fig. 2.3. Obviously, the same reasoning applies to hybrid constraints, in which case neither the feasibility set $\mathrm{F}(P_{\mathrm{t}}; P_1, \ldots, P_K)$ given by (2.13) nor its complement is a convex set in general. This immediately follows from (2.13) and the discussion above.

Part II

Applications and Algorithms

3

Introduction

Wireless networking has been a vibrant research area over the last two decades. During this time, we have observed the evolution of a number of different wireless communications standards that support a wide range of services. They include delay-sensitive applications such as voice and real-time video that usually have strict requirements with respect to quality-of-service (QoS) parameters such as data rate, delay and/or bit error rate. In such cases, a network designer must ensure that the QoS requirements are satisfied permanently. Data applications, however, may have fundamentally different QoS requirements and traffic characteristics than video or voice applications. In fact, most data applications are delay-insensitive, and therefore may tolerate larger transmission delays.

The principal contributor to many of the problems and limitations that beset wireless networks is the radio propagation channel or, simply, the wireless channel. Transmission signals can be severely distorted by the wireless channel whose parameters such as path delay, path amplitude, and carrier phase shifts may vary with time and frequency. Strict limitation on communication resources such as the power and the bandwidth is another major design criterion. As a consequence, the wireless channel is error-prone and highly unreliable being subject to several impairment factors that are of transient nature, such as those caused by co-channel interference or multipaths. In fact, a unique characteristic of wireless networks being absent in wired networks is that the channel behavior is a function of the interference level and location of the subscriber unit. Excessive interference can significantly deteriorate the network performance and waste scarce communication resources. For this reason, strategies for resource allocation and interference management are usually necessary in wireless networks to provide acceptable QoS levels to the users. The resource allocation problem is significantly aggravated when subscriber units self-configure to form a network without the aid of any established infrastructure. These so-called ad hoc wireless networks have a huge potential for many exciting applications, but also pose new technical challenges.

There are different mechanisms for resource allocation and interference management in wireless networks. The most important ones include congestion control, routing, link scheduling and power control [33]. Each of these components of the overall network design can be targeted separately, thereby ignoring important interdependencies between them. Exploiting these interdependencies through a joint optimization of these components may lead to significant performance gains, but it may otherwise be computationally prohibitive to be of any use in practice. In this book, we mainly focus on the power control problem and briefly discuss the possibility of combining power control with a node-by-node congestion control. Roughly speaking, the power control problem addresses the issue of coordinating transmit powers of links such that some aggregate utility function of link rates attains its maximum. We are convinced that power control will be of great importance for wireless ad hoc networks. Due to the lack of a central network controller in such networks, link scheduling strategies are notoriously difficult to implement. Therefore, a reasonable approach is to avoid only strong interference from neighboring links, and then use an appropriate power control policy to manage the remaining interference in a network.

Early work on power control focused on the problem of maximizing the minimum signal-to-interference ratio (SIR) [34, 35, 36, 37, 38, 39, 40]. A closely related approach aims at satisfying given target SIR levels with a minimum total transmit power [41, 42, 43]. In the latter case, optimal power allocations can be found by means of iterative algorithms that allow distributed implementation, provided that the SIR requirements are feasible [36, 37, 42]. However, the notion of being able to guarantee quality of service to applications is simply unrealistic in many ad-hoc wireless networks [33]. The channel and network dynamics of such networks coupled with multi-hop routing make it difficult to ensure some requirements permanently. Moreover, a number of (elastic) data applications such as file transfer or electronic mail do not have such permanent requirements. Here, link QoS is provided according to some link prices and low QoS levels are temporarily acceptable. Therefore, in such cases, best-effort power control strategies aiming at maximizing some aggregate utility function of link rates (or other quantities) appears to be a more appropriate approach. Such strategies implicitly use the relative delay tolerance of data applications as well as the network and channel dynamics to improve the network performance [44, 45, 46, 47, 48, 49, 50, 51, 52, 53, 54, 55] (and references therein). At the same time, the use of increasing and strictly concave utility functions ensures the desired degree of (link-layer and end-to-end) fairness [56, 57, 53].

Unfortunately, the power control problem is not a convex problem in general. Yet the convexity property is a crucial prerequisite for implementing power control algorithms in practical systems as this property opens the door to a widely developed theory and efficient solutions. Moreover, if the problem is convex, a global convergence of the algorithms can be guaranteed.

Based on the theory presented in Chapter 2, we will identify a class of utility functions for which the power control problem can be transformed into a convex optimization problem. The new utility functions differ from traditional ones but, under a standard rate model in wireless systems, they are still increasing and strictly concave functions of link rates.

This part of the book is structured as follows. Chapter 4 introduces the network and system model, which includes a brief description of the medium access control (MAC) layer and detailed information about the physical layer. We consider two examples of wireless networks to illustrate the definitions. Chapter 5 formulates the problem of resource allocation in communications networks. Based on some currently existing approaches for rate control in wired networks, we formulate the utility maximization problem for elastic traffic in wireless networks, which then gives rise to a utility-based power control problem. Finally, Chapter 6 presents and analyzes distributed gradient-based power control algorithms.

4

Network Model

4.1 Basic Definitions

A wireless communications network is a collection of nodes being capable of communicating with each other over wireless communications links. Let $\mathsf{N} := \{1, \ldots, N\}$ be the set of nodes, and let (n, m) with $n \neq m$ represent a wireless link from node $n \in \mathsf{N}$ to node $m \in \mathsf{N}$. We say that there is a wireless link (n, m) with $n \neq m$ if both

(i) node n is allowed to transmit data to node m, and
(ii) a minimum signal-to-noise ratio (SNR), being necessary for successful transmission, can be achieved on link (n, m), in the absence of interference and with transmit power on this link subject to some power constraints.

It is reasonable to assume that wireless links are bidirectional in the sense that (n, m) exists if and only if there exists (m, n). We label links (in any particular way) by the integers $1, 2, \ldots, L$ and use $\mathsf{L} = \{1, \ldots, L\}$ to denote a set of all wireless links.[1] The pair (N, L) is referred to as the network topology. With any network, we associate the topology graph, which is an undirected graph where a vertex corresponds to a node in the network, and an edge between two vertices represents a wireless link between the corresponding nodes.

Messages originate at source nodes where they are usually broken into shorter strings of bits called packets. The packets are passed from node to node to their destinations according to some routing protocol. We assume that no packets travel in a loop and that, for every flow, there is a at least one path (a sequence of connected links) from source node to destination node. All nodes (including source and destination nodes) may act as relays with packets being decoded and encoded at each relay. We use an on/off flow model by which messages are characterized by a sequence of bits flowing into the

[1] If it is necessary to specify which nodes are connected by link $l \in \mathsf{L}$, then we write $l = l(n, m)$ when l is a wireless link from node n to node m.

network at a given rate. Successive message arrivals are separated by random durations (inter arrival times) in which no flow enters the network. Assume that there are S flows (packet streams) represented by $\mathsf{S} = \{1, 2, \dots, S\}$, each flow having a unique origin and destination. On the way to a destination, packets of a single flow can take different routes (Fig. 4.1). The expected traffic, in nats per time unit, of flow $s \in \mathsf{S}$ is denoted by ν_s. There are no special demands on the arrival statistics, except that the traffic should not be very bursty. Indeed, some form of power control and link scheduling usually helps to improve the network performance in case of continuous data stream or long packet bursts.

The flows share wireless links by competing for access to wireless resources such as power, time and frequency. If routes are fixed, nodes along the flow paths maintain per flow queuing, thereby establishing a number of logical links on wireless links, each logical link associated with a flow. Without loss of generality, assume that there are K logical links labeled by $1, \dots, K$. Let $\mathsf{K} = \{1, \dots, K\}$ be the set of these links defined such that the set of logical links originating at node $n \in \mathsf{N}$ is $\mathsf{K}(n) = \{\sum_{j=1}^{n-1} |\mathsf{K}(j)| + 1, \dots, \sum_{j=1}^{n} |\mathsf{K}(j)|\}$ where $|\mathsf{K}(n)|$ denotes the cardinality of $\mathsf{K}(n) \subseteq \mathsf{K}$ with $|\mathsf{K}(0)| = 0$ (Fig. 4.1). Connections over logical links are referred to as MAC (medium access control) layer flows, being one-hop flows between neighboring nodes.

It is important to point out that L (and hence also the network topology) may change over time due to mobility of nodes or other time varying factors. However, these variations are usually on a much larger time scale than frame intervals, and therefore are neglected in this book. Actually, for the theory and algorithms presented here, it is essential that the radio propagation channel remains constant for the duration of a frame interval, with transitions between different channel states occurring at the frame boundaries. At the beginning of every frame, transmit powers are adjusted to changed channel and network conditions.

4.2 Medium Access Control

The purpose of data link control (DLC) is to provide reliable data transfer across the physical link. To this end, the DLC layer places some overhead control bits at the beginning of each packet and some more overhead bits at the end of each packet, resulting in a longer string of bits called a frame. These overhead bits determine whether an error has occurred during the transmission and, if errors occur, they require retransmissions. These bits also determine where one data frame ends and the next one starts (framing).

Another important component of the DLC layer is medium access control (MAC). It is often considered as the lower layer of the DLC layer. The MAC protocols dictate how different logical links (MAC layer flows) share available communication resources such as power and bandwidth. Methods for dividing the spectrum into different channels (the so-called channelization)

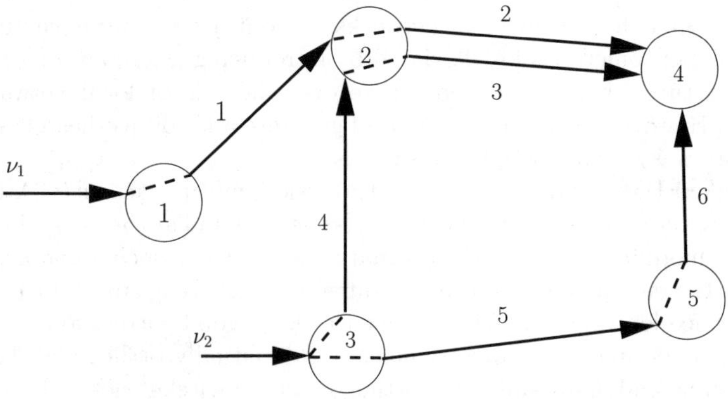

Fig. 4.1. There are five nodes represented by $\mathsf{N} = \{1,2,3,4,5\}$ and 10 wireless links: $(1,2),(2,1),(2,3),(3,2),(2,4),(4,2),(3,5),(5,3),(4,5),(5,4)$. The wireless links are not numbered in the figure. Two flows entering the network at nodes 1, 3 and destined for node 4 establish 6 logical links $\mathsf{K} = \{1,2,3,4,5,6\}$. For instance, logical links (or MAC layer flows) originating at node 2 are 2 and 3 so that we have $\mathsf{K}(2) = \{2,3\}$. These links share wireless link $(2,4)$. The flow rates are ν_1 and ν_2. Packets of flow 2 take two different routes to their destination that is node 4.

and assigning them to different links include time division multiple access (TDMA), frequency division multiple access (FDMA), code division multiple access (CDMA), and space division multiple access (SDMA) where the latter one is usually used in combination with TDMA or CDMA.[2] Obviously, hybrid combinations of all these methods are also possible.

FDMA is the oldest way for multiple radio transmitters to share the radio spectrum. Here each transmitter is assigned a distinct frequency channel so that receivers can discriminate among them by tuning to the desired channel. However, FDMA is inflexible and inefficient in handling flows with different bit rates. It would be necessary to modify FDMA so as to allocate frequency bands of different bandwidth to different logical links to accommodate differences in bit requirements. This requires simultaneous demodulation of multiple channels in different frequency bands, which is not a practicable solution. In case of TDMA, time is divided into nonoverlapping time slots. Each logical link is assigned one or multiple time slots such that there is only one link active at any time. TDMA is more flexible than FDMA in handling flows of various bit rates, but does not necessarily do this efficiently. The main difficulty with TDMA is the need for very accurate synchronization.

The efficiency of both FDMA and TDMA can be significantly improved by means of spatial reuse and dynamic allocation of bandwidth in terms of frequency or time. However, this requires a lot of coordination between nodes,

[2] Note that here the term "multiple access" refers to any situation where different logical transmitters (including those located at the same node) access the wireless channel.

which is difficult to achieve in networks without a fixed infrastructure. The problem can sometimes be alleviated by introducing a temporal hierarchical infrastructure where some nodes take over the role of local network controllers. However, such approaches still generate a lot of overhead traffic and therefore waste scarce wireless resources.

Unlike FDMA and TDMA, in code division multiple access (CDMA), the signal of every link occupies the entire frequency band at the same time. Each signal is modulated by a distinct signature sequence in such a manner that it enables the receivers to separate out different links. To ensure sufficiently low interference level, the signature sequences have good correlation properties. Often it is desired that some sequences are mutually orthogonal. However, establishing and maintaining the orthogonality in wireless networks is a quite tricky task. This is particularly true for fully asynchronous multipath channels in which case a complete elimination of multiple-access and intersymbol interference requires an allocation of signature sequences with zero aperiodic correlation side-lobes.[3] In fact, it was shown [59] that there are no such sequences in finite dimensional complex spaces. Yet the orthogonality can be established if all signals are at least coarsely synchronized at all receivers. One can also assign mutually orthogonal signature sequences to links originating at the same node. The problem is, however, that the number of mutually orthogonal sequences is strongly limited, making the reuse of sequences, and hence coordination between nodes necessary. Also maintenance of a coarse synchronization between different nodes may be a problem.

Consequently, the use of nonorthogonal sequences with relatively good correlation properties (semi-orthogonal sequences) appears to be a better strategy for some real world applications. The advantage of this approach is that there is no need for precise synchronization between logical links originating at different nodes. Moreover, little coordination is necessary if the number of sequences is relatively large. If the set is large enough, nodes can even pick up sequences randomly from a given set of sequences, with a low probability of choosing the same sequence. However, it should be noticed that there is a fundamental trade-off between the number of available signature sequences and their correlation properties. As a result, when semi-orthogonal sequences are used, the number of simultaneously active links is interference-limited. This means that the more links are active at the same time, the higher the level of interference, which in turn leads to the performance degradation of all links.

In the presence of interference, the network performance can be significantly improved by taking advantage of power control in combination with some link scheduling. These are two central mechanisms for resource allocation and interference management. Roughly speaking, whereas a link scheduling policy chooses groups of links that are to be activated at the same time,

[3] Aperiodic cross-correlations and autocorrelations not at the origin are referred to as aperiodic correlation side-lobes.

the power control part controls the interference level at the links by adjusting their transmit powers so as to achieve some sharing objectives. In order to achieve the best performance, power control and link scheduling should be optimized jointly. However, this problem is in general notoriously difficult to solve even in a centralized manner, not to mention the implementation of such policies in ad hoc wireless networks. To the best of our knowledge, there is no *efficient* distributed mechanism for assigning a number of time slots (subframes) to different wireless links. For these reasons, heuristics algorithms for link scheduling are quite common. A popular approach is to schedule neighboring links in different subframes, which is based on the common assumption that concurrent transmission of neighboring links will (with high probability) cause strong interference.

Our main focus in this book is on the power control problem for groups of interfering links that share the entire frequency band. We can assume any fixed link scheduling policy (in the time domain), including a suitable collision avoidance mechanism (see the remark in Sect. 4.3.1). Of course, pure TDMA is not of interest here since then the power control problem is trivial. However, it should be emphasized that the theory presented in this book does not necessarily apply to CDMA-based networks. For instance, interference may occur in any network with spatial reuse of resources. This is for instance true when multiple antennas are used to spatially separate different signals (see the example in Sect. 4.3.4).

4.3 Wireless Communication Channel

At the physical layer, frames are broken into shorter strings of bits and transmitted on logical links. The function of the physical layer is to provide logical links to the DLC layer while satisfying some quality of service requirements with respect to the bit loss and rate. To achieve this, there is a certain arrangement of several components on each side of a radio propagation channel (called a wireless channel) such as modulators, amplifiers, filters and mixers. The wireless channel distorts transmit signals in a way that can vary with time and system conditions. These distortions are usually of a random nature, and therefore cannot be exactly predicted. Still worse, wireless links share the available wireless channel, making each link prone to interference from other links. All this implies that the capacity of wireless links exhibits an ephemeral and dynamic nature, depending on both the wireless channel condition and transmit powers of all interfering links.

There is a huge amount of literature on physical layer methods for improving the overall network performance (see [60] and references therein). These methods include different multiuser and multiple antenna techniques whose purpose is to combat interference from other links as well as to mitigate the detrimental impact of the wireless channel. Such techniques make physical links robust against interference and channel variations, thereby increasing their capacity as well as simplifying the network design significantly. However, from the standpoint of practical design, there are some important disadvantages as well. First of all, most of these techniques may entail a significant increase of additional control traffic due to the increased demand on global information. So far, it is not clear whether the benefit of additional complexity outweighs these additional costs. Another important problem is an increased sensitivity of these methods against, for instance, erroneous channel state information. For these reasons, these techniques have not found wide usage in contemporary wireless networks.

Throughout this book, each link is a point-to-point communication link equipped with a linear receiver, followed by a single-user decoder. The interference at the output of each logical receiver is treated as noise. We adopt a block fading channel model in which the radio propagation channel holds its states for the duration of some frame interval, with transitions occurring on frame boundaries. This is a reasonable approximation for common radio propagation channels. For the sake of clarity, the channel is assumed to be flat, which, roughly speaking, means that each transmit signal is multiplied by some complex numbers, called the channel coefficients. Moreover, it is assumed that the frames of all links are perfectly synchronized in the sense that their beginnings and ends coincide. The common frame interval B of length $T_F = 1$ is partitioned into M disjoint intervals $\mathrm{T}(1), \ldots, \mathrm{T}(M)$ (called slots or symbol intervals) such that

$$\forall_{m \in \{1,\ldots,M\}}\ \mathrm{T}(m) = [(m-1)T, mT) \subset \mathrm{T}_F, \quad \bigcup_{m=1}^{M} \mathrm{T}(m) = \mathrm{B}\,.$$

The slot length T is chosen such that $M = T_F/T$ is large to ensure statistical significance (see also the following section). Each slot, say slot m, contains a certain number of information-bearing symbols $X_k(m), 1 \leq k \leq K$, transmitted on different links.[4] Since every symbol may carry one or several data bits, there is no loss in generality in assuming that every slot contains at most one symbol of link k. For simplicity, we also assume that baseband signals used to transmit the symbols on all links are square integrable functions on $[0, T]$. In particular, due to the perfect synchronization and the flat fading assumption, this implies that the information about $X_k(m), k \in \mathsf{K}$, at the kth logical transmitter is contained in time slot $\mathrm{T}(m)$.

Remark 4.1. If a baseband signal is time-limited (as assumed above), it cannot be band-limited. In such a case, one takes $W > 0$ as the bandwidth of a time-limited signal if "most" of its energy is contained in $(-W, W)$ [60].

Each logical receiver observes a superposition of all transmit signals corrupted by an additive white Gaussian noise. Let $y_k(t) \in \mathbb{C}, t \in \mathrm{T}(m)$, be an observation on link k in slot m, and let $c_k(t) \in \mathbb{C}, t \in [0, T)$, be a given square

[4] Throughout the book, we neglect the transmission of control symbols such as pilot or synchronization symbols.

integrable function associated with this link, which is called the kth logical receiver (or simply receiver). By far the most prominent example of $c_k(t)$ is the so-called matched-filter receiver [60, 61]. Given an arbitrary $1 \le m \le M$, the observation $y_k(t)$ in the interval $\mathrm{T}(m)$ is projected on $c_k(t-(m-1)T)$ to give

$$\hat{X}_k(m) = \int_{\mathrm{T}(m)} \overline{c_k(t-(m-1)T)} y_k(t) dt < +\infty, \quad 1 \le m \le M$$

which is bounded since

$$\forall_m \forall_k \int_{\mathrm{T}(m)} |y_k(t)|^2 dt < +\infty \quad \text{and} \quad \int_0^T |c_k(t)|^2 dt < +\infty .$$

The quantities $\hat{X}_k(1), \dots, \hat{X}_k(M)$ are referred to as soft-decision variables and are used to decode the information-bearing symbols $X_k(1), \dots, X_k(M)$ transmitted on link k. Consequently, it is desired that $\hat{X}_k(m)$ should be as close to $X_k(m)$ as possible (at least on average) with respect to a suitable performance measure. In all that follows, it is assumed that if $X_k(1), \dots, X_k(M)$ are chosen i.i.d. for every $1 \le k \le K$, then $y_k(t)$ is a realization of a stationary ergodic stochastic process.

4.3.1 Signal-to-Interference Ratio

A widespread and useful performance measure is the signal-to-interference+noise ratio (SIR). SIR gives the ratio of powers due to the desired link and due to all other interfering components at a soft-decision variable. To be precise, suppose that $X_k(1), \dots, X_k(M)$ are chosen i.i.d from some (finite) set and

$$E[X_k(m)] = 0 \quad \text{and} \quad E[|X_k(m)|^2] = p_k, \quad 1 \le m \le M, 1 \le k \le K . \tag{4.1}$$

Then, the *expected value* of SIR at the output of the kth logical receiver is independent of m (due to the stationarity) and is given by

$$\mathrm{SIR}_k(\mathbf{p}) := \frac{V_k p_k}{\sum_{l=1}^K p_l V_{k,l} + \sigma_k^2} = \frac{p_k}{\sum_{l=1}^K p_l \frac{V_{k,l}}{V_k} + \frac{\sigma_k^2}{V_k}}, \; 1 \le k \le K. \tag{4.2}$$

The notation in (4.2) is defined as follows.

- $p_k \ge 0$ is the transmit power on link k. We use

$$\mathbf{p} = (p_1, \dots, p_K) \in \mathbb{R}_+^K \tag{4.3}$$

  to denote a vector of transmit powers, referred to as power vector or power allocation.
- $V_k > 0$ is proportional to path gain on logical link k and depends on the spectrum allocation, various system parameters, as well as on the state of the wireless channel.

- $V_{k,l} \geq 0, l \neq k$, is a path gain (coupling factor) between link l and link k. In other words, if transmit power on link l is p_l, then the expected interference from this link on link $k \neq l$ is $p_l V_{k,l}$. Note that in general, we have $V_{k,l} \neq V_{l,k}$. If $V_{k,l} = 0$, then link k is said to be orthogonal to link l. As before, $V_{k,l}$ depends on the spectrum allocation, receiver structure, various system parameters, as well as the channel state. In general, there may be constraints in the simultaneous activation of links (e.g., nodes may not receive and transmit simultaneously). This situation can be captured by making both $V_{k,l}$ and $V_{l,k}$ sufficiently large for some $k \neq l$.
- $V_{k,k} \geq 0$ captures the effect of self- and intersymbol interference, which may occur, for instance, due to the time-dispersive nature of the wireless channel. In many cases of practical interest, it is reasonable to assume that $V_{k,k} = 0$ for every $1 \leq k \leq K$.
- $\sigma_k^2 > 0$ is the Gaussian noise variance at the output of the kth logical receiver.

Remark 4.2. In practice, strong interference is avoided using an appropriate collision avoidance protocol. If strong interference (collision) still occurs, then packets are lost and must often be retransmitted. Such protocols may be a part of a link scheduling policy (see Sect. 5.2.1). The design of collision avoidance protocols is beyond the scope of this book. In fact, we implicitly assume that collisions are perfectly avoided. For instance, when nodes may not receive and transmit simultaneously, this assumption means that wireless links constitute a matching in the topology graph. Recall that a matching is a set of edges such that no two of them have any common vertex [62].

It is convenient to write the quantities $V_k, V_{k,k}$ and $V_{k,l}$ in matrix form as follows

$$(\mathbf{V})_{k,l} = v_{k,l} = \begin{cases} \frac{V_{k,k}}{V_k} & k = l \\ \frac{V_{k,l}}{V_k} & k \neq l\,. \end{cases} \tag{4.4}$$

In a broader sense, the matrix $\mathbf{V}$ represents the effective state of the wireless channel and thus is referred to as the *channel state matrix*. The vector of effective noise variances

$$\mathbf{z} = \left(\frac{\sigma_1^2}{V_1}, \ldots, \frac{\sigma_K^2}{V_K}\right) \tag{4.5}$$

is called the (effective) noise vector. Note that this is actually the vector of normalized noise variances. We see from (4.2) that SIR depends on both $\mathbf{V}$ and $\mathbf{z}$.

Remark 4.3. Unless otherwise stated, it is assumed throughout the book that no self-interference is present, that is, $V_{k,k} = 0$ or, equivalently, trace($\mathbf{V}$) $= 0$. In fact, from the mathematical point of view, the self-interference presents no additional challenge. In particular, the algorithms presented in Chapt. 6 apply to systems with self-interference as well.

The primary message the reader should take away from this section is that SIR_k at the output of any receiver depends in general not only on transmit power on link k, but also on transmit powers of all other links. Furthermore, due to typically large values of M as well as due to the ergodicity of the received signal $y_k(t)$, it is reasonable to assume that the expected value of the signal-to-interference ratio SIR_k at every soft-decision variable is equal to the time-average SIR, with the average taken over all symbols transmitted in a given frame interval.

4.3.2 Power Constraints

In practice, there is a variety of system constraints. Most of them have only a marginal impact on the results presented in this book, and therefore are neglected for the sake of clarity. However, constraints on transmit powers must be incorporated into the system model since otherwise the results permit only crude insight into performance limits of practical networks.

Strict limitations on transmit powers in wireless networks result from a number of factors, including regulations, hardware costs and battery life. Most studies distinguish between two types of power constraints, namely peak constraint and average constraint. The first one is expressed in terms of the maximum crest factor (or peak-to-average power ratio (PAPR)) and is typically a result of some hardware constraints and regulations. Therefore, the peak power constraints usually pertain to individual physical communication links. In contrast, the average power constraint may be imposed on the overall transmit power in a network to reduce interference to adjacent networks as well as on individual nodes to prolong battery life.

The average transmit power is closely related to relevant performance measures such as data rate and bit error rate. As a consequence, some average transmit power on each logical link is necessary (but not always sufficient) to guarantee performance requirements of applications with regard to data rate and bit error rate. In this book, we assume individual power constraints on each node. To be precise, let $P_1, \ldots, P_N$ be positive real numbers, referred to as individual power constraints. Now we say that there are individual power constraints on each node if $\mathbf{p} \in \mathrm{P}$ where

$$\mathrm{P} := \mathrm{P}_1 \times \cdots \times \mathrm{P}_N, \quad \mathrm{P}_n := \left\{ \mathbf{x} \in \mathbb{R}_+^{|\mathsf{K}(n)|} : \sum_{k=1}^{|\mathsf{K}(n)|} x_k \leq P_n \right\}. \tag{4.6}$$

In other words, if $\mathbf{p}(n) = (p_k)_{k \in \mathsf{K}(n)}$ is the vector of average transmit powers on logical links originating at node $n \in \mathsf{N}$, then $\mathbf{p}(n) \in \mathrm{P}_n$. This model includes two types of power constraints often encountered in practice.

(i) Sum (or total) power constraint: A network is constrained only on total power so that $\|\mathbf{p}\|_1 = \sum_{k=1}^{K} p_k \leq P_{\mathrm{t}}$ for some given $P_{\mathrm{t}} > 0$. Constraints on total transmit power are typical for data transmission from a base station to mobile nodes. The best-known example is the down-link channel of a single-cell wireless cellular network. Total power constraints can also be imposed on transmit powers of mobile nodes to limit the radiation to other networks.

(ii) Individual power constraints on each link: This scenario corresponds to the situation where each node is a start point for exactly one logical link. Therefore, in this case, $p_k \leq P_k, k = 1, \dots, K$, for some given $P_1, \dots, P_K > 0$. A widely studied example is the up-link channel of a wireless cellular network.

Finally we point out that for the analysis presented in this book, the limitations on average transmit powers are of interest only when the noise variance σ_k^2 in (4.2) is relatively large in comparison with the interference factor $\sum_{l=1}^{K} p_l V_{k,l}$. Otherwise, if σ_k^2 is negligible when compared with the interference term and $\mathbf{V}$ is irreducible (Definition A.21), then it is justified to assume

$$\mathrm{SIR}_k(\mathbf{p}) \approx \mathrm{SIR}_k^0(\mathbf{p}) := \frac{p_k V_k}{\sum_{l=1}^{K} p_l V_{k,l}} .$$

This is a reasonable approximation for relatively large CDMA-based networks with pseudo-orthogonal spreading sequences. Due to the ray property

$$\forall_{c>0}\ \mathrm{SIR}_k^0(\mathbf{p}) = \mathrm{SIR}_k^0(c \cdot \mathbf{p}), \quad 1 \leq k \leq K$$

we see that if the Gaussian noise is neglected, the transmit power on each logical link can always be scaled down to satisfy given power constraints without influencing SIR values. Throughout the book, the effective noise vector $\mathbf{z}$ is assumed to be positive.

4.3.3 Data Rate Model

The data rate attainable on a wireless link is not fixed but depends in general on transmit powers, channel states and link scheduling policy involved. The data rate model under a link scheduling protocol is considered in Sect. 5.2.1. Now we assume that no link scheduling is involved which means that each link, say link $k \in \mathsf{K}$, is either active ($p_k > 0$) or idle ($p_k = 0$) during the whole frame interval. Then, given any channel matrix $\mathbf{V} \geq 0$ defined by (4.4), the data rate (in nats per channel use) on link k is a nonlinear function of the transmit power vector $\mathbf{p}$ and is given by

$$\nu_k(\mathbf{p}) = \Phi(\mathrm{SIR}_k(\mathbf{p})) \tag{4.7}$$

where $\mathrm{SIR}_k(\mathbf{p})$ is the signal-to-interference ratio defined by (4.2). Note that (4.7) is the data rate within a frame, and hence it may vary from frame to frame due to the changes of $\mathbf{V}$ and $\mathbf{p}$. Furthermore, note that the data rate on any link depends on transmit powers of other links. In other words, the data rates are interdependent since they are functions of global variables.

In this book, unless otherwise stated, it is assumed that

$$\nu_k(\mathbf{p}) = \Phi(\mathrm{SIR}_k(\mathbf{p})) = \kappa_1 \log(1 + \kappa_2 \mathrm{SIR}_k(\mathbf{p})), \quad k \in \mathsf{K} \tag{4.8}$$

where $\kappa_1, \kappa_2 > 0$ are some system constants, and $\log(x), x > 0$, is the natural logarithm.[5] The constant κ_1 depends primarily on the frequency bandwidth. Without loss of generality, it is assumed that $\kappa_1 = 1$. The constant $\kappa_2 > 0$ is dependent on a modulation scheme and the desired bit error rate. For simplicity, we also assume that $\kappa_2 = 1$.

It is important to emphasize that, for the analysis in this book, it is not necessary that the data rate on link k is exactly of the form given by (4.8). We will adhere to this common model for concreteness. In general, however, it is reasonable to assume that the rate-SIR relationship $\Phi : \mathbb{R}_+ \to \mathbb{R}_+$ satisfies the following conditions.

(i) Φ is a continuously differentiable and strictly increasing function.
(ii) $\Phi(x) \to 0$ as $x \to 0$ and $\Phi(x) \to +\infty$ as $x \to +\infty$.

Obviously, $\Phi(x) = \log(1 + x), x \geq 0$, satisfies both conditions. Note that due to the first condition, Φ is bijective (Definition B.4), and therefore there exists an inverse function $\Phi^{-1}(x) : \mathbb{R}_+ \to \mathbb{R}_+$ such that $\Phi(\Phi^{-1}(x)) = x, x \geq 0$ (Theorem B.5). In addition to these assumptions, in this book, the function Φ needs to be further restricted so as to guarantee that

$$U(x) = \Psi(\Phi^{-1}(x)), x > 0 \tag{4.9}$$

is increasing and strictly concave, where $\Psi : \mathbb{R}_{++} \to \mathbb{R}$ satisfies the Ψ-conditions 5.5. So if (4.8) holds, then $\Phi^{-1}(x) = e^x - 1, x \geq 0$ and $U(x)$ is increasing and strictly concave (see Sect. 5.2.5). But this requirement is also satisfied by the linear function $\Phi(x) = x, x \geq 0$.

4.3.4 Two Examples

Now we briefly illustrate the definitions introduced above by considering two examples of wireless communications networks.

A Cellular Network with Linear Beam-Forming Antennas

First consider a single-cell of a wireless cellular network with a multi-element antenna at the base station. No antenna arrays are considered for the mobiles. Such a network has a star topology with the base station acting as a central network controller. Due to the single-hop operation, no routing protocol is needed. Without loss of generality, we can assume that there is one logical

[5] In order to express data rate in bits per channel use, one should use the logarithm to the base 2.

link per wireless link. This in turn implies that there are as many down-links as flows (or users). This implies that there are $K = N - 1$ source destination pairs and no relays. If we assume that node 1 is the base station, then $l = l(n, m) \in \mathsf{L}$ is either a wireless link from the base station $n = 1$ to node $m \in \{2, \ldots, N\}$ or from node $n \in \{2, \ldots, N\}$ to the base station $m = 1$. The set $\mathsf{K}(1) \subset \mathsf{K}$ of wireless links originating at the base station (node 1) establishes the so-called down-link channel, whereas its complement $\mathsf{K} \setminus \mathsf{K}(1)$ constitutes the up-link channel from the mobile stations to the base station. In practice, down-links and up-links are used either in separate frame intervals (time division duplex (TDD) mode) or different frequency bands (frequency division duplex (FDD) mode). As a consequence the down-link and up-link channels can be considered separately with $\mathsf{K}(1)$ and $\mathsf{K} \setminus \mathsf{K}(1)$ as link sets for the down-link and up-link channel, respectively.

Let us first focus on the down-link scenario from the base station to $N - 1$ mobile nodes being arbitrarily distributed in a cell. As mentioned above, the base station is equipped with a multi-element antenna and each mobile station has an omnidirectional antenna (one antenna element). Suppose that there are $W \geq 1$ antenna elements at the base station. The data stream for each user, say user k, is spread over the antenna array by a vector $\mathbf{u}_k \in \mathbb{C}^W$ with $\|\mathbf{u}_k\|_2 = 1$, a so-called beam-forming vector. To be more precise, for an arbitrary slot m, consider the information-bearing symbols $X_1(m), \ldots, X_K(m)$ that are to be transmitted to the mobile nodes. The base station forms the vector $\mathbf{x}^T \mathbf{U}^H$ and transmits each element of this vector, say element j, over the jth antenna element, where $\mathbf{U} = (\mathbf{u}_1, \ldots, \mathbf{u}_K) \in \mathbb{C}^{W \times K}$ and $\mathbf{x} = (X_1(m), \ldots, X_K(m))$. The resulting transmit signals at each antenna element are distorted on their way to the mobile nodes. Here we focus on a multiplicative distortion meaning that the contribution of the jth antenna element to the received signal at node k is equal to $h_{j,k} \mathbf{x}^T \mathbf{U}^H \mathbf{e}_j$ where $h_{j,k} \in \mathbb{C}$ is usually referred to as the jth channel coefficient and $\mathbf{e}_j \in \{0, 1\}^W$ is the vector with 1 at the jth position and zeros elsewhere. The received signal at node k is a straightforward superposition of these contributions corrupted by a realization n_k of an independent Gaussian noise with the variance σ_k^2. As a result, the soft-decision variable $\hat{X}_k(m)$ is given by $\hat{X}_k(m) = \mathbf{x}^T \mathbf{U}^H \mathbf{h}_k + n_k$. The vector $\mathbf{h}_k = (h_{1,k}, \ldots, h_{W,k})$ is referred to as the channel signature of user k. It depends on channel and system parameters such as the array geometry, the relative position of a node to the base station, and the signal path attenuation. The soft-decision variable $\hat{X}_k(m)$ can be written as

$$\hat{X}_k(m) = \mathbf{x}^T \mathbf{U}^H \mathbf{h}_k + n_k = \underbrace{\mathbf{u}_k^H \mathbf{h}_k X_k(m)}_{\text{desired signal}} + \underbrace{\sum_{l \neq k} \mathbf{u}_l^H \mathbf{h}_k X_l(m) + n_k}_{\text{interference + noise}} .$$

Now if $X_k(m)$ are drawn i.i.d. from some zero-mean discrete probability distribution with $E[|X_k(m)|^2] = p_k$ (see also (4.1)), then the SIR measured at the antenna output of the kth logical receiver (over a sufficiently long frame interval) yields

$$\mathrm{SIR}_k(\mathbf{p},\mathbf{U}) = \frac{p_k \mathbf{u}_k^H \mathbf{R}_k \mathbf{u}_k}{\sum_{\substack{l=1 \\ l\neq k}}^K p_l \mathbf{u}_l^H \mathbf{R}_k \mathbf{u}_l + \sigma_k^2}, \qquad 1 \leq k \leq K \tag{4.10}$$

where the rank 1 matrix $\mathbf{R}_k = \mathbf{h}_k\mathbf{h}_k^H, 1 \leq k \leq K$, is called the spatial covariance matrix. We point out that if the channels are rapidly time-varying, the spatial covariance matrix $\mathbf{R}_k$ can be defined to be $\mathbf{R}_k = E[\mathbf{h}_k\mathbf{h}_k^H]$, in which case $\mathbf{R}_k$ may have full rank.

Now let us turn our attention to the up-link channel from mobile nodes to the base station. As before, it is assumed that there is one logical link per each wireless link, and that there are $K = N - 1$ users (flows) labeled by $1, \dots, K$. We use the same notation for the beam-forming vectors and channel signatures as in the case of the down-link channel. When compared with the down-link case, the roles in the up-link scenario are, in a sense, reversed with the antenna array acting as a linear receiver. Indeed, given an arbitrary slot m, the soft-decision variable is $\hat{X}_k(m) = \mathbf{u}_k^H\mathbf{y}$ where $\mathbf{y} \in \mathbb{C}^W$ is a vector whose jth entry is a sample of the received signal at the jth antenna element. As in the case of the down-link channel, each entry of $\mathbf{y}$ results from a superposition of different transmit signals corrupted by zero-mean Gaussian noise except that now each transmit signal is distorted by a user-specific channel signature. Thus, $\mathbf{y} = \sum_{l=1}^K \mathbf{h}_l X_l(m) + \mathbf{n}$ from which it follows that

$$\hat{X}_k(m) = \mathbf{u}_k^H \sum_{l=1}^K \mathbf{h}_l X_l(m) + \mathbf{u}_k^H\mathbf{n} = \underbrace{\mathbf{u}_k^H \mathbf{h}_k X_k(m)}_{\text{desired signal}} + \underbrace{\mathbf{u}_k^H \sum_{l\neq k} \mathbf{h}_l X_l(m) + n_k}_{\text{interference + noise}}$$

where $\mathbf{n} \in \mathbb{C}^W$ consists of the Gaussian noise samples at each antenna element and $n_k = \mathbf{u}_k^H\mathbf{n}$. Note that the interference term at the output of the kth logical receiver depends on the channel signatures of all other users but is independent of their beam-forming vectors. In the down-link channel, the situation is reversed with the interference term depending on the beam-forming vectors of all other users and being independent of their channel signatures. Thus, with the same assumptions on transmit symbols as before, we obtain

$$\mathrm{SIR}_k(\mathbf{p},\mathbf{U}) = \frac{p_k \mathbf{u}_k^H \mathbf{R}_k \mathbf{u}_k}{\sum_{\substack{l=1 \\ l\neq k}}^K p_l \mathbf{u}_k^H \mathbf{R}_l \mathbf{u}_k + \sigma_k^2}, \qquad 1 \leq k \leq K \tag{4.11}$$

where σ_k^2 is the noise variance. In fact, since $\|\mathbf{u}_k\|_2 = 1$ for each $1 \leq k \leq K$, we actually have $\sigma_1^2 = \cdots = \sigma_K^2$. To keep the model as general as possible though, the variances are allowed to be different. Note that in the down-link channel, the noise variances are in general different due to the existence of different receivers.

The best network performance can be achieved by jointly optimizing transmit powers and beam-forming vectors [63]. The theory presented in this book, however, targets networks with a classical approach of power control for fixed beam-formers. Due to its simplicity, this approach may be of interest in practice. A very simple and quite popular strategy is to choose $\mathbf{u}_k = \mathbf{h}_k, 1 \le k \le K$, in which case beam-forming vectors are said to be matched to channel signatures. As a consequence, if the channel signatures are fixed, so also are the beam-forming vectors. Moreover, both (4.10) and (4.11) are special cases of (4.2) with V_k and $V_{k,l} \ge 0$ given by

$$V_k = \mathbf{u}_k^H \mathbf{R}_k \mathbf{u}_k \qquad \text{and} \qquad V_{k,l} = \begin{cases} \mathbf{u}_l^H \mathbf{R}_k \mathbf{u}_l & \text{down-link}, k \neq l \\ \mathbf{u}_k^H \mathbf{R}_l \mathbf{u}_k & \text{up-link}, k \neq l \\ 0 & k = l\,. \end{cases}$$

From this, we can obtain the channel state matrix $\mathbf{V}$ defined by (4.4). It is interesting to point out that if $\mathbf{V}$ with $v_{k,l} = \mathbf{u}_k^H \mathbf{R}_l \mathbf{u}_k$ is the channel state matrix for the up-link channel,[6] then $\mathbf{V}^T$ is the channel state matrix for the down-link channel. This fact gives rise to the so-called duality theory for down-link and up-link multiuser beam-forming [63, 64]. This theory provides a framework for jointly optimizing power control and beam-forming in wireless networks.

A Distributed Network Based on Code Division Multiple Access

Finally we illustrate the definitions by considering a distributed wireless network based on code division multiple access (CDMA). We assume that all K logical links are perfectly synchronized as described in Sect. 4.3.

Let $J \ge 1$ be a common length of signature sequences, and suppose that logical link $k \in \mathsf{K}(n)$ is assigned a signature sequence $\mathbf{s}_k$ with $\|\mathbf{s}_k\|_2 = 1$, which is a vector in $\mathbb{C}^J$. In every time slot, say slot m, the logical transmitter on link k multiplies the signature sequence $\mathbf{s}_k$ by an information-bearing symbol $X_k(m)$ and transmits the resulting sequence elements at a rate of J/T. Note that the transmission rate is increased by the factor J, which is referred to as the spreading factor. Due to the perfect synchronization, we can drop the time index m and consider a discrete-time model where the logical receiver on link $k \in \mathsf{K}(n)$ originating at some node $n \in \mathsf{N}$ observes a vector of J samples $\mathbf{y}_k$ given by

$$\mathbf{y}_k = \underbrace{h_{k,k}\mathbf{s}_k X_k}_{\text{desired signal}} + \underbrace{h_{k,k} \sum_{l \in \mathsf{K}(n), l \neq k} \mathbf{s}_l X_l}_{\text{interference 1}} + \underbrace{\sum_{l \notin \mathsf{K}(n)} h_{k,l}\mathbf{s}_l X_l}_{\text{interference 2}} + \underbrace{\mathbf{n}}_{\text{noise}}, \quad k \in \mathsf{K}(n)\,.$$

Here, $\mathbf{n}$ is a zero-mean noise vector with $E[\mathbf{n}\mathbf{n}^H] = \sigma^2 \mathbf{I}$, $h_{k,l} \in \mathbb{C}$ with $|h_{k,k}| > 0$ is the channel coefficient between the logical transmitter of link

[6] This is true if $V_k = 1$ for each $1 \le k \le K$.

l and the logical receiver of link k, interference 1 is caused by other links originating at node n and interference 2 is due to all other links. Note that if $l \in \mathsf{K}(n)$ in the equation above, then $h_{k,l} = h_{k,k}$. In words, if logical link l originates at the same node as link $k \neq l$, then the impact of the transmit signal from link l at the output of the kth receiver is $h_{k,k}\mathbf{s}_l X_l$.

In the discrete-time domain, CDMA receivers are vectors in $\mathbb{C}^J$. Let $\mathbf{c}_k$ be the logical receiver of link k with $|\langle \mathbf{c}_k, \mathbf{s}_k \rangle| = 1$. Then, the soft-decision variable $\hat{X}_k = \langle \mathbf{c}_k, \mathbf{y}_k \rangle$ yields

$$\begin{aligned}\hat{X}_k &= \sum_{l \in \mathsf{K}} h_{k,l} \langle \mathbf{c}_k, \mathbf{s}_l \rangle X_l + n_k \\ &= h_{k,k} \langle \mathbf{c}_k, \mathbf{s}_k \rangle X_k + h_{k,k} \sum_{l \in \mathsf{K}(n), l \neq k} \langle \mathbf{c}_k, \mathbf{s}_l \rangle X_l + \sum_{l \notin \mathsf{K}(n)} h_{k,l} \langle \mathbf{c}_k, \mathbf{s}_l \rangle X_l + n_k\end{aligned}$$

where $n_k = \langle \mathbf{c}_k, \mathbf{n} \rangle$ and $k \in \mathsf{K}(n)$. Thus, $E[n_k] = 0$ and $E[|n_k|^2] = \|\mathbf{c}_k\|_2^2 \sigma^2 = \sigma_k^2$. Considering (4.1) and (4.2), we see that the signal-to-interference ratio at the soft-decision variable $\hat{X}_k$ is

$$\begin{aligned}\mathrm{SIR}_k(\mathbf{p}) &= \frac{|h_{k,k}|^2 p_k}{\sum_{l \neq k} |h_{k,l}|^2 p_l |\langle \mathbf{c}_k, \mathbf{s}_l \rangle|^2 + \sigma_k^2} \\ &= \frac{p_k}{\sum_{\substack{l \in \mathsf{K}(n) \\ l \neq k}} p_l |\langle \mathbf{c}_k, \mathbf{s}_l \rangle|^2 + \sum_{l \notin \mathsf{K}(n)} \frac{|h_{k,l}|^2}{|h_{k,k}|^2} p_l |\langle \mathbf{c}_k, \mathbf{s}_l \rangle|^2 + \frac{\sigma_k^2}{|h_{k,k}|^2}}, \quad k \in \mathsf{K}(n).\end{aligned}$$

Therefore, the channel state matrix $\mathbf{V}$ defined by (4.4) is given by

$$(\mathbf{V})_{k,l} = v_{k,l} = \begin{cases} \frac{|h_{k,l}|^2}{|h_{k,k}|^2} |\langle \mathbf{c}_k, \mathbf{s}_l \rangle|^2 & l \neq k \\ 0 & l = k \end{cases}$$

and the effective noise vector is $\mathbf{z} = (\frac{\sigma_1^2}{|h_{1,1}|^2}, \ldots, \frac{\sigma_K^2}{|h_{K,K}|^2})$.

5

Resource Allocation Problem in Communications Networks

This chapter formulates the resource allocation problem for wireless networks. Before that, however, we briefly discuss the fundamental trade off between efficiency and fairness in wired networks. This trade off eventually led researchers to consider the problem of maximizing the sum of increasing and strictly concave utility functions of source rates. We review some existing solutions to this problem and explain the insufficiency of these solutions in case of wireless networks. Section 5.2 reformulates the problem to better capture the situation encountered in wireless networks. We will argue in favor of MAC-layer fair policies that have already been used in wired networks as a basis to achieve end-to-end fairness. We precisely define the concept of joint power control and link scheduling as well as introduce the notion of the feasible rate region. It is shown that this set is not convex in general, which makes the optimization of wireless networks a fairly tricky task. The utility-based power control problem is formulated in Sect. 5.2.4. In particular, we introduce a class of increasing and strictly concave utility functions of link rates for which the power control problem can be converted into a convex optimization problem. The reader will realize a strong connection to the results of the first part of the book because the inverse functions of the considered utility functions are log-convex functions. Finally, we will utilize some results of Chapt. 2 to obtain valuable insights into the problem of joint power control and link scheduling.

5.1 End-to-End Rate Control in Wired Networks

A standard problem in network design concerns how the available bandwidth should be shared between competing flows to meet some share objectives [56, 65, 58, 57, 49] (and references therein). One possible objective is to allocate rates to the set of flows so as to maximize the total throughput subject to link capacity constraints. The main drawback of this strategy is that it may be quite unfair in the sense that some flows (users) may be denied access to the links [65]. Therefore, any rate control scheme must address the issue of fairness. One of the most common ideas of fairness is max-min fairness. The idea behind the max-min approach is to treat all users as fairly as possible by making their rates as equal as possible [65]. The main drawback of this approach is that such "perfect fairness" is usually achieved at the expense of a considerable drop in efficiency expressed in terms of total throughput. Indeed, there seems to be a fundamental trade off between throughput and fairness, with the throughput-optimal policy and max-min fair policy being two extremes of this trade off [58]. A common approach to balance the issue of fairness and efficiency is to maximize the aggregate (overall) utility of rate allocations represented through continuously differentiable, increasing, and strictly concave functions (the law of diminishing returns) [56, 65, 57].

In this section, we briefly discuss the utility maximization problem in wired networks and summarize some interesting results. In the next section, we build on these results to formulate the utility maximization problem in wireless networks.

5.1.1 Fairness Criteria

Consider a network with an established topology (N, L) and fixed routes for each flow. Let $\phi_l(s)$ with $\phi_l(s) = 0$ if $l \notin \mathsf{L}$ (no traffic is routed over nonexistent links) be a routing variable so that the product $\phi_l(s)\nu_s$ is the expected data rate of flow $s \in \mathsf{S}$ going through link $l \in \mathsf{L}$. Notice that in the special case of single-path routing, $\phi_l(s) = 1$ if flow s goes through link l and $\phi_l(s) = 0$ otherwise. Let $\boldsymbol{\nu} = (\nu_1, \ldots, \nu_S)$ be a vector of source rates. Then, the problem of end-to-end network utility maximization can be stated as follows

$$\max_{\boldsymbol{\nu} \geq 0} U(\boldsymbol{\nu}) \qquad \text{subject to} \qquad \forall_{l \in \mathsf{L}} \sum_{s \in \mathsf{S}} \phi_l(s)\nu_s \leq C_l \tag{5.1}$$

where C_l denotes a fixed capacity of wired link $l \in \mathsf{L}$ and $U : \mathbb{R}_+^S \to \mathbb{R}$ is a continuous, concave (strictly increasing in each entry) function representing the total utility of all flows. It is important to notice that link capacities are fixed and that flows can share wired links over both time and frequency. Furthermore, notice that (5.1) deals with the expected traffic and thus does not preclude the existence of traffic queues at the nodes. Any vector of source rates $\boldsymbol{\nu} \geq 0$ satisfying the link capacity constraints in (5.1) is called *feasible*.

The standard formulation of network utility maximization for elastic traffic is to maximize the sum of individual sources' utilities subject to the link capacity constraints [56]. In this case, $U(\boldsymbol{\nu})$ is of the form

$$U(\boldsymbol{\nu}) = \sum_{s \in \mathsf{S}} U_s(\nu_s), \quad U_s : \mathbb{R}_+ \to \mathbb{R} \tag{5.2}$$

where U_s is a continuously differentiable, strictly increasing, and concave function. Choosing $U_s(x) = x, x \geq 0$, for every $s \in \mathsf{S}$ turns (5.1) into the problem of maximizing the total end-to-end throughput. In general there are infinitely many throughput-optimal allocations for a given network topology. However, simple examples show that a necessary condition for attaining the maximum is that, roughly speaking, relatively long flows are allocated zero source rates. Therefore, throughput optimal policies are said to be unfair. The problem can be illustrated by means of a network with three flows and two links as depicted in Fig. 5.1.

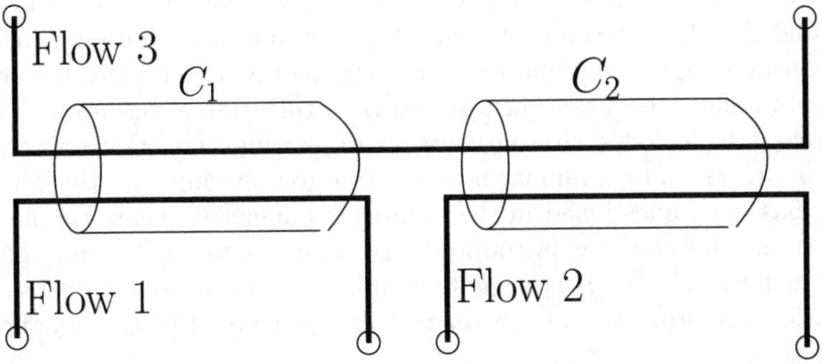

Fig. 5.1. Three flows compete for access to two links [58, 1]. Whereas flows 1 and 2 are one-link flows going through links 1 and 2, respectively, flow 3 uses both links. The links have fixed capacities C_1 and C_2, respectively. Clearly, the maximum total throughput is $C_1 + C_2$ and, in the maximum, the longer flow must be shut off ($\nu_3 = 0$) so that the one-link flows can be allocated rates of $\nu_1 = C_1$ and $\nu_2 = C_2$. In contrast, if $C_1 \leq C_2$, the max-min fair allocation is $\nu_1 = C_1/2, \nu_2 = C_2 - C_1/2$ and $\nu_3 = C_1/2$. Thus, the total throughput is $C_2 + C_1/2$ which is strictly smaller than $C_1 + C_2$. Note that if $C_1 = C_2$, then all source rates are equal under the max-min fair solution.

A contrary approach is to make the flow rates as equal as possible, which leads us to the concept of *max-min fairness*: A feasible flow rate vector $\boldsymbol{\nu}$ is defined to be max-min fair if any rate ν_s cannot be increased without decreasing some $\nu_r, r \neq s$, which is smaller than or equal to ν_s [1]. It is well known that the max-min fair rate allocation is unique when the number of resources and the number of flows are both finite. Furthermore, it can be derived using the following simple procedure: Starting from a zero rate allocation, increase uniformly the rate of each flow until the capacity constraint of some link is achieved. Freeze the rate allocations of the flows going through this link and continue the procedure for the remaining flows until all flows are constrained (e.g., [1, p. 526]).

The max-min fair approach is a *user-centric* approach in the sense that all users are treated fairly. In fact, we see from the definition above that, under the max-min fair policy, no flow is allocated a higher rate at the expense of other flows. This corresponds to an ideal "social" network, where all flows (users) are provided with data rates that are as close to each other as possible, regardless of how much resources each flow needs. As a result, a significant drop in the overall throughput should be expected, especially if there exist long flows going through many bottleneck links. Recall that a link $l \in \mathsf{L}$ is a bottleneck link with respect to a rate vector $\boldsymbol{\nu}$ for a flow $s \in \mathsf{S}$ if $\sum_{s \in \mathsf{S}} \phi_l(s)\nu_s = C_l$ and $\nu_s \geq \nu_{s'}$ for all flows s' going through link l [1]. For instance, considering the example in Fig. 5.1 shows with $C_2 > C_1$ that the bottleneck links of flows $1, 2$ and 3 with respect to the max-min fair allocation are links $1, 2$ and 1, respectively.

It is a matter of controversy whether the max-min fair rate allocation is desirable. As illustrated above, under the max-min fair solution, some flows may consume significantly more resources than others. Generally, the problem is how to balance between fairness and the utilization of resources. This led researchers to look for alternative ways of sharing network resources. The appropriateness of max-min fairness as a resource sharing objective for elastic traffic has been questioned in the landmark paper [56] where the notion of proportional fairness was introduced. A vector of rates $\boldsymbol{\nu}^*$ is proportionally fair if it is feasible ($\boldsymbol{\nu}^*$ is nonnegative and satisfies the capacity constraints) and if for any other feasible vector $\boldsymbol{\nu}$, the aggregate of proportional change is negative:

$$\sum_{s \in \mathsf{S}} \frac{\nu_s - \nu_s^*}{\nu_s^*} \leq 0\,.$$

Considering the Karush–Kuhn–Tucker [11] conditions for problem (5.1) with (5.2), it may be shown that $\boldsymbol{\nu}^*$ is a proportional fair rate allocation if and only if $\boldsymbol{\nu}^*$ solves (5.1) with $U(\boldsymbol{\nu}) = \sum_s \log \nu_s$. Thus, since $\log(x), x > 0$, is a strictly concave function, it may be inferred that proportional fair rates are unique. Reference [56] has also considered a weighted version of the proportional fairness criterion in which case flow rates ν_s are chosen so as to maximize $U(\boldsymbol{\nu}) = \sum_s w_s \log \nu_s$. The use of the weights has been advocated as a way for each user (associated with a flow) to choose the charge per unit time that the user is willing to pay. The user's rate as a result of optimization increases as the charge the user is willing to pay increases.

Instead of linear utility functions for throughput maximization, we have logarithmic functions in case of proportional fairness. Since $\log(x) \to -\infty$ as $x \to 0$, it is easy to see that each source rate is strictly positive under proportionally fair allocation. But this is actually the strict concavity property of the logarithm which forces fairness between sources. Indeed, whereas the rate of increase of $U_s(x) = x$ is the same for all $x \geq 0$, the rate of increase of $U_s(x) = \log(x)$ is decreasing in x (the law of diminishing returns), and hence smaller source rates are favored in the latter case. On the other hand, if the rate of increase does not decrease too rapidly, then the total throughput is improved in comparison with the max-min fair allocation. For instance, consider the network in Fig. 5.1. It may be easily verified that the proportional fair rate allocation satisfies the following set of equations:

$$\begin{cases} \nu_1 + \nu_3 = C_1 \\ \nu_2 + \nu_3 = C_2 \\ \nu_3 = \frac{\nu_1 \nu_2}{\nu_1 + \nu_2} \end{cases} \qquad \nu_1, \nu_2, \nu_3 \geq 0 .$$

Under the assumption of equal link capacities $C_1 = C_2 = C$, the solution to the above set of equations is given by $\boldsymbol{\nu}^* = (2C/3, 2C/3, C/3)$. The total throughput is $5C/3$, which is smaller than $2C$ (maximum throughput) but greater than $3C/2$ (max-min fair throughput). So, in this example, the introduction of a strictly concave utility function provides some balance between efficiency and fairness.

The notion of proportional fairness has been generalized by [57]. This generalization includes arbitrarily close approximation of max-min fairness. To be more precise, let $\mathbf{w} = (w_1, \ldots, w_S)$ be a positive vector, and let α be a nonnegative constant. Then a vector of rates $\boldsymbol{\nu}^*$ is said to be $(\mathbf{w}, \alpha)$-proportionally fair if it is feasible and for any other feasible vector $\boldsymbol{\nu}$:

$$\sum_{s \in \mathsf{S}} w_s \frac{\nu_s - \nu_s^*}{\nu_s^{*\alpha}} \leq 0 . \tag{5.3}$$

Obviously, if $\alpha = 1$ and $\mathbf{w} = \mathbf{1}$, $\boldsymbol{\nu}^*$ is a proportionally fair rate vector. Further examination reveals that (5.3) is equal to $\nabla U(\boldsymbol{\nu}^*)^T(\boldsymbol{\nu} - \boldsymbol{\nu}^*)$ with $U(\boldsymbol{\nu}) = \sum_s U_s(\nu_s)$ and $U_s : \mathbb{R}_{++} \to \mathbb{R}$ given by

$$U_s(x) = \begin{cases} w_s \frac{x^{1-\alpha}}{1-\alpha} & \alpha > 1 \\ w_s \log x & \alpha = 1 . \end{cases} \tag{5.4}$$

Consequently, since $U : \mathbb{R}^S_{++} \to \mathbb{R}$ is a strictly concave function and $U(\boldsymbol{\nu}) = U(\boldsymbol{\nu}^*) + \nabla U(\boldsymbol{\nu}^*)^T(\boldsymbol{\nu} - \boldsymbol{\nu}^*) + \frac{1}{2}(\boldsymbol{\nu} - \boldsymbol{\nu}^*)^T \nabla^2 U(\boldsymbol{\nu}^*)(\boldsymbol{\nu} - \boldsymbol{\nu}^*) + o(\|\boldsymbol{\nu} - \boldsymbol{\nu}^*\|_2^2)$, we have $U(\boldsymbol{\nu}) \leq U(\boldsymbol{\nu}^*) + o(\|\boldsymbol{\nu} - \boldsymbol{\nu}^*\|_2^2)$ for every feasible rate vector $\boldsymbol{\nu}$. So $\boldsymbol{\nu}^*$ is a local maximum of U. However, as $U(\boldsymbol{\nu}) = \sum_s U_s(\nu_s)$ with (5.4) has a unique global maximum, it follows that the $(\mathbf{w}, \alpha)$-proportionally fair rate vector maximizes $U(\boldsymbol{\nu}) = \sum_s U_s(\nu_s)$ over the set of all feasible rate vectors. The converse holds as well, which can be deduced from the associated Karush–Kuhn–Tucker conditions [57]. Summarizing, we can say that $\boldsymbol{\nu}^*$ is $(\mathbf{w}, \alpha)$-proportionally fair if and only if $\boldsymbol{\nu}^*$ solves (5.1) with $U(\nu_1, \ldots, \nu_S) = \sum_s U_s(\nu_s)$ and $U_s(x)$ given by (5.4). Furthermore, it is shown in [57] that the $(\mathbf{1}, \alpha)$-proportionally fair rate vector approaches the max-min rate vector as $\alpha \to \infty$ (see also Observation 5.8).

5.1.2 Algorithms

Given a network with fixed link capacities and a fixed number of sources (flows), the max-min fair rates for these sources can be easily computed by employing the filling procedure described in the previous section. Such a solution may be appropriate for small networks with an omniscient network controller that could easily compute the max-min fair rates and update them as the number of flows changes. Since this is impractical for moderately large networks, there are many publications on distributed max-min fair algorithms that dynamically adjust the source rates as the number of flows changes (see [1], pp. 528–530 and references therein). Most of those algorithms require some coordination and exchange of information between network nodes. An interesting exception is an approach suggested by [66, 67], where the authors show that max-min fairness can be achieved by performing per-flow *fair queuing* on all network links. More precisely, in this approach, each link offers a transmission slot to its flows by polling them in round-robin order. In addition, node-by-node window flow control is performed to prevent excessive packet queues at the network nodes. Now as the window size increases, the source rates approach the max-min fair rates. Finally, we mention reference [68]. This paper provides an asynchronous distributed algorithm that converges to the exact max-min fair rate allocation. In the proposed scheme, each source progressively discovers its rate allocation by comparing it with the "advertised rate" of the links on its route.

Note that the max-min fair utility function is not differentiable so that some standard optimization methods cannot be applied in this case. In contrast, proportionally fair objectives are continuously differentiable, increasing and strictly concave, therewith admitting a convex optimization formulation with zero duality gap. In [56], the authors proposed two algorithms (primal and dual) that arbitrarily closely approximate the $(\mathbf{w}, 1)$-proportionally fair rates. The primal algorithm changes the rate of flow $s \in \mathsf{S}$ according to the following system of differential equations:

$$\frac{d}{dt}\nu_s(t) = \kappa\Big(w_s - \nu_s(t)\sum_{l\in\mathsf{L}}\phi_l(s)\mu_l(t)\Big) \qquad \mu_l(t) = p_l\Big(\sum_{s\in\mathsf{S}}\phi_l(s)\nu_s(t)\Big)$$

with $\phi_l(s) \in \{0, 1\}$ (single-path routing) where κ is a positive constant and $p_l(x) = (x - C_l + \epsilon)^+/\epsilon^2, \epsilon > 0$, is a nonnegative, continuous, and increasing function. In words, each source, say source s, gets feedback $\mu_l(t)$ (related to residual capacity on link l) from the links and gradually changes its rate as follows: Increase the rate linearly proportional to w_s and decrease it multiplicatively proportional to total feedback. In the dual algorithm, instead of rates, the Lagrange multipliers (shadow prices) $\mu_l(t)$ are adjusted gradually, with rates given as functions of the shadow prices. Algorithms for computing $(\mathbf{w}, \alpha)$-proportionally fair rates have been developed in [57]. Here, each source adjusts its window size based on the total delay. This stands in contrast to [56] where flow rates are calculated explicitly.

Another interesting work is [65], where the static regime of a network with perfectly fluid flows is considered. Given a fixed end-to-end window control, the authors have showed that different fair rate allocation objectives can be met by implementing different queuing disciplines in network nodes, provided that the network is not too congested. For instance, it turns out that if round trip delays are negligible, then the static rates under the FIFO (first in first out) queuing discipline are $(\mathbf{w}, 1)$-proportionally fair rates with the weights being equal to the window sizes. In contrast, the maximum throughput allocation is achieved with *longest queue first* policy if the round trip delay is small. There is a similar conclusion for $(\mathbf{w}, 2)$-proportional fair rates if each node maintains per-flow queuing with service rate for each queue proportional to the square root of the queue size. These results as well as the work of [66, 67] show that network-wide (end-to-end) fairness can be also achieved if each node executes an appropriate contention resolution algorithm. These results may serve as a motivation for MAC layer fair power control algorithms for wireless networks presented later in this book.

5.2 Problem Formulation for Wireless Networks

In the previous section, we have briefly outlined the problem of rate control in wired networks. In what follows we turn our attention to wireless networks. The first question which may arise is the following: *Is there something fundamental about the nature of wireless networks that prevents us from reusing the well-developed techniques for wired networks?* In fact, one of the most important unique features was already mentioned in Sect. 4.3.3, namely that the data rate achievable on any link is a nonlinear function of global variables such as transmit powers and channel states of all links. Moreover, the channel state matrix $\mathbf{V} \geq 0$ can be only partially influenced (if at all) since it depends on relative locations of nodes and other objects (scatters) in the vicinity of a network. Therefore, even if the nodes are stationary and the network topology is fixed, $\mathbf{V}$ is not known in advance as the channel states can vary due to the mobility of these objects. In a mobile network environment with nodes changing their positions permanently, the network topology is not known in advance either and the process of route discovery and maintenance may consume a lot of wireless resources.

Due to the variation of the wireless radio environment, the capacity of wireless links exhibits an ephemeral and dynamic nature. This stands in clear contrast to wired networks where the capacity of any link is fixed and independent of the transmission rate on other links. In wireless networks, nodes in general do not even know the exact capacities of their own links because, as mentioned above, the capacity of each link depends on some global network variables. Due to this mutual dependence, it is clear that the scope of the utility maximization problem (5.1) is limited for wireless networks. Furthermore, when designing protocols and algorithms for wireless networks, coordination between nodes should be reduced to a minimum in order to save wireless resources. This suggests the development of smart strategies for resource allocation and interference management that achieve network-wide fairness with minimum global coordination. In this book, we argue in favor of power control and link scheduling policies designed to ensure fairness at the MAC layer (MAC layer fairness; see Sect. 5.2.4 for more details). This provides a better utilization of scarce resources. In a sense, such an approach can be viewed as an extension of the work done by [66, 67, 65] to wireless networks.

Remark 5.1. It must be emphasized that the problem of maximizing the aggregate utility of flow (or link) rates is not appropriate for all scenarios. For instance, if the power supply is a bottleneck (like in sensor networks), then the throughput and fairness performance should be balanced against power consumption to prolong battery life [69, 45, 46]. Considering only the throughput performance would discharge the batteries after a relatively short time. Therefore, the rate control strategies presented in this book are not applicable to wireless networks where mobile devices are equipped with low-capacity batteries.

The following section introduces the notion of joint power control and link scheduling. This model is used in Sect. 5.4.1 to provide some interesting insights into the design of throughput-optimal MAC policies. However, as mentioned before, the main focus is on the power control problem for a given link scheduling policy. In the face of implementation constraints, this seems to be a reasonable approach in many cases of practical interest.

5.2.1 Joint Power Control and Link Scheduling

Now let us introduce the notion of joint power control and link scheduling (JPCLS). Our definition is tailored to better illustrate throughput-optimal MAC policies discussed in Sect. 5.4.1.

Roughly speaking, a JPCLS policy is a (distributed or centralized) mechanism of the MAC layer that divides every frame into a finite number of perfectly synchronized subframe intervals, assigns a group of logical links to each subframe, and allocates transmit powers to them. In what follows, we formalize these ideas. To this end, let B be a bounded interval on the real line, and let $\Lambda := \{1, 2, \dots, |\Lambda|\}$, where $|\Lambda|$ is usually significantly smaller than the number of time slots M (symbol intervals) in each frame. Assume that $\mathsf{A} = \{\mathrm{B}_n : n \in \Lambda\}$ is a given system of subsets of B with

$$\bigcup_{n\in\Lambda} \mathrm{B}_n = \mathrm{B} \qquad\qquad \forall_{\substack{n,m\in\Lambda \\ n\neq m}}\ \mathrm{B}_n \cap \mathrm{B}_m = \emptyset\,.$$

In words, A partitions B into a finite number of disjoint sets B_n. We use $\mu : \mathsf{A} \to [0, 1]$ to represent any real (set) function such that

$$\forall_{n\in\Lambda}\ \mu(\mathrm{B}_n) \geq 0 \quad \mu(\emptyset) = 0 \quad \mu\Big(\bigcup_{n\in\Lambda} \mathrm{B}_n\Big) = \sum_{n\in\Lambda} \mu(\mathrm{B}_n) = \mu(\mathrm{B}) = 1\,. \tag{5.5}$$

Furthermore, with each logical link $k \in \mathsf{K}$, we associate a set function $p_k : \mathsf{A} \to \mathbb{R}_+$. For any given B and A, any functions $\mu : \mathsf{A} \to [0, 1]$ (satisfying

(5.5)) and $\mathbf{p} = (p_1, \ldots, p_K) : \mathsf{A} \to \mathbb{R}_+^K$ have the following interpretation: The expected data rate (in nats per channel use) on link $k \in \mathsf{K}$ is equal to

$$\nu_k(\mathbf{p}, \mu) = \sum_{n\in\Lambda} \mu(\mathrm{B}_n)\Phi(\mathrm{SIR}_k(\mathbf{p}(\mathrm{B}_n)))$$

where

$$\mathrm{SIR}_k(\mathbf{p}(\mathrm{B}_n)) = \frac{p_k(\mathrm{B}_n)V_k}{\sum_{l=1}^K p_l(\mathrm{B}_n)V_{k,l} + \sigma_k^2}, \qquad n \in \Lambda\,. \tag{5.6}$$

Note that in the special case when $\mu(\mathrm{B}_1) = \mu(\mathrm{B}) = 1$ and $\mu(\mathrm{B}_n) = 0$ for all $n > 1$, we have $\mathrm{SIR}_k(\mathbf{p}(\mathrm{B}_1)) = \mathrm{SIR}_k(\mathbf{p})$ where $\mathrm{SIR}_k(\mathbf{p})$ is defined by (4.2).

Definition 5.2. *Given* B *and* A, *link scheduling refers to the operation of choosing* $\mu : \mathsf{A} \to [0, 1]$ *satisfying (5.5), while power control determines* $\mathbf{p}$: $\mathsf{A} \to \mathbb{R}_+^K$. *A mechanism that* jointly *determines the pair*

$$(\mathbf{p}, \mu) : \mathsf{A} \times \mathsf{A} \to \mathbb{R}_+^K \times [0, 1] \tag{5.7}$$

is called joint power control and link scheduling (JPCLS). If $\mu(\mathrm{B}_1) = \mu(\mathrm{B}) = 1$, *then we say that there is no link scheduling involved.*

Throughout the book, we adopt the following assumptions and interpretations of the above definitions.

(i) B is referred to as a frame, while $\mathrm{B}_n, n \in \Lambda$, is the nth subframe. The subframes are ordered in any particular way.
(ii) $\mathbf{p}(\mathrm{B}_n)$ is a vector of transmit powers allocated to links in subframe B_n. If $p_k(\mathrm{B}_n) > 0$, we say that link k is active in B_n, otherwise it is said to be idle.
(iii) $\mu(\mathrm{B}_n)$ is the fraction of the frame occupied by subframe B_n. Also, $\mu(\mathrm{B}_n)$ can be viewed as the relative frequency at which the power vector $\mathbf{p}(\mathrm{B}_n)$ is utilized. If $\mu(\mathrm{B}_n) = 0$, then the power vector $\mathbf{p}(\mathrm{B}_n)$ is not utilized.
(iv) $\mathrm{SIR}_k(\mathbf{p}(\mathrm{B}_n))$ is the signal-to-interference ratio at each soft-decision variable in subframe B_n. It is assumed that every *nonempty* subframe is large enough (in terms of the number of transmitted symbols) to ensure that SIR defined by (5.6) is close to the time average SIR. If $\mu(\mathrm{B}_n) = 0$, then the average SIR in B_n is equal to zero.

As a consequence, $\Phi(\mathrm{SIR}_k(\mathbf{p}(\mathrm{B}_n)))$ is equal to the time average rate on link k in subframe $\mathrm{B}_n \in \mathsf{A}$. Choosing μ in Definition 5.2 is equivalent to determining the lengths of the subframes, and therefore this operation can be viewed as time slot management, where groups of symbol intervals are merged to form subframes. In practice, there is usually a fixed division of a frame into subframes whose lengths are multiples of T (the length of a single symbol interval). Link scheduling then refers to the operation of assigning links to the subframes. Power control in turn allocates transmit powers to links in each subframe. Link scheduling may be implemented in a centralized or decentralized manner. In the first case, there is a central scheduler that coordinates time slot management and link assignment across the network. A distributed implementation requires coordination between local link schedulers at every node.

Remark 5.3. In this book, the definition of link scheduling is slightly different from that described above. According to Definition 5.2, link scheduling determines the lengths of all subframes and assigns all links to each of them. Power control determines transmit powers of all links in each subframe. In other words, the actual task of assigning links to subframes is carried out by power control if we assume that a link is assigned to some subframe if and only if it is active in this subframe.

Power Constraints under Link Scheduling

In Sect. 4.3.2, we have specified constraints on transmit powers that can be dissipated over the duration of the frame interval when no link scheduling is involved. The average transmit power on each link is assumed to be approximately equal to the expected transmit power in any symbol interval. Thus, if the expected transmit power in every symbol interval is kept constant, then the expected transmit power over a frame period decreases when the link active time decreases. So the question arises whether the links can compensate the power loss by increasing their (expected) transmit powers over the active time periods. Formally, the question is which of the following should hold:

$$\sum_{n\in\Lambda} \mathbf{p}(\mathrm{B}_n)\mu(\mathrm{B}_n) \in \mathrm{P} \tag{5.8}$$

or

$$\forall_{n\in\Lambda}\ \mathbf{p}(\mathrm{B}_n) \in \mathrm{P} \tag{5.9}$$

where $\mathbf{p}(\mathrm{B}_n)$ is a vector of the expected transmit powers used in subframe B_n. Hence, if subframes are sufficiently long, condition (5.8) limits the average transmit powers that can be dissipated over a frame interval. In contrast, there are no constraints on the entries of $\mathbf{p}(\mathrm{B}_n)$ which may become arbitrarily large as $\mu(\mathrm{B}_n) \to 0$. This implicitly requires amplifiers with an ideal linear transfer characteristic. However, practical amplifiers have nonlinear characteristics and will (hopefully) go into saturation beyond a certain limit. Therefore, from a practical point of view, it is reasonable to assume that (5.9) holds. In this case, we say that transmit powers are subject to (MAC layer) peak power constraints.

Note that if there are peak power constraints, the maximum average transmit power decreases as time occupied by a link decreases. Thus, link scheduling is less attractive in the case of peak power constraints. With these constraints, it follows from the definitions above that the data rate on link k under a JPCLS policy $(\mathbf{p}, \mu)$ is

$$\nu_k(\mathbf{p}, \mu) = \sum_{n \in \Lambda} \mu(\mathrm{B}_n) \Phi(\mathrm{SIR}_k(\mathbf{p}(\mathrm{B}_n))) \tag{5.10}$$

where $\mathbf{p}(\mathrm{B}_n) \in \mathrm{P}$ for each $n \in \Lambda$. Note that whereas $\mathbf{p}$ in (4.7) is a vector of transmit powers, $\mathbf{p}$ in (5.10) is a vector of set functions defined on A, each of which is subject to the (MAC-layer) peak power constraints.

5.2.2 Feasible Rate Region

The set of all achievable data rate vectors $\boldsymbol{\nu}(\mathbf{p}) = (\nu_1(\mathbf{p}), \ldots, \nu_K(\mathbf{p})) \in \mathbb{R}_+^K$ is called the feasible rate region. It is a set of all data rates that are achievable on wireless links under a given coding strategy. Hence, this notion is distinct from the information theoretic capacity region, which includes optimization over all possible coding schemes. When no link scheduling is involved ($\mu(\mathrm{B}_1) = 1$ and $\Lambda = \{1\}$), the feasible rate region $\mathrm{C} \subset \mathbb{R}_+^K$ is given by

$$\mathrm{C} := \{\boldsymbol{\omega} \in \mathbb{R}_+^K : \boldsymbol{\omega} \le \boldsymbol{\nu}(\mathbf{p}), \mathbf{p} \in \mathrm{P}\} \tag{5.11}$$

where P is the set of feasible power vectors defined by (4.6).

When dealing with the utility maximization problem, one of the main difficulties stems from the fact that the feasible rate region is not a convex set in general. This stands in clear contrast to wired networks where the capacity region is a box. To see that C is not convex in general, let $\boldsymbol{\omega} \in \mathrm{C}$ be arbitrary. Hence, by the definition of C, it follows that there exists $\mathbf{p} \in \mathrm{P}$ such that $\omega_k \le \Phi(\mathrm{SIR}_k(\mathbf{p}))$ for each $k \in \mathsf{K}$. Since $\Phi : \mathbb{R}_+ \to \mathbb{R}_+$ is a bijection with $\phi(0) = 0$ (Definition B.4), we can rewrite this set of inequalities using the inverse function $\Phi^{-1}(x)$ as

$$\Phi^{-1}(\omega_k)\Big(\sum_{l=1}^{K} v_{k,l} p_l + z_k\Big) \le p_k, \quad k \in \mathsf{K}\,.$$

In vector form, this becomes

$$\boldsymbol{\Gamma}(\boldsymbol{\omega})\mathbf{z} \le (\mathbf{I} - \boldsymbol{\Gamma}(\boldsymbol{\omega})\mathbf{V})\mathbf{p} \tag{5.12}$$

where $\boldsymbol{\Gamma}(\boldsymbol{\omega}) = \mathrm{diag}(\Phi^{-1}(\omega_1), \ldots, \Phi^{-1}(\omega_K))$ and $\mathbf{z} > 0$ is the noise vector defined by (4.5). Now Theorem A.35 implies that if $\rho(\boldsymbol{\Gamma}(\boldsymbol{\omega})\mathbf{V}) < 1$, there exists a unique vector $\mathbf{p}(\boldsymbol{\omega}) = (p_1(\boldsymbol{\omega}), \ldots, p_K(\boldsymbol{\omega})) \ge 0$ given by[1]

$$\mathbf{p}(\boldsymbol{\omega}) := (\mathbf{I} - \boldsymbol{\Gamma}(\boldsymbol{\omega})\mathbf{V})^{-1}\boldsymbol{\Gamma}(\boldsymbol{\omega})\mathbf{z}\,. \tag{5.13}$$

Conversely, if $\mathbf{p}(\boldsymbol{\omega}) \ge 0$ exists, then $\boldsymbol{\omega} \ge 0$ is unique and $\rho(\boldsymbol{\Gamma}(\boldsymbol{\omega})\mathbf{V}) < 1$. In other words (see also Lemma 2.10), there exists a *bijective* function from

[1] Note that the kth coordinate of $\mathbf{p}(\boldsymbol{\omega})$ is zero if and only if $\omega_k = 0$. Thus, for every $\boldsymbol{\omega} > 0$ with $\rho(\boldsymbol{\Gamma}(\boldsymbol{\omega})\mathbf{V}) < 1$, there is a unique *positive* vector $\mathbf{p}(\boldsymbol{\omega}) > 0$.

C onto P such that for every $\mathbf{p} \in \mathrm{P}$, there is exactly one $\boldsymbol{\omega} \in \mathrm{C}$ such that $\mathbf{p} = \mathbf{p}(\boldsymbol{\omega})$. Considering this, it follows from (5.12) that $\boldsymbol{\omega} \in \mathrm{C}$ if and only if $\mathbf{p}(\boldsymbol{\omega}) \in \mathrm{P}$, and hence one has

$$\mathrm{C} = \{\boldsymbol{\omega} \in \mathbb{R}_+^K : \mathbf{p}(\boldsymbol{\omega}) \in \mathrm{P}\} . \tag{5.14}$$

Comparing this with the results of Chapt. 2 reveals that $\mathbf{p}(\boldsymbol{\omega})$ is a special form of (2.4) with $\mathbf{X}(\boldsymbol{\omega}) = \boldsymbol{\Gamma}(\boldsymbol{\omega})\mathbf{V}$ and $\mathbf{b}(\boldsymbol{\omega}) = \boldsymbol{\Gamma}(\boldsymbol{\omega})\mathbf{z}$ (see also Sects. 2.3 and 2.3.3). Using the terminology of the first part of the book, the feasible rate region is nothing but the feasibility set when the parameter vector $\boldsymbol{\omega}$ is chosen to be a vector of link rates.[2] The nonnegative solution in (2.4) is a unique power vector for which the data rate allocation is equal to $\boldsymbol{\omega} \geq 0$.

It follows from Theorem 2.5 that each element of $\mathbf{p}(\boldsymbol{\omega})$ would be a log-convex function of $\boldsymbol{\omega}$ if $\Phi^{-1}(x)$ was log-convex. Unfortunately, for $\Phi(x) = \log(1+x), x \geq 0$, we have $\Phi^{-1}(x) = e^x - 1, x \geq 0$, which is not log-convex on any interval $\mathrm{I} \subseteq \mathbb{R}_+$. This is because

$$\frac{d^2\theta}{dx^2}(x) = -\frac{e^x}{(e^x-1)^2} < 0, \quad x > 0,\ \theta(x) = \log(e^x - 1) .$$

Although the log-convexity property is not necessary for the feasible rate region to be a convex set, Example 2.4 shows that the feasible rate region is indeed not convex in general. The non-convexity of C makes the utility maximization problem over a joint space of transmit powers and link schedulers significantly more challenging and, in general, difficult to solve. Indeed, if C is not a convex set, then a throughput-optimal MAC policy (involving joint power control and link scheduling (JPCLS) under some peak power constraints as discussed in Sect. 5.2.1) is related to the problem of computing the points of the convex hull of C. Recall that the convex hull of a set of points is the intersection of all convex sets containing this set. Therefore,

$$\mathrm{C} \subseteq \tilde{\mathrm{C}} := \mathrm{ConvexHull}(\mathrm{C}) \tag{5.15}$$

where $\tilde{\mathrm{C}}$ is the set of data rates being achievable by means of some feasible MAC policy. This immediately follows from (5.10), which says that data rates under any JPCLS policy is equal to a convex combination of some points in C. In Sect. 5.4.1, we define throughput-optimal MAC policies as those policies that achieve some points on the boundary of $\tilde{\mathrm{C}}$. Consequently, the problem of finding throughput-optimal policies is in general a non-convex problem of combinatorial nature that is difficult to solve [70].

Finally we point out that if each entry of the channel state matrix $\mathbf{V}$ (note that $\mathbf{V}$ represents the current state of the channel) follows an ergodic stochastic process, taking values on a finite state space V, with time average

[2] More precisely, C is the closure of the feasibility set with $\gamma(x) = e^x - 1$ as the latter set does not contain the boundary vectors with zero entries.

probabilities $p_{\mathbf{V}}$, then the set of all average rate vectors (averaged over all channel state matrices) is given by [71, 72]:

$$\overline{\mathsf{C}} = \sum_{\mathbf{V} \in \mathsf{V}} p_{\mathbf{V}} \tilde{\mathsf{C}}(\mathbf{V})$$

where $\tilde{\mathsf{C}}(\mathbf{V})$ is used to denote the convex hull of the feasible rate region when the channel state matrix $\mathbf{V}$ is given. However, notice that the feasible rate region C (and not $\overline{\mathsf{C}}$) serves as a basis for the development of dynamic throughput-optimal policies.

5.2.3 End-to-End Window-Based Rate Control for Wireless Networks

Having introduced the notion of the feasible rate region, we briefly discuss the end-to-end window-based rate control problem for wireless networks. This discussion should be primarily considered as a motivation for the power control problem formulated in the next section. For a more detailed presentation of this approach, the reader is referred to [49, 50].

One of the major difficulties in achieving end-to-end fairness in wired networks is that, in the optimum, any source rate is a function of not only routing variables but also of other source rates. In wireless networks, an additional problem is that link capacities are not fixed but depend on the interference levels, which in turn depend on all transmit powers. We may therefore conjecture that the network performance can be improved by implementing a cross-layer protocol that couples an end-to-end window flow control with power control in the lower layers of the protocol stack. Such a cross-layer protocol could work as follows: Each source gets implicit feedback from the network such as round-trip delay or throughput and regulates the source rate by adjusting its window size that limits the maximum number of packets to be transmitted but not yet acknowledged. Then, a power control protocol utilizes some information from the transport layer to determine the pair $(\boldsymbol{\mu}, \mathbf{p})$ defined by (5.7). Note that each logical link is associated with a flow and that there is a per flow queuing at every node.

Let $\mathbf{A} = (a_{k,s}) \in \mathbb{R}_+^{K \times S}$ be a matrix such that $a_{k,s}\nu_s$ is the expected traffic of flow $s \in \mathsf{S}$ going through logical link $k \in \mathsf{K}$. Thus, $a_{k,s} = \phi_l(s)$ (see Sect. 5.1) if logical link k shares wireless link l and $a_{k,s} = 0$ otherwise. Note that each row of $\mathbf{A}$ has exactly one positive entry since flows cannot share logical links. As before, assume that the objective is to maximize the sum of sources' utilities subject to link capacity constraints given a channel state matrix $\mathbf{V}$. A formal problem formulation is given by (5.1) except that now the vector of logical link rates $\mathbf{A}\boldsymbol{\nu}$ must lie in the convex hull of the feasible rate region. Therefore,

$$\boldsymbol{\nu}^* = \arg\max_{\boldsymbol{\nu}} \sum_{s \in \mathsf{S}} U_s(\nu_s) \quad \text{subject to} \quad \mathbf{A}\boldsymbol{\nu} \in \tilde{\mathsf{C}} \tag{5.16}$$

where $\tilde{\mathrm{C}}$ is defined by (5.15) and the maximum is assumed to exist.

Now suppose for the moment that $\Phi(\mathrm{SIR}_k(\mathbf{p}))$ is a concave function of $\mathbf{p} \geq 0$ for each $k \in \mathsf{K}$ and[3] that Slater's condition holds [11]. By (4.8) and (5.11), the concavity property implies that $\mathrm{C} = \tilde{\mathrm{C}}$, from which it follows that power control with all logical links being active concurrently is an optimal policy. Under these assumptions, the problem in (5.16) is convex without performing the convex hull operation and the Karush–Kuhn–Tucker conditions provide necessary and sufficient conditions for optimality. Therefore, solving (5.16) is equivalent to satisfying the complementary slackness condition and finding a stationary point of the Lagrangian function [11, 50].

In what follows, let us assume that the complementary slackness conditions are satisfied for any primal and dual optimal solutions. We see that the link capacity constraints in (5.16) can be written as $\forall_{s \in \mathsf{S}} \forall_{k \in \mathsf{K}_s} \nu_s \leq \Phi(\mathrm{SIR}_k(\mathbf{p})), \mathbf{p} \in \mathrm{P}$, where $\mathsf{K}_s \subseteq \mathsf{K}$ is a set of those logical links through which flow $s \in \mathsf{S}$ passes. So, the Lagrangian function associated with the problem is

$$L(\boldsymbol{\nu}, \mathbf{p}, \boldsymbol{\lambda}) = \sum_{s \in \mathsf{S}} U_s(\nu_s) - \boldsymbol{\lambda}^T \mathbf{A} \boldsymbol{\nu} + \sum_{k \in \mathsf{K}} \lambda_k \Phi(\mathrm{SIR}_k(\mathbf{p}))$$

where $\boldsymbol{\lambda} = (\lambda_1, \ldots, \lambda_K) \geq 0$ are dual variables, $\mathbf{p} \in \mathrm{P}$ and $\boldsymbol{\nu} \geq 0$. We see that whereas the last addend on the right-hand side depends on $\mathbf{p}$, the first two addends are independent of transmit powers. Thus, by linearity of the differentiation operator, the problem of finding a stationary point of the Lagrangian function can be decomposed into two problems coupled by the optimal Lagrange multiplier vector $\boldsymbol{\lambda}^*$:

$$\begin{aligned} \boldsymbol{\nu}^* &= \arg\max_{\boldsymbol{\nu} \in \mathbb{R}_+^S} \sum_{s \in \mathsf{S}} U_s(\nu_s) - \boldsymbol{\lambda}^{*T} \mathbf{A} \boldsymbol{\nu} \\ \mathbf{p}^* &= \arg\max_{\mathbf{p} \in \mathrm{P}} \sum_{k \in \mathsf{K}} \lambda_k^* \Phi(\mathrm{SIR}_k(\mathbf{p})) . \end{aligned} \tag{5.17}$$

The first subproblem can be implicitly solved by employing an appropriate end-to-end window-based congestion control algorithm for different functions $U_s(x)$, such as those given by (5.4) [57, 50, 73]. For example, TCP Vegas has been shown to solve the first subproblem for logarithmic utility functions with the associated dual variable λ_k being the queuing delay along link k. Therefore, the Vegas source rates are $(\mathbf{w}, 1)$-fair where the weight w_s linearly increases with the round-trip propagation delay for source $s \in \mathsf{S}$. The equilibrium backlogs at the links provide the optimal Lagrange multipliers. The second subproblem in (5.17) is to allocate transmit powers to interfering

[3] If $\Phi(x) = \log(1 + x), x \geq 0$, the concavity requirement is actually never satisfied unless $\mathbf{V} = 0$, that is, unless all links are mutually orthogonal. Therefore, for the rest of this section, the reader can think of other functions that could satisfy the concavity requirement.

links so as to maximize a weighted sum of link rates. Under the assumption of concavity, the problem is mathematically tractable and can be solved using methods of convex optimization such as gradient projection methods. Unfortunately, if $\Phi(x) = \log(1+x), x \geq 0$, the problem is not convex in general. In order to make it tractable, it is a common practice [49] to assume that $\Phi(x) = \log(x), x > 0$, which is equivalent to assuming the high SIR regime (Sect. 5.4.2). The constraint set is then a convex set since the inverse of $\log(x), x > 0$, is log-convex on $\mathbb{R}$ (see the previous section as well as Sect. 5.3). Furthermore, as shown in Chapt. 6, $\log(\mathrm{SIR}_k(e^{\mathbf{s}}))$ is a concave function of $\mathbf{s} = \log \mathbf{p}$.

In [49], the stationary point of the Lagrange function is found iteratively by a simultaneous application of a congestion control mechanism and a gradient projection algorithm for the second subproblem, with the weights being equal to the dual variables associated with the problem (the queuing delays in case of TCP Vegas).

5.2.4 MAC Layer Fair Rate Control for Wireless Networks

The previous section illustrates how traditional TCP protocols may be coupled with power control and link scheduling algorithms to enhance the network performance in terms of some aggregate utility function. Upon receiving information about queuing delays (the dual variables in case of TCP Vegas), each source node updates its window size to adjust the source rate. At the same time, a distributed MAC (medium access control) protocol assigns groups of links to subframes and allocates transmit powers to them so as to maximize a weighted sum of link rates. This is however still an end-to-end control scheme, and therefore it has some important disadvantages common to such schemes. The rates are adjusted with a period proportional to the end-to-end round-trip delays, which are usually large in wireless networks. As a result, such schemes can be expected to have slow convergence and extensive rate oscillations. The latter one may cause large queues (excessive memory) or data loss on congested links when intermediate nodes have no means to limit the traffic generated by other nodes in their vicinity. In fact, because of slow convergence, it is justified to claim that the determination of correct rates would not be affordable if such a scheme was implemented in dynamic wireless environments. Finally, the commonly raised argument in favor of end-to-end schemes that they keep the network simple and scalable by placing the complexity in the hosts hardly applies to wireless networks where the number of flows per node is of a much smaller order than in the Internet. Moreover, wireless networks have per-flow queuing for reasons of scheduling and power control.

For these reasons (see also the discussion at the beginning of Sect. 5.2), we argue in favor of MAC layer (also called per link) fairness with some kind of node-by-node (or link-by-link) congestion control as described, for instance, in [67, 1, 74, 75]. MAC layer fair mechanisms such as weighted fair queuing have already served as a basis for achieving end-to-end fairness in wired networks. Due to the unique characteristics of wireless networks, however, it is clear that MAC layer policies for wired networks cannot be simply reused in the wireless environment. Actually, as pointed out by [76], MAC layer flows (one-hop flows between neighboring nodes) in wireless networks have location-dependent contention for resource allocation, and therefore have some commonalities with network layer flows in wired networks in the sense that they experience different contention.

The notion of MAC layer fairness is defined along similar lines to end-to-end fairness in Sect. 5.2.3, except that instead of network flows, MAC layer flows are considered. A MAC layer policy with node-by-node (or hop-by-hop) congestion control may work as follows: At the beginning of every frame, a (distributed) MAC controller chooses link rates $\boldsymbol{\nu}^* \in \mathbb{R}_+^K$ such that[4]

$$\boldsymbol{\nu}^* = \arg\max_{\boldsymbol{\nu}} \sum_{k \in \mathsf{K}} U_k(\nu_k) \quad \text{subject to} \quad \boldsymbol{\nu} \in \tilde{\mathsf{C}} \tag{5.18}$$

where U_k is a continuously differentiable, increasing and strictly concave function, and $\tilde{\mathsf{C}}$ is the convex hull of C defined by (5.11). The utility function is usually of the form $U_k(x) = w_k U(x)$ where w_k is a nonnegative weight that couples the MAC layer with a congestion control protocol for each link, and therefore usually depends on the current queue states. In addition, at source nodes, a simple window-based end-to-end congestion control may be used to prevent excessive queues at the network nodes. A possible strategy for choosing the weights inspired by the so-called back-pressure policy [77, 78] is

$$w_k = \max\{u_s^{(k)} - u_d^{(k)}, 0\} \tag{5.19}$$

where $u_s^{(k)} \geq 0$ and $u_d^{(k)} \geq 0$ are buffer occupancies at the source and destination of link k, respectively. Thus, w_k is zero if the queue occupancy at the source node is smaller than the queue occupancy at the destination node. The choice of the weights is beyond the scope of this book. We simply assume that the weights are provided at the beginning of every frame according to some strategy. Furthermore, it is assumed that all weights are positive, which does not impact the generality of the analysis. However, we point out that the process of determining the weights is an important issue and has a decisive impact on the overall network behavior.

As mentioned in Sect. 5.2.2, optimal joint power and link scheduling policies are difficult to determine even in a centralized manner. For this reason, practical MAC protocols are usually based on heuristic approaches that attempt to avoid strong interference by activating neighboring links in different subframes. In particular, when nodes cannot transmit and receive simultaneously, wireless links assigned to the same subframe must constitute

[4] Note that, beginning with this section, $\boldsymbol{\nu}$ denotes a K-dimensional nonnegative vector of link rates.

a matching in the corresponding topology graph [62]. This can be achieved by a suitable collision avoidance mechanism (see Sect. 4.2) to avoid strong interference.

In this book, we neglect the problem of link scheduling and focus on the power control problem. The only exception is Sect. 5.4 where some interesting consequences of the results presented in Chapt. 2 on throughput-optimal policies are discussed. For simplicity, it is assumed that no link scheduling is involved, which means that $\Lambda = \{1\}$ and $\mu(\mathrm{B}_1) = \mu(\mathrm{B}) = 1$. This implies that all links can be activated simultaneously without having a collision or, equivalently, without causing strong interference. In particular, each node can transmit and receive simultaneously over all its logical links.

Remark 5.4. Although this is not a particularly realistic scenario, we again emphasize that the assumption does not impact the generality of the analysis presented in this book. In fact, the extension to an arbitrary link scheduling policy is straightforward. Due to the MAC layer peak power constraints (5.9), it is clear that the power control problem under some link scheduling policy decomposes into separate problems of the same type, each for one subframe. The data rate on each logical link follows from (5.10) to be a linear combination of data rates achieved in each subframe, with the coefficients being equal to the length of the subframes. In real networks, the computation of an optimal power vector for each subframe makes sense only if the number of subframes is relatively small. Alternatively, groups of links can be scheduled in different frames which may be a reasonable approach when larger delays are acceptable. Finally it is interesting to point out that if we had the average power constraints in (5.8), the power control problem does not decompose into separate problems for each subframe since then the power vectors for different subframes are subject to a common power constraint.

5.2.5 Utility-Based Power Control

Under the above assumptions, the vector of data rates $\boldsymbol{\nu}$ is confined to be an element of the feasible rate region C defined by (5.11). Moreover, by (5.14) and the discussion in Sect. 5.2.2, we know that there exists a bijective map from the feasible rate region onto the set of feasible power vectors P. As an immediate consequence of this, we can change the optimization domain in (5.18) so as to arrive at an equivalent power control problem subject to the power constraints:

$$\mathbf{p}^* = \arg\max_{\mathbf{p}\in\mathsf{P}} \sum_{k\in\mathsf{K}} U_k(\nu_k(\mathbf{p})) = \arg\max_{\mathbf{p}\in\mathsf{P}} \sum_{k\in\mathsf{K}} U_k(\Phi(\mathrm{SIR}_k(\mathbf{p}))) \tag{5.20}$$

where it is assumed that the maximum exists (see also below). Recall that, unless otherwise stated, $\Phi(x) = \log(1+x), x \geq 0$. Having found $\mathbf{p}^*$, the MAC layer fair rate on link k is given by $\Phi(\mathrm{SIR}_k(\mathbf{p}^*))$. Note that in comparison with the original problem, we have only changed the optimization domain from C to P. Although other domains could be considered by using different bijective mappings (see also the next section), the power domain appears to be the most natural choice in case of wireless networks. If $U_k(x)$ is of the form given by (5.4) for some $\alpha > 0$, the power vector $\mathbf{p}^*$ defined by (5.20) is referred to as a $(\mathbf{w}, \alpha)$-fair power allocation.

Although the functions in (5.4) are strictly concave, it may be easily verified by computing the Hessian matrix for $K = 2$ that $U_k(\nu_k(\mathbf{p}))$ is in general not concave with respect to the power vector $\mathbf{p}$. Thus, the power control problem in (5.20) is not convex. General-type nonconvex problems are too difficult for numerical solution. The computational effort required to solve such a problem by the best-known numerical methods may grow prohibitively fast with the dimension of the problem, and there are serious theoretical reasons to conjecture that this is the intrinsic feature of non-convex problems rather than a drawback of the existing optimization techniques. The situation is further complicated when decentralized algorithms are desired. Therefore, we slightly modify the traditional utility criteria, thereby preserving monotonicity and strict concavity with respect to data rate. To be precise, given a weight vector $\mathbf{w} \in \mathbb{R}^K_{++}$, we consider the following class of utility functions [53]

$$U_k(x) = w_k \Psi(\Phi^{-1}(x)),\ x > 0,\ w_k > 0 \quad k \in \mathsf{K} \tag{5.21}$$

and assume that the following conditions on $\Psi(x)$ hold:

Definition 5.5 (Ψ-Conditions).

(i) $\Psi : \mathbb{R}_{++} \to \mathsf{Q}$ *is a twice continuously differentiable and strictly increasing function where* Q *is some open interval on the real line.*

(ii) There holds

$$\lim_{x\to 0} \Psi(x) := -\infty \qquad \Rightarrow \qquad \lim_{x\to 0} \Psi'(x) = \lim_{x\to 0} \frac{d\Psi}{dx}(x) = +\infty \tag{5.22}$$

where the second limit follows by considering condition (i). This requirement guarantees that $\mathbf{p}^*$ *given by (5.20) is a positive vector, and hence all links are assigned positive transmit powers in the maximum.*

(iii) $\Psi_e(x) = \Psi(e^x)$ *is concave on* $\mathbb{R}$. *Since* Ψ *is twice differentiable and* e^x *is positive on* $\mathbb{R}$, *this is equivalent to*

$$\Psi''_e(x) = \frac{d^2\Psi_e}{dx^2}(x) \le 0, x \in \mathbb{R} \quad \text{and} \quad \Psi'(x) + x\Psi''(x) \le 0, x > 0\,. \tag{5.23}$$

We point out that twice differentiability of Ψ could be replaced by continuous differentiability and the Lipschitz continuity condition on the gradient of the aggregate utility function (see also Chapt. 6). The second condition ensures that, in the maximum, each link is assigned a nonzero data rate. However, the most important condition is the third one since it enables us to convert the utility optimization problem into a convex problem (Chapt. 6). It is pointed out that a class of functions satisfying the Ψ-conditions forms a proper superset of functions considered in [51, 51]. Indeed, these papers considered a set of twice continuously differentiable and strictly increasing functions $\Psi : \mathbb{R}_{++} \to \mathrm{Q}$ satisfying

$$-\frac{\Psi''(x)x}{\Psi'(x)} \in [1,2]\,.$$

Now it is easy to see that this condition follows from (5.23). The converse, however, does not hold as the examples below will show.

It is important to notice that, with the Ψ-conditions and $\Phi(x) = \log(1+x), x > 0$, the utility function $U_k(x)$ given by (5.21) is increasing and strictly concave, and hence satisfies the fundamental properties of the traditional utility functions defined in [57].

Observation 5.6. *Let $\Phi(x) = \log(1+x), x > 0$, and let Ψ satisfy the Ψ-conditions. Then, $U_k(x), x > 0$, defined by (5.21) is an increasing and strictly concave function.*

Proof. It is clear that U_k is increasing so that we only need to show strict concavity. To this end, let $\hat{x}, \check{x} \in \mathbb{R}_{++}$ be arbitrary. By concavity of Ψ_e, we have $(1-\mu)\Psi(e^{\hat{x}}-1)+\mu\Psi(e^{\check{x}}-1) \leq \Psi\big((e^{\hat{x}}-1)^{1-\mu}(e^{\check{x}}-1)^{\mu}\big) < \Psi(e^{(1-\mu)\hat{x}+\mu\check{x}}-1)$ for all $\mu \in (0,1)$ where the last inequality follows since Ψ is strictly increasing and $(e^{\hat{x}}-1)^{1-\mu}(e^{\check{x}}-1)^{\mu} < e^{(1-\mu)\hat{x}}e^{\mu\check{x}}-1$ for all $\hat{x},\check{x} > 0$ with $\hat{x} \neq \check{x}$ and $\mu \in (0,1)$. Since this holds for any $\hat{x},\check{x} \in \mathbb{R}_{++}$, we deduce that U_k is strictly concave on $\mathbb{R}_{++}$.

As pointed out in Sect. 4.3.3, $\Phi(x) = \log(1+x), x \geq 0$, could be replaced by any increasing and strictly concave rate-SIR function $\Phi(x)$, provided that $\Psi(\Phi^{-1}(x)), x > 0$, is strictly concave (see also the remark in Sect. 4.3.3).

Remark 5.7. Sometimes it is desired to guarantee certain quality-of-service (QoS) requirements, for instance, in terms of some maximum delay. In such cases, it is convenient to take $\mathbf{dom}(\Psi) = [a, +\infty)$ with $a > 0$ chosen such that the QoS requirements are guaranteed, and define the value of Ψ outside of its original domain to be $-\infty$. See also the example in the following section.

It may be easily verified that the Ψ-conditions are satisfied by the traditional utility functions defined in (5.4) for all $\alpha \geq 1$. Thus, potential choices of the function Ψ are

$$\Psi(x) = \Psi_\alpha(x) := \begin{cases} \frac{x^{1-\alpha}}{1-\alpha} & \alpha > 1 \\ \log x & \alpha = 1 \end{cases} \quad x > 0\,. \tag{5.24}$$

Another interesting family of functions is

$$\Psi(x) = \tilde{\Psi}_\alpha(x) = \begin{cases} \log \frac{x}{1+x} & \alpha = 2 \\ \log \frac{x}{1+x} + \sum_{j=1}^{\alpha-2} \frac{1}{j(1+x)^j} & \alpha > 2 \end{cases} \quad x > 0\,. \tag{5.25}$$

Fig. 5.2 depicts the utility function (5.21) for different choices of Ψ and compares them to the traditional utility functions corresponding to proportional fairness and total potential delay utility criteria ($\alpha = 2$) defined in Sect. 5.1.1. We see that the "new" class of utility functions is obtained by com-

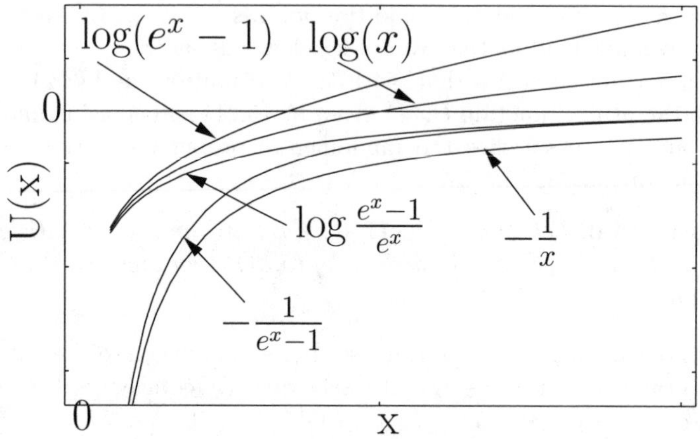

Fig. 5.2. Assuming $\Phi^{-1}(x) = e^x - 1, x \in \mathbb{R}$, the figure compares the modified utilities $U(x) = \Psi(\Phi^{-1}(x)), x > 0$, with the traditional ones $U(x) = \Psi(x), x > 0$, for $\Psi(x) = \log(x), \Psi(x) - 1/x, x > 0$, and $\Psi(x) = \log x/(1+x), x > 0$.

posing the traditional utilities with the function $\Phi^{-1}(x) = e^x - 1, x > 0$. The effect of this is a linearization of the logarithmic rate-SIR curve $\Phi(x)$. Indeed, substituting (4.7) into $U_k(\nu_k(\mathbf{p}))$ with U_k defined by (5.21) yields

$$U_k(\nu_k(\mathbf{p})) = w_k \Psi(\mathrm{SIR}_k(\mathbf{p})) = w_k \Psi\Big(\frac{p_k}{(\mathbf{V}\mathbf{p} + \mathbf{z})_k}\Big) \tag{5.26}$$

so that the utility maximization problem in (5.20) becomes

$$\mathbf{p}^* = \arg\max_{\mathbf{p} \in \mathsf{P}} \sum_{k \in \mathsf{K}} w_k \Psi(\mathrm{SIR}_k(\mathbf{p})) = \arg\max_{\mathbf{p} \in \mathsf{P}} \sum_{k \in \mathsf{K}} w_k \Psi\Big(\frac{p_k}{(\mathbf{V}\mathbf{p} + \mathbf{z})_k}\Big)\,. \tag{5.27}$$

So, in analogy to the wired case (see (5.4)), if $\Psi(x)$ has the form given by (5.24), $\mathbf{p}^*$ in (5.27) can be regarded as being $(\mathbf{w}, \alpha)$-fair with respect to the SIR. For example, according to this view, $\mathbf{p}^* \in \mathsf{P}$ with $\alpha = 1$ and $\mathbf{w} = \mathbf{1}$ is an SIR proportional fair power allocation since it maximizes the sum of logarithmic SIR values. In this book, we adopt this interpretation and refer to power allocations given by (5.27) as SIR fair or, more specifically, as $(\mathbf{w}, \Psi)$-fair power allocations.

SIR proportional fair policies are close to throughput-optimal ones if $\Phi(\mathrm{SIR}_k(\mathbf{p})) \approx \log(\mathrm{SIR}_k(\mathbf{p}))$ for each $k \in \mathsf{K}$ (high SIR regime), since then $\Psi(\mathrm{SIR}_k(\mathbf{p})) \approx \nu_k(\mathbf{p})$. On the other hand, if $\mathrm{SIR}_k(\mathbf{p}) \ll 1$ for each $k \in \mathsf{K}$, then $\nu_k(\mathbf{p}) = \log(1 + \mathrm{SIR}_k(\mathbf{p})) \approx \mathrm{SIR}_k(\mathbf{p})$. Hence, at low SIR values, the modified utility functions are good approximations of the traditional ones, that is, we have $\Psi(\mathrm{SIR}_k(\mathbf{p})) \approx \Psi(\nu_k(\mathbf{p}))$. This can be seen in Fig. 5.2 where, for values of $x > 0$ close to zero, the modified and traditional utility functions almost coincide. Another interesting observation is the following.

Observation 5.8. *Let* $\mathbf{w} > 0$ *be arbitrary, and let* U_k *be given by (5.21) with* $\Psi(x) = \Psi_\alpha(x)$ *where* Ψ_α *is defined by (5.24). Suppose that* $\boldsymbol{\nu}_\alpha^*$ *is a solution to (5.18) with* $\mathrm{C} = \tilde{\mathrm{C}}$*. Then,* $\boldsymbol{\nu}_\alpha^*$ *converges to a max-min fair rate allocation as* $\alpha \to \infty$*.*

Proof. For $\alpha \geq 2$, we have $U(x) = \Psi_\alpha(\Phi^{-1}(x)) = -\big(-h(x)\big)^{\alpha-1}/(\alpha-1), x > 0$, where $h(x) = -1/(e^x - 1), x > 0$, is a continuously differentiable, increasing, negative, and strictly concave function. So, since C in (5.18) (with $\tilde{\mathrm{C}}$ replaced by C) is a compact set, the observation follows from [57, Lemma 3 and Corollary 2].

As explained in Sect. 5.2.2, $\mathbf{p}(\boldsymbol{\omega})$ defined by (5.13) is a continuous bijective map from the feasible rate region C onto P. Therefore, if $\boldsymbol{\nu}_\alpha^* \in \mathrm{C}$ tends to a max-min fair rate allocation as $\alpha \to \infty$, then $\mathbf{p}(\boldsymbol{\nu}_\alpha^*)$ exists for every $\alpha \geq 2$ and goes to a max-min fair power allocation $\tilde{\mathbf{p}} \in \mathrm{P}$. If $\mathbf{V}$ is an irreducible matrix (Definition A.4.1), then

$$\tilde{\mathbf{p}} = \arg\max_{\mathbf{p}\in\mathrm{P}} \min_{1\leq k\leq K} \nu_k(\mathbf{p}) = \arg\max_{\mathbf{p}\in\mathrm{P}} \min_{1\leq k\leq K} \Phi(\mathrm{SIR}_k(\mathbf{p}))\,.$$

The irreducibility of the matrix $\mathbf{V}$ ensures that all links are coupled, which in turn admits the max-min representation above.

We complete this section by showing that the maximum in (5.27) exists.

Lemma 5.9. *There exists some* $\mathbf{p}^* \in \mathrm{P}$ *such that*

$$\sup_{\mathbf{p}\in\mathrm{P}} \sum_{k\in\mathsf{K}} w_k \Psi(\mathrm{SIR}_k(\mathbf{p})) = \sum_{k\in\mathsf{K}} w_k \Psi(\mathrm{SIR}_k(\mathbf{p}^*))\,. \tag{5.28}$$

Proof. A standard method for showing that a function $f : \mathbb{R}^K \to \mathbb{R}$ has a maximum value on a compact set is to argue that f is continuous on this set (see Theorem B.8). The set P is closed and bounded so that, by Theorem B.3, P is compact. So the only problem is that the objective function is discontinuous on P due to the zero components in $\mathbf{p} \in \mathrm{P}$. However, this can be easily fixed. To this end, let $F(\mathbf{V}, \mathbf{p}) = \sum_k w_k \Psi(\frac{p_k}{(\mathbf{V}\mathbf{p}+\mathbf{z})_k})$, and let $\overline{\mathbf{p}} \in \mathrm{P}$ be arbitrary and fixed. Define $\overline{\mathrm{P}} = \{\mathbf{p} \in \mathrm{P} : F(\mathbf{V}, \overline{\mathbf{p}}) \leq F(\mathbf{V}, \mathbf{p})\} \subseteq \mathrm{P}$. Clearly, $\overline{\mathrm{P}}$ is a nonempty compact set. Moreover, $F(\mathbf{V}, \mathbf{p})$ is continuous on $\overline{\mathrm{P}}$ and

$$\sup_{\mathbf{p}\in\mathrm{P}} \sum_{k\in\mathsf{K}} w_k \Psi(\mathrm{SIR}_k(\mathbf{p})) = \sup_{\mathbf{p}\in\overline{\mathrm{P}}} \sum_{k\in\mathsf{K}} w_k \Psi(\mathrm{SIR}_k(\mathbf{p}))\,.$$

Consequently, since $\overline{\mathrm{P}}$ is compact, the supremum is attained.

5.3 Interpretation in the QoS Domain

The function $\Psi(x)$ introduced in the previous section can be interpreted as a SIR-QoS mapping if $\Psi(\text{SIR}_k(\mathbf{p}))$ is a QoS value for link flow k. In this case, Ψ can be either strictly increasing or strictly decreasing depending on whether a larger value of $\Psi(\text{SIR}_k(\mathbf{p}))$ implies a better QoS for flow k (as in the case of data rate) or smaller values of $\Psi(\text{SIR}_k(\mathbf{p}))$ are desired (as in the case of delay). Widely considered QoS parameters are delay, bit error rate and data rate but other quantities such as effective bandwidth [79] and effective spreading gain [26] have been also considered in the literature. In this section, we are going to formulate the power control problem in the QoS domain. In fact, note that (5.18) can be viewed as an equivalent formulation of the problem (5.20) in the QoS domain when the QoS parameter of interest is data rate.

Let $\text{Q} \subseteq \mathbb{R}$ be an interval on the real line and suppose that $U : \mathbb{R}_{++} \to \text{Q}$ is a twice continuously differentiable and strictly monotone function. In contrast to the preceding section, the function $U(x)$ can be either strictly increasing or strictly decreasing. The function value at $x \in \mathbb{R}_{++}$ can be interpreted as the degree of satisfaction of a link flow to the service quality if the link rate is equal to $x > 0$. Now suppose that $\omega_k \in \text{Q}$ is a QoS parameter value of flow k, and let $\boldsymbol{\omega} = (\omega_1, \ldots, \omega_K) \in \text{Q}^K$ be a QoS vector. This can be a vector of delays, data rates or other QoS parameter values of interest.[5]

Definition 5.10. *We say that a QoS vector* $\boldsymbol{\omega}$ *is feasible if there exists a power vector* $\mathbf{p} \in \text{P}$ *such that*

$$\begin{aligned} \omega_k &\leq U(\Phi(\text{SIR}_k(\mathbf{p}))) \quad && U \textit{ strictly increasing} \\ \omega_k &\geq U(\Phi(\text{SIR}_k(\mathbf{p}))) \quad && U \textit{ strictly decreasing}. \end{aligned} \tag{5.29}$$

Note that this definition implicitly implies a one-to-one relationship between the QoS parameter value of interest and the signal-to-interference ratio at the receiver output. Let us characterize the set of all feasible QoS parameter values. To this end, define a function $\gamma(x)$ with $\mathbf{dom}(\gamma) = \text{Q}$ as follows

$$\gamma(U(x)) = \Phi^{-1}(x), \; x \in \mathbb{R}_{++}\,. \tag{5.30}$$

So $\gamma(x)$ is positive for all $x \in \text{Q}$, which is in fact equivalent to assuming that each link rate is positive. Since $\Phi^{-1}(x)$ is strictly increasing, it is clear that if U is strictly increasing (decreasing), then γ is strictly increasing (decreasing) as well. Now combining (5.29) with (5.30), and then proceeding essentially as in Sect. 5.2.2 shows that $\boldsymbol{\omega} \in \text{Q}^K$ is feasible if and only if

$$\forall_{1\leq k\leq K}\, \gamma(\omega_k) \leq \text{SIR}_k(\mathbf{p}) \qquad \Leftrightarrow \qquad \mathbf{\Gamma}(\boldsymbol{\omega})\mathbf{z} \leq (\mathbf{I} - \mathbf{\Gamma}(\boldsymbol{\omega})\mathbf{V})\mathbf{p}$$

[5] Notice that QoS parameter values are relative values and are not expressed in any absolute units like, for instance, seconds (delay) or bits per second (data rate).

where $\mathbf{\Gamma}(\boldsymbol{\omega}) = \text{diag}(\gamma(\omega_1), \ldots, \gamma(\omega_K))$ represents the minimum signal-to-interference ratios that are necessary to provide the QoS vector $\boldsymbol{\omega}$ to the flows. Reasoning further along these lines shows that $\boldsymbol{\omega}$ is feasible if and only if $\boldsymbol{\omega} \in \mathrm{F}_\gamma(\mathrm{P})$ where $\mathrm{F}_\gamma(\mathrm{P})$, referred to as the *feasible QoS region*, is given by

$$\begin{aligned} \mathrm{F}_\gamma(\mathrm{P}) &= \{\boldsymbol{\omega} \in \mathrm{Q}^K : \mathbf{p}(\boldsymbol{\omega}) \in \mathrm{P}\} \\ \text{with} \quad \mathbf{p}(\boldsymbol{\omega}) &= (\mathbf{I} - \mathbf{\Gamma}(\boldsymbol{\omega})\mathbf{V})^{-1}\mathbf{\Gamma}(\boldsymbol{\omega})\mathbf{z} \,. \end{aligned} \tag{5.31}$$

By Theorem A.35, $\mathbf{p}(\boldsymbol{\omega}) \in \mathbb{R}^K_{++}$ exists (and is unique) if and only if the spectral radius of the matrix $\mathbf{\Gamma}(\boldsymbol{\omega})\mathbf{V}$ satisfies $\rho(\mathbf{\Gamma}(\boldsymbol{\omega})\mathbf{V}) < 1$. Therefore,

$$\mathrm{F}_\gamma = \{\boldsymbol{\omega} \in \mathrm{Q}^K : \rho(\mathbf{\Gamma}(\boldsymbol{\omega})\mathbf{V}) < 1\} \tag{5.32}$$

can be interpreted as the feasible QoS region when there are no constraints on transmit powers. Note that if we choose $\gamma(x) = e^x - 1, x > 0$, (or, equivalently, $U(x) = x$), the closure of $\mathrm{F}_\gamma(\mathrm{P})$ is the feasible rate region defined by (5.14).

Remark 5.11. Considering Chapt. 2 and, in particular, Sect. 2.2 reveals that $\mathrm{F}_\gamma(\mathrm{P})$ is closely related to the feasibility set $\mathrm{F}(P_{\mathrm{t}}; P_1, \ldots, P_K)$ defined by (2.13). Indeed, if $\mathbf{X}(\boldsymbol{\omega}) = \mathbf{\Gamma}(\boldsymbol{\omega})\mathbf{V}$ and $\mathbf{z} = \mathbf{1}$, then

$$\mathrm{F}(P_{\mathrm{t}}; P_1, \ldots, P_K) = \mathrm{F}_\gamma(\mathrm{P}_t \cap \mathrm{P}_i)$$

where $\mathrm{P}_t := \{\mathbf{x} \in \mathbb{R}^K_+ : \|\mathbf{x}\|_1 \leq P_{\mathrm{t}}\}$ and $\mathrm{P}_i := \{\mathbf{x} \in \mathbb{R}^K_+ : \forall_{k \in \mathsf{K}} x_k \leq P_k\}$. In special cases of no power constraints, total power constraint and individual power constraints on each logical link, we have (respectively)

$$\begin{aligned} \mathrm{F}_\gamma &= \mathrm{F} && (\mathrm{F} \text{ defined by (2.5)}) \\ \mathrm{F}_\gamma(\mathrm{P}_t) &= \mathrm{F}(P_{\mathrm{t}}) && (\mathrm{F}(P_{\mathrm{t}}) \text{ defined by (2.9)}) \\ \mathrm{F}_\gamma(\mathrm{P}_i) &= \mathrm{F}(P_1, \ldots, P_K) && (\mathrm{F}(P_1, \ldots, P_K) \text{ defined by (2.11)}) \end{aligned} \tag{5.33}$$

where it is assumed that $\mathbf{z} = \mathbf{1}$.

Now we are in a position to state the power control problem in the QoS domain. First assume that a *larger* value of ω_k implies a better QoS for link $k \in \mathsf{K}$. Then, the problem is to find a QoS vector $\boldsymbol{\omega}^* \in \mathrm{F}_\gamma(\mathrm{P})$ such that

$$\boldsymbol{\omega}^* = \arg\max_{\boldsymbol{\omega} \in \mathrm{F}_\gamma(\mathrm{P})} \mathbf{w}^T \boldsymbol{\omega}, \quad \mathbf{w} \in \mathbb{R}^K_{++} \,.$$

In contrast, if a *smaller* value of ω_k implies a better QoS performance, then

$$\boldsymbol{\omega}^* = \arg\min_{\boldsymbol{\omega} \in \mathrm{F}_\gamma(\mathrm{P})} \mathbf{w}^T \boldsymbol{\omega} \,.$$

Now it becomes obvious why convexity of the feasible QoS region is a highly desired property. Indeed, if $\mathrm{F}_\gamma(\mathrm{P})$ is a convex set, the problem in the QoS domain simply reduces to finding a vector $\boldsymbol{\omega}^*$ (if exists) at the boundary of $\mathrm{F}_\gamma(\mathrm{P})$ where the hyperplane with the normal vector $\mathbf{w}$ supports the feasible QoS region. The corresponding power vector is then $\mathbf{p}(\boldsymbol{\omega}^*) \in \mathrm{P}$.

Remark 5.12. For convenience, in this book, "the boundary $\partial \mathrm{F}_\gamma(\mathrm{P})$ of $\mathrm{F}_\gamma(\mathrm{P})$" always refers to points $\boldsymbol{\omega} \in \mathrm{Q}^K$ such that $\mathbf{p}(\boldsymbol{\omega})$ satisfies some power constraints with equality. Formally, we have

$$\partial \mathrm{F}_\gamma(\mathrm{P}) = \Big\{\boldsymbol{\omega} \in \mathrm{Q}^K : \rho(\boldsymbol{\Gamma}(\boldsymbol{\omega})\mathbf{V}) < 1 \text{ and } \exists_{n \in \mathsf{N}} \sum_{k \in \mathsf{K}(n)} p_k(\boldsymbol{\omega}) = P_n\Big\}. \quad (5.34)$$

According to this convention, $\mathrm{F}_\gamma(\mathrm{P})$ is strictly convex (see Definitions 2.15 and 2.15) if every boundary point of $\mathrm{F}_\gamma(\mathrm{P})$ cannot be written as a convex combination of any two other points of $\mathrm{F}_\gamma(\mathrm{P})$.

Since P is a convex set, it follows from (5.31) that $\mathrm{F}_\gamma(\mathrm{P})$ is convex if $p_k(\boldsymbol{\omega})$ is convex for each $k \in \mathsf{K}$. Hence, by Corollary 1.39 and Theorem 2.5 (see also Corollary 2.8), we can conclude that both F_γ and $\mathrm{F}_\gamma(\mathrm{P})$ are convex sets if $\gamma(x)$ is log-convex on Q. This raises the question of whether $\gamma(x)$ is log-convex when the Ψ-conditions (Definition 5.5) are satisfied. To answer this question, we combine the identity $\Psi(\Phi^{-1}(x)) = U(x), x > 0$, with (5.30) to obtain

$$\begin{aligned} \Psi(\gamma(x)) &\equiv x,\ x > 0 \quad \gamma \text{ strictly increasing} \\ \psi(\gamma(x)) &\equiv x,\ x > 0 \quad \gamma \text{ strictly decreasing} \end{aligned} \quad (5.35)$$

where $\psi : \mathbb{R}_{++} \to \mathrm{Q}$ is used to denote a negative version of Ψ, that is,

$$\psi(x) := -\Psi(x), \quad x > 0\,. \quad (5.36)$$

By (5.35), if $\gamma(x)$ is strictly increasing (decreasing), then $\Psi(x)$ ($\psi(x)$) is its inverse function. Now Theorem B.28 in Appendix B.3.1 asserts that $\gamma(x)$ is log-convex if and only if $\Psi_e(x) = \Psi(e^x)$ is concave or if and only if $\psi_e(x) = \psi(e^x)$ is convex depending on whether $\gamma(x)$ is a strictly increasing or decreasing function. Consequently, if the Ψ-conditions are satisfied, then $\gamma(x)$ is log-convex. This in turn implies that the corresponding feasible QoS region is a convex set.

The functions $\Psi(x)$ and $\psi(x)$ relate a QoS parameter of interest and the signal-to-interference ratio at the output of a linear receiver, and therefore these functions can be referred to as SIR-QoS mappings. Note that because of the strict monotonicity property, the increase of SIR always leads to a better quality-of-service. Below we present two interesting examples of strict monotone functions whose inverse functions are log-convex.

(i) Data rate in the high SIR regime: When $\mathrm{SIR}_k(\mathbf{p}) \gg 1$, we have $\log(1+\mathrm{SIR}_k(\mathbf{p})) \approx \log(\mathrm{SIR}_k(\mathbf{p}))$. Thus, at high SIR values, the relationship between the data rate and the signal-to-interference ratio is well approximated by $\Psi(x) = \log(x), x > 0$. The inverse function $\gamma(x) = e^x$ is log-convex on $x \in \mathbb{R}$.
(ii) Average customer time for a $M/M/1$ queuing system in the low SIR regime: If $\mathrm{SIR}_k \ll 1$, then the data rate is linear in SIR since then

$\log(1+\mathrm{SIR}_k) \approx \mathrm{SIR}_k$. On the other hand, the average customer time is $1/(\nu-\lambda)$ where ν and λ denote service rate and arrival rate, respectively [1]. Thus, in the low SIR regime, the average customer time in a $M/M/1$ queuing system is the inverse function of SIR. Therefore, provided that $\lambda > 0$ is not too large, our power control strategy with $\Psi(x) = -\psi(x) = -1/x, x > 0$, is a good approximation for minimizing the total delay. Of course, the inverse function of $\psi(x) = 1/x, x > 0$, is $\gamma(x) = 1/x, x > 0$, which is a log-convex function implying that $\mathrm{F}_\gamma(\mathrm{P})$ is a convex set. Note that the convexity property holds even if $\psi(x) = 1/(x - \lambda)$ for some $\lambda > 0$. This, however, requires that data rates greater than λ can be guaranteed on every link. If $(\lambda, \ldots, \lambda)$ does not lie in the feasible rate region C, the problem could be resolved by an appropriate link scheduling, provided that $(\lambda, \ldots, \lambda)$ is in the convex hull of C.

5.4 Remarks on Joint Power Control and Link Scheduling

The objective of this section is to discuss some potential consequences of the results from Chapt. 2 on throughput-optimal MAC policies. Our definition of the MAC layer includes two mechanisms for resource allocation and interference management, namely link scheduling and power control. The operation of dividing a frame into a number of shorter subframes and assigning links to each subframe is referred to as link scheduling. The power control protocol determines transmit powers of the links in each subframe. The process of jointly optimizing these two mechanisms is called joint power control and link scheduling (JPCLS) (see Sect. 5.2.1). We say that a MAC policy does not involve any link scheduling if each link is either active or idle during the whole frame interval. In other words, there is no time sharing protocol between different points in the feasible rate region that prevents some links from being active concurrently.

5.4.1 Optimal Joint Power Control and Link Scheduling

We are interested in throughput-optimal strategies defined as follows.

Definition 5.13. *Let* $\mathbf{w} \geq 0$ *be a given weight vector, and let* $\mathbf{p}(\mathrm{B}_n) \in \mathrm{P}$ *for every* $n \in \Lambda$. *We say that* $(\mathbf{p}^*, \mu^*)$ *is throughput-optimal if*

$$(\mathbf{p}^*, \mu^*) = \arg\max_{(\mathbf{p},\mu)} \sum_{k \in \mathsf{K}} w_k \nu_k(\mathbf{p}, \mu)$$

where $\nu_k(\mathbf{p}, \mu)$ *is given by (5.10). The corresponding JPCLS is referred to as throughput-optimal.*

As mentioned in Sect. 5.2.1, link rates under an optimal JPCLS correspond to some point on the boundary of the convex hull of the feasible rate region defined by (5.15). Thus, the problem of determining $(\mathbf{p}^*, \mu^*)$ is related to the computation of the points of the convex hull of C. In this section, our main concern is the question of when (if at all) a concurrent transmission of links should be preferred. This is of interest as sophisticated link scheduling policies can be too prohibitive to be implemented in a distributed manner.

As already mentioned in Sect. 5.2.2, the question is directly linked to the geometry of the feasible rate region $\mathrm{C} \subset \mathbb{R}_+^K$. Recall that C includes all data rates being achievable by means of power control when link scheduling is not implemented. Thus, since data rates under any JPCLS policy is a convex combination of some points in C, pure geometrical reasoning shows that:

(i) If C is a convex set, all boundary points of the convex hull of C can be achieved without resorting to link scheduling. More precisely, there exists an optimal JPCLS policy with $\mu^*(\mathrm{B}_1) = \mu^*(\mathrm{B}) = 1$ and some power vector $\mathbf{p}^* = \mathbf{p}^*(\mathrm{B}_1) \in \mathrm{P}$.

(ii) If C is strictly convex (see the remark in the previous section and Definition 2.15), an optimal strategy does not involve any link scheduling. Using the definitions of Sect. 5.2.1, this means that in the optimum, we must have $\mu^*(\mathrm{B}_1) = 1$.

Summarizing, we can say that the JPCLS problem becomes a pure power control problem if C is a convex set. To illustrate this, consider a network with mutually orthogonal links or, equivalently, with the channel state matrix $\mathbf{V} = 0$. It may be easily verified that in this case, the feasible rate region C is a convex set (Example 2.3). Figure 5.3 depicts the feasible rate region for two mutually orthogonal links subject to a sum power constraint $(p_1 + p_2 \le P_\mathrm{t})$. Note that if $\mathbf{V} = \mathbf{0}$ and $\sum_k p_k \le P_\mathrm{t}$, C is a strictly convex set since $\Phi(x) = \log(1+x)$ is strictly concave on $\mathbb{R}_+$. In case of $K \ge 2$ mutually orthogonal links, every point $\boldsymbol{\nu}^*$ on the boundary of C (see the remark on the boundary of $\mathrm{F}_\gamma(\mathrm{P})$ in the previous section) is given by $\nu_k^* = w_k \Phi(p_k^*/z_k)$, $k \in \mathsf{K}$, where $\mathbf{w} \in \mathbb{R}_{++}^K$ is a fixed weight vector and

$$\mathbf{p}^* = \arg\max_{\substack{\mathbf{p}\in\mathbb{R}_+^K \\ \|\mathbf{p}\|_1 = P_\mathrm{t}}} \sum_{k=1}^K w_k \Phi\Big(\frac{p_k}{z_k}\Big) = \arg\max_{\substack{\mathbf{p}\in\mathbb{R}_+^K \\ \|\mathbf{p}\|_1 = P_\mathrm{t}}} \sum_{k=1}^K w_k \log\Big(1 + \frac{p_k}{z_k}\Big).$$

This is a weighted version of the standard water-filling problem [11] for which a closed form solution can be easily found and is given by

$$p_k^* = \max\Big\{0, \frac{w_k}{\lambda} - z_k\Big\},\ z_k > 0$$

where the dual variable λ is chosen to satisfy $\sum_{k=1}^K \max\{0, \frac{w_k}{\lambda} - z_k\} = P_\mathrm{t}$. In a special case, when $p_k^* > 0$ for each $k \in \mathsf{K}$ (no link is idle), one obtains

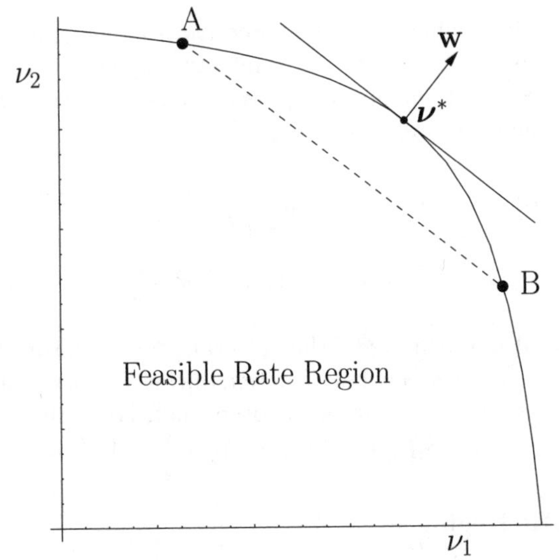

Fig. 5.3. The feasible rate region for two mutually orthogonal links subject to a sum power constraint. The region is a strictly convex set so that link scheduling between arbitrary points on the boundary of the feasible rate region is suboptimal.

$$p_k^* = w_k \frac{P_t + \sum_{l=1}^K z_l}{\sum_{l=1}^K w_l} - z_k > 0\,.$$

From this equation, we see that every optimal *positive* power vector is associated with a unique weight vector $\mathbf{w}$ normalized to be $\|\mathbf{w}\|_1 = 1$. Thus, except for the points on the boundary where at least one of the links is idle ($\nu_k(\mathbf{p}) = 0$ for some $k \in \mathsf{K}$), every rate vector $\boldsymbol{\nu}^*$ is associated with a unique (up to a scaling factor) positive weight vector $\mathbf{w}$ which is normal to the hyperplane supporting the feasible rate region at $\boldsymbol{\nu}^*$.

Due to the strict convexity property of C in the example above, the points on the boundary of C cannot be achieved when link scheduling is involved. Indeed, if there are at least two subframes B_1 and B_2 with $\mu(B_1) > 0$ and $\mu(B_2) > 0$, and associated power vectors $\mathbf{p}(B_1) \neq \mathbf{p}(B_2)$, then the resulting data rate $\mu(B_1)\boldsymbol{\nu}(B_1) + \mu(B_2)\boldsymbol{\nu}(B_2)$ with $\mu(B_1) + \mu(B_2) = 1$ is interior to C where $\nu_k(B_n) = \Phi(\mathrm{SIR}_k(\mathbf{p}(B_n))), k \in \mathsf{K}$. If C is convex (but not strictly convex), an optimal MAC policy may involve link scheduling. To see this, consider two links with $p_1 + p_2 \leq P_t$ and

$$\nu_1(\mathbf{p}) = \log\Big(1 + \frac{p_1}{p_2\varrho + 1}\Big) \qquad \nu_2(\mathbf{p}) = \log\Big(1 + \frac{p_2}{p_1\varrho + 1}\Big)$$

for some $\varrho > 0$. In this case, the feasible rate region is equal to the closure of the feasibility set in Example 2.4. It follows that if $\varrho = (\sqrt{1 + P_t} - 1)/P_t$, the feasible rate region C is convex but not strictly convex. Now suppose that $w_1 = w_2 = 1$, in which case the objective is to maximize $\nu_1(\mathbf{p}) + \nu_2(\mathbf{p})$ subject to $p_1 + p_2 \le P_t$. Due to the symmetry of the objective function, it may be easily seen that the optimal power allocation is $p_1 = p_2 = P_t/2$ and the corresponding data rates are

$$\nu_1(\mathbf{p}) = \nu_2(\mathbf{p}) = \log\Big(1 + \frac{P_t}{\sqrt{1+P_t}+1}\Big) = \log\Big(1 + \frac{P_t(\sqrt{1+P_t}-1)}{P_t}\Big)$$
$$= \log\sqrt{1+P_t} = \frac{1}{2}\log(1+P_t)\,.$$

These rates can also be achieved if the links take turns transmitting at powers equal to P_t, such that the one link is active during the first half of the frame interval and the second link during the other half. Formally, this means that $\mu(\mathrm{B}_1) = \mu(\mathrm{B}_2) = 1/2$ and $\mathbf{p}(\mathrm{B}_1) = (P_t, 0), \mathbf{p}(\mathrm{B}_2) = (0, P_t)$.

5.4.2 High SIR Regime

We say that link k operates in the high SIR regime if

$$\nu_k(\mathbf{p}) = \Phi(\mathrm{SIR}_k(\mathbf{p})) = \log(1 + \mathrm{SIR}_k(\mathbf{p})) \approx \log(\mathrm{SIR}_k(\mathbf{p}))\,.$$

Thus, in the high SIR regime, the data rate behaves like a logarithmic function of SIR. As a result, a linear increase of SIR results only in a logarithmic increase of data rate. In this section, we discuss how this impacts optimal MAC strategies.

To this end, define the feasible rate region with a common rate requirement $\alpha \ge 0$ as follows

$$\mathrm{C}(\alpha) = \{\boldsymbol{\omega} \in \mathbb{R}_+^K : \alpha \le \boldsymbol{\omega} \le \boldsymbol{\nu}(\mathbf{p}), \mathbf{p} \in \mathrm{P}\}\,.$$

Clearly, $\mathrm{C}(\alpha) \subseteq \mathrm{C}$ for all $\alpha \ge 0$ with $\mathrm{C}(0) = \mathrm{C}$ and $\mathrm{C}(\alpha) = \emptyset$ for some sufficiently large α. Now suppose that $\alpha > 0$ is chosen such that both[6]

(i) $\mathrm{C}(\alpha) \ne \emptyset$, and
(ii) for every $\boldsymbol{\nu} \in \mathrm{C}(\alpha)$, there holds $\nu_k \le \nu_k(\mathbf{p}) \approx \log(\mathrm{SIR}_k(\mathbf{p}))$ for each $k \in \mathsf{K}$ and some $\mathbf{p} \in \mathrm{P}$.

Under this assumption, $\mathrm{C}(\alpha)$ is well approximated by the feasible QoS region $\mathrm{F}_\gamma(\mathrm{P})$ defined by (5.31) with $\gamma(x) = e^x, x \in [\alpha, \infty)$. Now since the exponential function is log-convex, Corollary 2.8 implies that $\mathrm{F}_\gamma(\mathrm{P})$ is a convex set. This in turn implies that $\mathrm{C}(\alpha)$ is a convex set under the assumption of the logarithmic relationship between data rate and SIR. Moreover, if $\mathbf{V}$ is irreducible, which is usually the case in practice, Corollary 2.16 ensures that $\mathrm{F}_\gamma(\mathrm{P})$ is a strictly convex set. Thus, by the discussion above, we can conclude that in the high

[6] Note that if there is no such α, then the network cannot operate in the high SIR regime.

SIR regime, no link scheduling should be involved. This is quite intuitive since a logarithmic increase in data rate due to a higher value of SIR cannot compensate a linear decrease due to a shorter transmission time (by virtue of partitioning a frame into subframes).

5.4.3 Low SIR Regime

Now let us assume that a network operates in the low SIR regime. In this case, the relationship between data rate and SIR can be well approximated by a linear function:

$$\nu_k(\mathbf{p}) = \Phi(\mathrm{SIR}_k(\mathbf{p})) = \log(1 + \mathrm{SIR}_k(\mathbf{p})) \approx \mathrm{SIR}_k(\mathbf{p}) .$$

Thus, when compared with the high SIR regime, we have an entirely different situation here since a linear increase of SIR entails a linear increase in data rate. This has a tremendous impact on the design of optimal MAC policies. The main driving force behind the operation in the low SIR regime is the ability to transmit at an energy per information bit close to the minimum [80, 81]. For instance, in wireless sensor networks, the energy consumption (rather than the spectral efficiency) is one of the major design criteria.

Comparing (5.14) with (5.31) reveals that the feasible rate region in the low SIR regime is equal to the closure of the feasible QoS region $\mathrm{F}_\gamma(\mathrm{P})$ with $\gamma(x) = x, x > 0$. In all that follows, let us assume that[7] $\gamma(x) = x, x > 0$. We refer to $\mathrm{cl}(\mathrm{F}_\gamma(\mathrm{P}))$ as the *feasible SIR region*, while its complement $\mathrm{F}^c_\gamma(\mathrm{P}) = \mathbb{R}^K_+ \setminus \mathrm{cl}(\mathrm{F}_\gamma(\mathrm{P}))$ is called the *infeasible SIR region*. Since the linear function is not log-convex, the results of Sect. 2.3 cannot be applied to the low SIR regime. Furthermore, it follows from Sect. 2.4 that the feasible SIR region is not a convex set. However, for $K = 2$, the convex hull of the feasible SIR region is a convex polygon, regardless of whether the links are subject to individual power constraints or are constrained on total power. This is easy to see from (2.16) and the following discussion in Sect. 2.4, which shows that $p_1(\boldsymbol{\omega})$ and $p_2(\boldsymbol{\omega})$ are both concave on $\mathrm{F}_\gamma = \mathrm{F}$. Thus, in the case of two links subject to a total power P_t, the convex hull of $\mathrm{cl}(\mathrm{F}_\gamma(\mathrm{P}))$ is a triangle with the vertices given by $(0, 0)$, $(P_\mathrm{t}/z_1, 0)$ and $(0, P_\mathrm{t}/z_2)$. The nonzero vertices are the points E and F in Fig. 5.4. Clearly, if $\mathbf{V} \neq \mathbf{0}$, a throughput-optimal MAC policy is then a simple link scheduling (or time sharing) protocol between the vertices $(P_\mathrm{t}/z_1, 0)$ and $(0, P_\mathrm{t}/z_2)$, which correspond to the power vectors $\mathbf{p} = (P_\mathrm{t}, 0)$ and $\mathbf{p} = (0, P_\mathrm{t})$, respectively. In other words, the links take turns transmitting at powers equal to P_t, such that only one link is active at any time. Using the definitions of Sect. 5.2.1, this means that each frame B is divided into (at most) two subframes B_1 and B_2 such that $\mu(\mathrm{B}_1) +$

[7] The only reason for excluding zero from the definition of $\gamma(x)$ is the compatibility with the previous definitions. Because of this, we have to take the closure of $\mathrm{F}_\gamma(\mathrm{P})$ to obtain the feasible rate region.

$\mu(\mathrm{B}_2) = \mu(\mathrm{B}) = 1$, with the corresponding power vectors being equal to $\mathbf{p}(\mathrm{B}_1) \in \{(P_\mathrm{t}, 0), (0, P_\mathrm{t})\}$ and $\mathbf{p}(\mathrm{B}_2) = (P_\mathrm{t}, P_\mathrm{t}) - \mathbf{p}(\mathrm{B}_1)$. The set function $\mu : \mathrm{A} \to [0, 1]$ with $\mathrm{A} = \{\mathrm{B}_1, \mathrm{B}_2\}$ indicates at which relative frequencies each of the power vectors is utilized in an optimal MAC policy. This then determines a family of optimal MAC policies parameterized by the function μ.

The situation is slightly more complicated in the case of two links subject to individual power constraints $P_1 > 0$ and $P_2 > 0$. In this case, the convex hull of the feasible SIR region can be either a triangle spanned by the points $(0, 0), (P_1/z_1, 0)$ and $(0, P_2/z_2)$ or a convex quadrilateral whose vertices are $(0, 0), (P_1/z_1, 0), (0, P_2/z_2)$ and (ω_1^*, ω_2^*) satisfying $p_1(\boldsymbol{\omega}^*) = p_2(\boldsymbol{\omega}^*)$. Clearly, if the convex hull is the triangle ((ω_1^*, ω_2^*) is then a member of this triangle) and $\mathbf{V} \neq 0$, then an optimal MAC policy is similar to that for a total power constraint except that now it involves a time sharing protocol between the points $(P_1/z_1, 0)$ and $(0, P_2/z_2)$ (D and A in Fig. 5.4), with the corresponding power vectors being $(P_1, 0)$ and $(0, P_2)$. This again implies two subframes B_1 and B_2 in each frame such that $\mu(\mathrm{B}_1) + \mu(\mathrm{B}_2) = 1$, with the power vectors being given by $\mathbf{p}(\mathrm{B}_1) \in \{(P_1, 0), (0, P_2)\}$ and $\mathbf{p}(\mathrm{B}_2) = (P_1, P_2) - \mathbf{p}(\mathrm{B}_1)$. When the convex hull is a convex quadrilateral, a time sharing protocol either between $(P_1/z_1, 0)$ and (ω_1^*, ω_2^*) (D and G in Fig. 5.4) or between $(0, P_2/z_2)$ and (ω_1^*, ω_2^*) (A and G) is optimal. Again, each frame is divided into two subframes B_1 and B_2 such that $\mu(\mathrm{B}_1) + \mu(\mathrm{B}_2) = 1$. However, the power vectors can be either

$$\begin{aligned} \mathbf{p}(\mathrm{B}_1) &\in \{(P_1, 0), (p_1(\boldsymbol{\omega}^*), p_2(\boldsymbol{\omega}^*))\} \\ \mathbf{p}(\mathrm{B}_2) &= \{(P_1, 0), (p_1(\boldsymbol{\omega}^*), p_2(\boldsymbol{\omega}^*))\} \setminus \{\mathbf{p}(\mathrm{B}_1)\} \end{aligned}$$

or

$$\begin{aligned} \mathbf{p}(\mathrm{B}_1) &\in \{(0, P_2), (p_1(\boldsymbol{\omega}^*), p_2(\boldsymbol{\omega}^*))\} \\ \mathbf{p}(\mathrm{B}_2) &= \{(0, P_2), (p_1(\boldsymbol{\omega}^*), p_2(\boldsymbol{\omega}^*))\} \setminus \{\mathbf{p}(\mathrm{B}_1)\}\,. \end{aligned}$$

In the point (ω_1^*, ω_2^*), both links are active and transmit at powers specified by the power vector $\mathbf{p}(\boldsymbol{\omega}^*)$. Again, the set function μ determines at which relative frequencies the power vectors are utilized in an optimal strategy.

Now the question is whether it is possible to generalize these observations to a network with K links and transmit powers subject to some constraints? More precisely, we are interested in the following problem.

Problem 5.14. Is the convex hull of the feasible SIR region a convex polytope in $\mathbb{R}_+^K$, regardless of the type of power constraints?

If this was true, then a MAC policy based on a time sharing protocol between the polytope vertices similar to that for the two dimensional case would be optimal. However, this cannot be true in *full generality*. To see this, consider the feasible SIR region given by $\mathrm{cl}\big(\mathrm{F}_\gamma(\mathrm{P}_t) \cap (\mathrm{F}_\gamma(\mathrm{P}_i)\big)$ where $\mathrm{F}_\gamma(\mathrm{P}_t)$ and $\mathrm{F}_\gamma(\mathrm{P}_i)$

are defined by (5.33) for $\mathbf{z} = \mathbf{1}$. In words, while $\text{cl}(\text{F}_\gamma(\text{P}_t))$ is the feasible SIR region in a network constrained on total power, $\text{cl}(\text{F}_\gamma(\text{P}_i))$ denotes the feasible SIR region under individual power constraints on each logical link. The situation is illustrated in Fig. 5.4 for $K = 2$. In this two dimensional

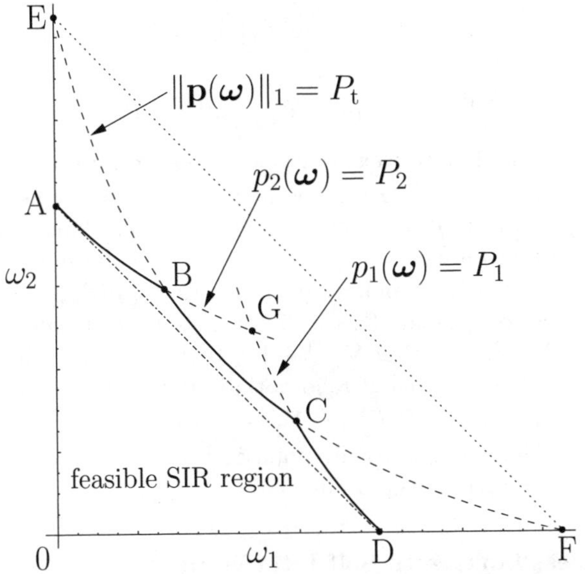

Fig. 5.4. The feasible SIR region for two users under total power constraint P_t and individual power constraints on each link $P_1 < P_t$ and $P_2 < P_t$. If there were no individual power constraints, a MAC policy involving a time sharing protocol between the points E and F, corresponding to power vectors $(0, P_t)$ and $(P_t, 0)$, respectively, would be optimal. In contrast, when in addition individual power constraints are imposed, a time sharing protocol between A and D (that correspond to power vectors $(0, P_2)$ and $(P_1, 0)$, respectively) is suboptimal. In this case, it is better to schedule either between A and B or between B and C or between C and D depending on the target signal-to-interference ratios.

example, the convex hull of the feasible SIR region is a convex pentagon generated by the points $(0, 0), A, B, C$ and D. So, in this case, a time sharing protocol between some of the pentagon vertices A, B, C and D is optimal. While in A and D only one of the links is active, both links are active in B and C. However, Theorem 2.17 asserts that $\text{F}^c_\gamma(\text{P}_t)$ is not a convex set in general. Therefore, the convex hull of $\text{F}_\gamma(\text{P}_t) \cap \text{F}_\gamma(\text{P}_i)$ does not need to be a convex polytope, which may require alternative MAC policies to achieve points on the boundary of the convex hull. For instance, if $\text{F}_\gamma(\text{P}_t)$ was not convex in the two-user case, a time sharing protocol between the points B and C (Fig. 5.4) would not be necessarily the best strategy since the boundary of $\text{F}_\gamma(\text{P}_t)$ may intersect the straight line connecting B and C, and hence requiring another policy to achieve the boundary points.

It is interesting to point out that in a network constrained only on total power, the convex hull of the feasible SIR region $\mathrm{cl}(\mathrm{F}_\gamma(\mathrm{P}_t))$ is a convex polytope despite the fact that $\mathrm{F}^c_\gamma(\mathrm{P}_t)$ is not a convex set in general. Indeed, we have

$$\tilde{\mathrm{F}}_\gamma(\mathrm{P}_t) = \mathrm{ConvexHull}\big(\mathrm{F}_\gamma(\mathrm{P}_t)\big) \tag{5.37}$$

where

$$\tilde{\mathrm{F}}_\gamma(\mathrm{P}_t) = \{\boldsymbol{\omega} \in \mathbb{R}^K_{++} : \boldsymbol{\omega}^T \mathbf{z} \leq P_\mathrm{t}\} .$$

Clearly, $\mathrm{cl}(\tilde{\mathrm{F}}_\gamma(\mathrm{P}_t))$ is a convex polytope whose vertices are $(0, \dots, 0)$ and $P_\mathrm{t}/z_k\, \mathbf{e}_k$ with $k = 1, \dots, K$. Every point in $\tilde{\mathrm{F}}_\gamma(\mathrm{P}_t)$ can be achieved using a TDMA-like protocol that alternates between different links transmitting at power P_t, such that only one link is active in every subframe. More formally, given B and A, any JPCLS policy $(\mathbf{p}^*, \mu^*)$ with $\sum_k \mu(\mathrm{B}_k) = 1$ and $\mathbf{p}(\mathrm{B}_k) = P_\mathrm{t}\mathbf{e}_k, 1 \leq k \leq K$, is optimal. To see (5.37), note that, for all $\boldsymbol{\omega} \geq 0$ $(\boldsymbol{\omega} \neq 0)$ with $\rho(\mathbf{\Gamma}(\boldsymbol{\omega})\mathbf{V}) < 1$, $(\mathbf{I} - \mathbf{\Gamma}(\boldsymbol{\omega})\mathbf{V})^{-1} - \mathbf{I} \geq 0$ with equality (in all entries) if and only if $\mathbf{V} = 0$. From this, it follows that $\boldsymbol{\omega}^T\mathbf{z} < \|\mathbf{p}(\boldsymbol{\omega})\|_1$ for any $\boldsymbol{\omega} \in \mathrm{F}_\gamma$ and $\mathbf{V} \neq \mathbf{0}$, and therefore $\mathrm{F}_\gamma(\mathrm{P}_t) \subset \mathrm{F}_\gamma$ is a proper subset of $\tilde{\mathrm{F}}_\gamma(\mathrm{P}_t)$ for any matrix $\mathbf{V} \neq 0$. Moreover, since any point in $\tilde{\mathrm{F}}_\gamma(\mathrm{P}_t)$ is a convex combination of some points in $\mathrm{F}_\gamma(\mathrm{P}_t)$, we obtain (5.37).

5.4.4 Wireless Links with Self-Interference

Up to now, we have assumed that wireless links are only exposed to interference from other links, and therefore no link interferes with itself whenever it is active. The assumption is reasonable when multiple-access interference is a dominant factor.

Nevertheless, self-interference is usually present in wireless networks, mainly due to the time-dispersive nature of the radio propagation channel, but also due to the nonlinear characteristics of deployed components. Whatever the reason is, it is reasonable to assume that the self-interference is proportional to transmit power. In case of a multipath propagation channel, the received signal may be composed of a strong signal path and some (usually weaker) delayed paths. When the time dispersion of the channel is sufficiently large and only the strong path is used to decode transmitted symbols, other paths may cause relatively strong intersymbol interference. It is important to emphasize that it is not the absolute value of self-interference that matters but rather its relation to multiple access interference. Indeed, link scheduling can entail a noteworthy performance improvement only if multiple access interference is dominant. On the contrary, if self-interference is a dominant factor on each link, concurrent transmission may be preferable.

But when exactly is self-interference dominant? Although it is difficult to give a definite answer to this question in a general context, Theorem 1.48 in

Sect. 1.4.2 suggests some useful guidelines for the design of MAC strategies in the presence of relatively strong self-interference. This theorem asserts that the Perron root is convex if both the channel state matrix $\mathbf{V} \in \mathbb{R}_+^{K\times K}$ is symmetric positive semidefinite and $\gamma : \mathrm{Q} \to \mathbb{R}_+$ defined by (5.30) is a convex function. This in turn implies (Corollary 1.49) that the feasible QoS region F_γ defined by (5.32) is a convex set if both $\mathbf{V}$ is symmetric positive semidefinite and γ is convex. Consequently, under the assumption of positive semidefiniteness of $\mathbf{V}$, convexity of $\gamma : \mathrm{Q} \to \mathbb{R}_+$ is sufficient for F_γ to be a convex set, which is a significantly weaker requirement than log-convexity. In particular, $\gamma(x) = e^x - 1, x \geq 0$ is a convex function implying that the feasible rate region is a convex set when the channel state matrix is symmetric positive semidefinite.

If there is no self-interference or, equivalently, if $\mathrm{trace}(\mathbf{V}) = 0$, the channel state matrix $\mathbf{V}$ cannot be positive semidefinite, which immediately follows from the non-negativeness of $\mathbf{V}$. Roughly speaking, for the matrix $\mathbf{V}$ to be positive semidefinite, self-interference must be dominant on each link. However, the results mentioned above can be applied only if $\mathbf{V}$ is symmetric (or approximately symmetric). In these cases, all points on the boundary of the convex hull of the feasible rate region C can be achieved by power control with all links being active concurrently. Hence, throughput-optimal MAC policies do not need to involve link scheduling.

For the case that $\mathbf{V}$ is not symmetric but its diagonal elements are dominant in the sense that

$$\mathbf{V}_s = (\mathbf{V} + \mathbf{V}^T)/2$$

is positive semidefinite, numerical experiments suggest that similar conclusions may be possible. However, we have no proof for that to be true in general. If the symmetric part $\mathbf{V}_s$ of $\mathbf{V}$ is positive definite, then it is known that $\rho(\mathbf{V}) \leq \rho(\mathbf{V}_s)$ for any nonnegative matrix $\mathbf{V}$ [7].

Note that each quadratic matrix can be uniquely written as the sum of a symmetric matrix and a skew-symmetric one. If we view $\mathbb{R}^{K\times K}$ as a Hilbert space with the inner product given by $\langle \mathbf{A}, \mathbf{B}\rangle = \mathrm{trace}(\mathbf{A}^T\mathbf{B})$, then the sets of symmetric matrices and skew-symmetric ones are orthogonal complementary subspaces in $\mathbb{R}^{K\times K}$. Moreover, $\mathbf{V}_s$ is the orthogonal projection of $\mathbf{V}$ onto the space of $K \times K$ symmetric matrices. Thus, in this sense, $\mathbf{V}_s$ is the closest symmetric matrix to $\mathbf{V}$. This suggests that F_γ is a "nearly convex set" when both $\mathbf{V}_s$ is positive semidefinite (self-interference dominant) and the distance between $\mathbf{V}$ and $\mathbf{V}_s$ is not too large. In these cases, no link scheduling should be preferred when maximizing total throughput.

5.5 Remarks on the Efficiency–Fairness Trade Off

In Sect. 5.1.1, we briefly described the fundamental efficiency–fairness trade off in wired communications networks. It was pointed out that throughput-optimal policies may lead to significant rate deviations among competing flows. It is even possible that some flows are denied access to the links. For instance, the example in Fig. 5.1 shows a simple scenario where the longer flow is allocated zero source rate under a throughput-optimal rate allocation. Because this is in general not tolerable, the network designers are forced to address the issue of fairness. The most common understanding of fairness is the max-min fairness, in which case all source rates are made as equal as possible. However, it is quite intuitive that, for instance, an allocation of equal source rates to all flows is suboptimal in terms of throughput or some other (suitably chosen) aggregate utility function. Since the value of the utility function can be identified as some measure of overall efficiency of the network, the outlined trade off situation is often referred to as the efficiency–fairness trade off. It is important to notice that the discussion below aims at highlighting the potential of incompatibility between efficiency and fairness issues. In fact, in the case of wired networks, this trade off depends on the network topology. Simple examples show that there exist wired network topologies where an "ideal" combination of fairness and efficiency is possible [58].

The wired network topology together with the fixed link capacities is, in the case of wireless networks, replaced by some constraints on transmit powers and the channel state matrix $\mathbf{V} \geq 0$, which describes the crosstalk between different links. As described in Sect. 5.2.4, in wireless networks, the utility maximization problem becomes a power control problem of the form

$$\mathbf{p}^* = \arg\max_{\mathbf{p}\in\mathsf{P}} \sum_{k\in\mathsf{K}} w_k \Psi(\mathrm{SIR}_k(\mathbf{p})) \tag{5.38}$$

for some given weight $\mathbf{w} > 0$, where the function $\Psi : \mathbb{R}_{++} \to \mathsf{Q} \subseteq \mathbb{R}$ is assumed to satisfy the Ψ-conditions (Definition 5.5) and $\mathbf{p}^*$ is referred to as a $(\mathbf{w}, \Psi)$-fair power allocation (see Sect. 5.2.5). Under such an allocation of transmit powers, the data rate for link k is $\nu_k(\mathbf{p}^*) = \Phi(\mathrm{SIR}_k(\mathbf{p}^*))$ and, of course, the data rates are in general different for different links. This raises the following interesting question: Under which conditions (if at all) is a $(\mathbf{w}, \Psi)$-fair power allocation $\mathbf{p}^*$ a max-min fair power allocation $\overline{\mathbf{p}}$ defined to be

$$\overline{\mathbf{p}} = \arg\max_{\mathbf{p}\in\mathsf{P}} \min_{1\leq k\leq K} \mathrm{SIR}_k(\mathbf{p}) \,. \tag{5.39}$$

In other words, we are asking whether there exists a power vector $\mathbf{p}$ that is both $(\mathbf{w}, \Psi)$-fair and max-min fair in the sense of (5.39). In the following section, we use the results of Sects. 1.2.3 and 1.2.4 to answer this question. In doing so, we assume that

(i) $\mathbf{z} = 0$, from which it follows that $\mathrm{SIR}_k(\mathbf{p}) = \mathrm{SIR}_k(\alpha\mathbf{p}), \alpha > 0$. This assumption is justified when the noise is negligible in comparison with multiple access interference.
(ii) $\mathbf{V} \geq 0$ is irreducible, that is, $\mathbf{V} \in \mathrm{X}_K$. Furthermore, $\mathbf{V}$ is chosen such that the maximum in (5.38) with $\mathbf{z} = 0$ exists. As discussed in Sect.

1.2.3, if $\mathbf{z} = 0$ and $\Psi(x) = \log(x), x > 0$, irreducibility of $\mathbf{V} \geq 0$ is not sufficient for the maximum in (5.38) to exist. At the end of the section, we make some remarks on general nonnegative matrices.

Remark 5.15. Irreducibility of $\mathbf{V}$ can be explained by means of the directed graph $\mathrm{G}(\mathbf{V})$ associated with $\mathbf{V}$. Indeed, it is well known (see Appendix A.4.1) that $\mathbf{V}$ is irreducible if and only if $\mathrm{G}(\mathbf{V})$ is strongly connected. In terms of interference, we can interpret strong connectivity of $\mathrm{G}(\mathbf{V})$ as a kind of interference coupling spanning the entire network. Therefore, when $\mathbf{V}$ is irreducible, a network is said to be *entirely coupled.*

As an immediate consequence of these assumptions, we have

$$\max_{\mathbf{p}\in\mathrm{P}} \min_{1\leq k\leq K} \mathrm{SIR}_k(\mathbf{p}) = \max_{\mathbf{p}\in\mathbb{R}^K_{++}} \min_{1\leq k\leq K} \mathrm{SIR}_k(\mathbf{p}) .$$

Since $\mathbf{V} \geq 0$ is irreducible, the existence of the maximum on $\mathbb{R}^K_{++}$ immediately follows from the Collatz–Wielandt formula (Theorem A.27). Furthermore, it follows from this theorem that the maximum is attained if and only if $\overline{\mathbf{p}}$ is a positive right eigenvector of $\mathbf{V}$. Thus, any positive right eigenvector of $\mathbf{V} \in \mathrm{X}_K$ is a max-min fair power allocation. Finally, we must have

$$\mathrm{SIR}_1(\overline{\mathbf{p}}) = \ldots = \mathrm{SIR}_K(\overline{\mathbf{p}}) = 1/\rho(\mathbf{V}) \tag{5.40}$$

which immediately follows from the fact that $\overline{\mathbf{p}} > 0$ is a right eigenvector of $\mathbf{V}$ associated with $\rho(\mathbf{V})$. Obviously, (5.40) implies that $\nu_1(\overline{\mathbf{p}}) = \ldots = \nu_K(\overline{\mathbf{p}})$. In what follows, $\overline{\mathbf{p}}$ is used to denote both a positive right eigenvector of $\mathbf{V}$ associated with $\rho(\mathbf{V})$ and a max-min fair power allocation given by (5.39).

5.5.1 Efficiency of the Max-Min Fair Power Allocation

Due to the above assumptions, the optimization domain $\mathrm{P} \subset \mathbb{R}^K_+$ in (5.38) and (5.39) can be replaced by $\mathbb{R}^K_{++}$. Moreover, to conform with the results presented in Sects. 1.2.3 and 1.2.4, we rewrite the utility-based power control problem as an equivalent minimization problem:

$$\begin{aligned}\mathbf{p}^* &= \arg\min_{\mathbf{p}\in\mathbb{R}^K_{++}} \sum_{k\in\mathsf{K}} w_k \psi(\mathrm{SIR}_k(\mathbf{p})) = \arg\min_{\mathbf{p}\in\mathbb{R}^K_{++}} \sum_{k\in\mathsf{K}} w_k F\Big(\frac{1}{\mathrm{SIR}_k(\mathbf{p})}\Big) \\ &= \arg\min_{\mathbf{p}\in\mathbb{R}^K_{++}} G(\mathbf{p},\mathbf{w})\end{aligned} \tag{5.41}$$

where $F(x) = \psi(1/x) = -\Psi(1/x)$ and $G : \mathbb{R}^K_{++} \times \mathbb{R}^K_{++} \to \mathbb{R}$ is defined to be

$$G(\mathbf{p},\mathbf{w}) := \sum_{k\in\mathsf{K}} w_k F\Big(\frac{1}{\mathrm{SIR}_k(\mathbf{p})}\Big) = \sum_{k\in\mathsf{K}} w_k F\Big(\frac{(\mathbf{V}\mathbf{p})_k}{p_k}\Big) .$$

Now since Ψ-conditions are satisfied and the minimum exists (by assumption), it is clear by Sect. 6.2 that F belongs to the function class $\mathcal{G}(\mathbf{V})$ (Definition 1.22). Indeed, based on the results of Sect. 6.2, it may be seen that the problem can be converted into an equivalent convex optimization problem. This in turn implies that every local minimum is a global one.

Since $\mathbf{V} \in \mathrm{X}_K$, the problem falls into the theoretical framework presented in Sects. 1.2.3 and 1.2.4. So let us apply these results to our problem. To this end, suppose that $\overline{\mathbf{q}} \in \mathbb{R}_+^K$ is a left eigenvector of $\mathbf{V}$ associated with $\rho(\mathbf{V})$. Recall that since $\mathbf{V}$ is irreducible, $\rho(\mathbf{V})$ is an eigenvalue of $\mathbf{V}$ and $\overline{\mathbf{q}}, \overline{\mathbf{p}} > 0$ (Theorem A.25). For the purposes in this section, the most important results are provided by Theorems 1.24 and 1.31. By the first theorem, we know that if $\mathbf{w} = \overline{\mathbf{q}} \circ \overline{\mathbf{p}} > 0$, then[8]

$$G(\mathbf{p}, \mathbf{w}) = \sum_{k \in \mathsf{K}} w_k F\Big(\frac{(\mathbf{V}\mathbf{p})_k}{p_k}\Big) \geq F\big(\rho(\mathbf{V})\big)\|\mathbf{w}\|_1$$

for all $\mathbf{p} \in \mathbb{R}_{++}^K$. Equality holds for $\mathbf{p} = \overline{\mathbf{p}} \in \mathbb{R}_{++}^K$ if and only if $\mathbf{w} = \overline{\mathbf{q}} \circ \overline{\mathbf{p}}$.[9] This means that if $\mathbf{w} > 0$ happens to be equal to the Hadamard product of $\overline{\mathbf{q}}$ and $\overline{\mathbf{p}}$, then the objective function in (5.41) attains its minimum if the power vector is equal to a positive right eigenvector of $\mathbf{V}$. Thus, at the minimum in (5.41), we have

$$\mathrm{SIR}_1(\mathbf{p}^*) = \cdots = \mathrm{SIR}_K(\mathbf{p}^*) = 1/\rho(\mathbf{V})\,.$$

Consequently, if $\mathbf{w} = \overline{\mathbf{q}} \circ \overline{\mathbf{p}}$, then $\mathbf{p}^* = \overline{\mathbf{p}}$, that is, $\mathbf{p}^*$ is both $(\mathbf{w}, \Psi)$-fair and max-min fair power allocation. By Theorem 1.31, $(\overline{\mathbf{q}} \circ \overline{\mathbf{p}}, \overline{\mathbf{p}})$ is a saddle point of $G(\mathbf{p}, \mathbf{w})$. Moreover, both $\overline{\mathbf{q}} \circ \overline{\mathbf{p}}$ and $\overline{\mathbf{p}}$ are unique up to constant multiples. This gives rise to the following theorem.

Theorem 5.16. *Let $\mathbf{V} \in \mathrm{X}_K$ and $F \in \mathcal{G}(\mathbf{V})$ be given. Then, a $(\mathbf{w}, \Psi)$-fair power allocation is max-min fair if and only if $\mathbf{w} = \overline{\mathbf{q}} \circ \overline{\mathbf{p}} > 0$, which is unique up to a constant multiple.*

The theorem asserts that, in the case of entirely coupled networks ($\mathbf{V} \in \mathrm{X}_K$), $(\mathbf{w}, \Psi)$-fair power allocations and max-min fair power allocations coincide if and only if $\mathbf{w} = \overline{\mathbf{q}} \circ \overline{\mathbf{p}} > 0$. The corresponding power vectors are positive and unique up to constant multiples.

The result can be extended to more general nonnegative matrices by considering the results in Sects. 1.6.2 and 1.6.3. However, it is emphasized that some restrictions on $\mathbf{V} \neq \mathbf{0}$ are necessary. The reason is that if $\mathbf{V} \geq 0$ is reducible (Theorem A.21), a max-min fair power allocation in the sense of (5.39) with (5.40) does not need to exist. Indeed, by Theorem A.30, we know that a reducible matrix $\mathbf{V} \geq 0$ may have no positive right eigenvectors associated with $\rho(\mathbf{V}) \geq 0$. Moreover, as explained in Sect. 1.6.3, we may have

[8] If $\mathbf{w} \in \Pi_K^+$, then the lower bound is equal to that in (1.32).

[9] Note that in the first part of the book, $\mathbf{p} > 0$ is a right eigenvector of $\mathbf{X} \in \mathrm{X}_K$.

$$\sup_{\mathbf{p} \in \mathbb{R}^K_{++}} \min_{1 \le k \le K} \frac{1}{\mathrm{SIR}_k(\mathbf{p})} < \rho(\mathbf{V}) .$$

So, the question is under which conditions the above supremum is attained and the left-hand side is equal to the spectral radius of $\mathbf{V}$. This is equivalent to asking under which conditions a positive right eigenvector of $\mathbf{V}$ associated with $\rho(\mathbf{V}) > 0$ can be constructed, which is actually necessary and sufficient for a max-min fair power allocation in the sense of (5.39) with (5.40) to exist. In fact, in Sect. 1.6.2, we completely characterized the set of such matrices. By these results, we know that a max-min fair power allocation exists if and only if $\mathbf{V} \in \mathrm{B}_K$. Note that the set B_K is larger than X_K. In particular, if $\mathbf{V} \in \mathrm{B}_K$, there may be an entirely coupled subnetwork with links being orthogonal to all other links outside the subnetwork. Such a subnetwork is called isolated. Now the condition $\mathbf{V} \in \mathrm{B}_K$ means that each isolated subnetwork is maximal (see Sect. 1.6.2 and Definition A.31) and all other (nonisolated) entirely coupled subnetworks are not maximal.

Now let us consider the existence of a $(\mathbf{w}, \Psi)$-fair power allocation given by (5.41). First of all, if $\mathbf{V} \ge 0$ is reducible, then the function $G : \mathbb{R}^K_{++} \times \mathbb{R}^K_{++} \to \mathbb{R}$ in (5.41) may not be well defined. As discussed in Sect. 1.6.2, for the function to be well defined, it is necessary to assume that $\mathbf{V} \in \mathrm{N}^+_K$. Furthermore, we have $F \in \mathcal{G}(\mathbf{V})$ where $\mathcal{G}(\mathbf{V})$ is specified by Definition 1.22 except that now $G(\mathbf{p}, \mathbf{w})$ does not need to have a minimum on $\mathbb{R}^K_{++}$ for every $\mathbf{w} > 0$ but only a finite infimum. Under these assumptions, Theorem 1.60 implies that if $\mathbf{w} = \overline{\mathbf{q}} \circ \overline{\mathbf{p}} \ge 0$, then, for any $\mathbf{V} \in \mathrm{N}^+_K$,

$$\inf_{\mathbf{p} \in \mathbb{R}^K_{++}} G(\mathbf{p}, \mathbf{w}) = \inf_{\mathbf{p} \in \mathbb{R}^K_{++}} \sum_{k \in \mathsf{K}} w_k F\Big(\frac{(\mathbf{V}\mathbf{p})_k}{p_k}\Big) = F\big(\rho(\mathbf{V})\big) \|\mathbf{w}\|_1 .$$

Moreover, according to the discussion above, the infimum is attained for $\mathbf{p} = \overline{\mathbf{p}} > 0$ if $\mathbf{V} \in \mathrm{B}_K$. However, note that even if $\mathbf{V} \in \mathrm{B}_K$, it might be impossible to construct a positive weight vector $\mathbf{w}$ such that $\mathbf{w} = \overline{\mathbf{q}} \circ \overline{\mathbf{p}}$. This is simply because the existence of a positive left eigenvector of $\mathbf{V} \in \mathrm{B}_K$ associated with $\rho(\mathbf{V})$ is not guaranteed.

From a practical point of view, allocating positive transmit powers to links with zero weights is a waste of resources, and therefore does not make sense. For this reason, we are interested in the characterization of channel state matrices $\mathbf{V} \in \mathrm{N}^+_K$ for which positive left and right eigenvectors can be constructed. Such a characterization is provided by Theorem A.34 (see also Sect. 1.6.2). This theorem asserts that $\mathbf{V}$ has positive left and right eigenvectors associated with $\rho(\mathbf{V})$ if and only if $\mathbf{V} \in \overline{\mathrm{B}}_K$, that is, if and only if both $\mathbf{V}$ is block-irreducible in accordance with Definition A.33 and each diagonal block in the normal form (A.20) is maximal. Note that if $\mathbf{V}$ is block-irreducible, the network consists of entirely coupled subnetworks each of which is isolated. As a consequence, Theorem 5.16 essentially extends to the set $\overline{\mathrm{B}}_K$ except that $\overline{\mathbf{p}} > 0$ and $\mathbf{w} = \overline{\mathbf{q}} \circ \overline{\mathbf{p}} > 0$ are not unique up to constant multiples. In fact, $(\overline{\mathbf{q}}, \overline{\mathbf{p}})$ can be any member of $\mathrm{E}_K^+(\mathbf{V})$, the eigenmanifold of $\mathbf{V} \in \overline{\mathrm{B}}_K$ defined in Sect. A.4.3. Finally, considering Theorem 1.62 shows that if $\mathbf{V} \in \overline{\mathrm{B}}_K$, then $(\overline{\mathbf{p}}, \overline{\mathbf{q}} \circ \overline{\mathbf{p}})$ is a saddle point of $G(\mathbf{p}, \mathbf{w})$ for any $(\overline{\mathbf{q}}, \overline{\mathbf{p}}) \in \mathrm{E}_K^+(\mathbf{V})$.

5.5.2 Axiom-Based Interference Model

In this section, we point out the work of [82] (see also other references therein) where the authors go one step further in assuming that interference not only depends on the power allocation, but also on some adaptive *receive strategies*, like interference filtering or channel assignment. The additional optimization of the receive strategy adds new degrees of freedom to the problem of resource allocation. Thus, new concepts and algorithms are required.

More precisely, if $\mathbf{z} = 0$, [10] then the SIR is of the form $\mathrm{SIR}_k(\mathbf{p}) = p_k / I_k(\mathbf{p})$ where $I_k : \mathbb{R}_+^K \to \mathbb{R}_+$ is an interference function characterized by the following axioms.

A1 $I_k(\mathbf{p}) \geq 0$ (non-negativity).
A2 $I_k(\mu \mathbf{p}) = \mu I_k(\mathbf{p})$ for all $\mu \geq 0$ (scalability).
A3 $I_k(\mathbf{p}^{(1)}) \geq I_k(\mathbf{p}^{(2)})$ if $\mathbf{p}^{(1)} \geq \mathbf{p}^{(2)}$ (monotonicity).

Obviously, the interference function $I_k(\mathbf{p}) = (\mathbf{V}\mathbf{p})_k = \sum_l v_{k,l} p_l$ considered in this book satisfies the axioms. This linear mapping is however only one possible choice of an interference function. Indeed, I_k may be nonlinear and can also model the impact of adaptive receiver designs, such as minimum mean square error (MMSE) filtering or interference cancellation, as well as other system aspects. Common examples are:

- $I_k(\mathbf{p}) = \min_{z_k \in \mathrm{Z}_k} \sum_l v_{k,l}(z_k) p_l$, where the adjustable receive strategy z_k (taken from some compact set of possible strategies Z_k) has impact on the effective interference on link k. This specific model holds, for instance, for linear multi-antenna MMSE filtering or MMSE detection in code division multiple access systems.
- $I_k(\mathbf{p}) = \max_c f_k(\mathbf{p}, c)$, where $f_k(\mathbf{p}, c)$ is the interference for a given power allocation $\mathbf{p}$ under some interference uncertainty c. This definition can be used to model worst-case interference under imperfect channel knowledge.

---

10 The case of positive noise powers can be incorporated by considering an extended power vector [82].

6

Power Control Algorithm

This chapter presents algorithmic solutions to the power control problem as stated in the previous chapter. We focus on recursive gradient-based algorithms with a constant step size [83, 11]. Although much more powerful algorithms can be devised to solve the problem, we are going to confine our attention to such methods because of their simplicity. In particular, there is no need for step size control, which is difficult to realize in practice.

The significance of simple iterative algorithms that allow an *efficient* distributed implementation cannot be emphasized enough in the case of wireless networks where the judicious assessment of the complexity–performance trade off is particularly important. Given the limited and costly nature of wireless resources, minimizing the control message overhead for each iteration step must be a high priority. In case of gradient-based algorithms, one of the major challenges is the computation of the gradient vector of the aggregate utility function in a distributed manner. In general, due to the mutual dependence of logical links, this computation involves coordination and exchange of global information between all network nodes. Therefore, the use of classical flooding protocols to exchange this information results in a relatively high cost in terms of wireless resources.

In this book, we present a scheme based on the use of an adjoint network to efficiently distribute some locally measurable quantities to all other logical transmitters. A network is said to be adjoint to a given (primal) network with the channel state matrix $\mathbf{V}$ if it has the same network topology and its channel state matrix is $\mathbf{V}^T$. We call the procedure *cooperative flooding* as nodes cooperate by transmitting its local information to other nodes. More precisely, instead of each node sending its message separately as in the case of classical flooding protocols, nodes cooperate by transmitting simultaneously over the adjoint network in such a way that each node can estimate its gradient component based on some local measurements. The disadvantage of this scheme is that some coarse synchronization is necessary.

6.1 Some Basic Definitions

Throughout this chapter, the utility function is assumed to have the form given by (5.21) where $\Psi : \mathbb{R}_{++} \to \mathsf{Q}$ satisfies the Ψ-conditions defined right below (5.21). For this class of utility functions, we prove global convergence of the algorithm. The rate of convergence is shown to be geometric (or linear) under some mild conditions on the channel state matrix $\mathbf{V}$. In order to conform to the classical formulation as a minimization problem, we define $\psi : \mathbb{R}_{++} \to \mathsf{Q}$ to be[1]

$$\psi(x) := -\Psi(x), \quad x \in \mathbb{R}_{++}. \tag{6.1}$$

Thus, using

$$F(\mathbf{p}) := \sum_{k \in \mathsf{K}} w_k \psi(\mathrm{SIR}_k(\mathbf{p})), \quad \forall_{k \in \mathsf{K}}\, w_k > 0 \tag{6.2}$$

with the signal-to-interference ratio (SIR) defined by (4.2), the power control problem in (5.20) can be rewritten in an equivalent form as

$$\mathbf{p}^* = \arg\min_{\mathbf{p} \in \mathsf{P}} F(\mathbf{p}). \tag{6.3}$$

Note that due to Lemma 5.9 and $\psi(x) = -\Psi(x)$, the minimum exists. Throughout this chapter, it is assumed that

$$\mathrm{trace}(\mathbf{V}) = 0 \tag{6.4}$$

which implies that there is no self-interference on each link (see also Sect. 5.4). However, we point out that this requirement does not impact the generality of the analysis and could be easily dropped. It is convenient to define the interference function $I(\mathbf{p}) > 0$ as

$$I_k(\mathbf{p}) := (\mathbf{V}\mathbf{p} + \mathbf{z})_k = \sum_{l \in \mathsf{K}} v_{k,l} p_l + z_k = \sum_{\substack{l \in \mathsf{K} \\ l \neq k}} v_{k,l} p_l + z_k, \; k \in \mathsf{K} \tag{6.5}$$

where the last equality follows from (6.4). Hence,

$$\mathrm{SIR}_k(\mathbf{p}) = \frac{p_k}{I_k(\mathbf{p})}. \tag{6.6}$$

For completeness, the definition below summarizes key properties of the function ψ, which are an immediate consequence of (6.1) and the Ψ-conditions (Definition 5.5).

[1] The reader is also referred to Sect. 5.3 for definitions and other interpretations of the functions $\Psi(x)$ and $\psi(x) = -\Psi(x)$

Definition 6.1 (ψ-Conditions).

(i) $\psi : \mathbb{R}_{++} \to \mathrm{Q}$ *is a twice continuously differentiable and strictly decreasing function.*

(ii) We have

$$\lim_{x\to 0}\psi(x) := +\infty \qquad \Rightarrow \qquad \lim_{x\to 0}\psi'(x) = \lim_{x\to 0}\frac{d\psi}{dx}(x) = -\infty\,. \tag{6.7}$$

This requirement guarantees that $\mathbf{p}^*$ *given by (6.3) is positive.*

(iii) $\psi_e(x) := \psi(e^x)$ *is convex on* $\mathbb{R}$. *Since* ψ *is twice continuously differentiable and* $e^x > 0$ *is strictly monotone on* $\mathbb{R}$, *Theorem B.18 implies that this is equivalent to*

$$\psi_e''(x) = \frac{d^2\psi_e}{dx^2}(x) \geq 0, \quad x \in \mathbb{R}\,. \tag{6.8}$$

It is worth pointing out that the last condition implies that ψ is strictly convex. To see this, let $\hat{x}, \check{x} \in \mathbb{R}$ with $\hat{x} \neq \check{x}$ be arbitrary, and let $x(\mu) = (1-\mu)\hat{x} + \mu\check{x}$ be their convex combination. By convexity of ψ_e, we have

$$\psi_e(x(\mu)) = \psi(e^{x(\mu)}) \leq (1-\mu)\psi(e^{\hat{x}}) + \mu\psi(e^{\check{x}})$$

for all $\mu \in [0,1]$. On the other hand, it is a well-known fact [84] that the arithmetic mean bounds above the geometric one. Hence, we have

$$e^{x(\mu)} = \left(e^{\hat{x}}\right)^{1-\mu}\left(e^{\check{x}}\right)^{\mu} \leq (1-\mu)\left(e^{\hat{x}}\right) + \mu\left(e^{\check{x}}\right) \tag{6.9}$$

for all $\mu \in (0,1)$. Equality holds if and only if $e^{\hat{x}} = e^{\check{x}}$ or, equivalently, if and only if $\hat{x} = \check{x}$. Thus, combining this and (6.9) with the previous inequality as well as taking into account that ψ is strictly decreasing yields

$$\psi((1-\mu)\hat{z} + \mu\check{z}) < (1-\mu)\psi(\hat{z}) + \mu\psi(\check{z}), \quad \hat{z}, \check{z} > 0, \hat{z} \neq \check{z}$$

for all $\mu \in (0,1)$ where we used $\hat{z} = e^{\hat{x}} > 0$ and $\check{z} = e^{\check{x}} > 0$. This proves the claim. This observation, however, should not tempt the reader to conclude that $F(\mathbf{p})$ is a convex function of $\mathbf{p}$.

6.2 Convex Statement of the Problem

The main objective of this section is to show that the power control problem in (6.3) can be transformed into a convex problem, provided that the ψ-conditions are satisfied. A key ingredient in this formulation is the fact that $\mathrm{SIR}_k(e^{\mathbf{s}})$ is a log-concave function of the *logarithmic power vector* [85]

$$\mathbf{s} := \log \mathbf{p}\,, \qquad \mathbf{p} \in \mathrm{P}_+ := \mathrm{P} \cap \mathbb{R}^K_{++} \tag{6.10}$$

where the logarithm is taken elementwise. There is no loss in generality in assuming that $\mathbf{p} \in \mathrm{P}_+$ (a positive vector) since, in the minimum, every link must be assigned a positive transmit power (equation (6.7)). Therefore, we have

$$\min_{\mathbf{p}\in\mathrm{P}} F(\mathbf{p}) = \inf_{\mathbf{p}\in\mathrm{P}_+} F(\mathbf{p}) = \min_{\mathbf{p}\in\mathrm{P}_+} F(\mathbf{p}) .$$

By strict monotonicity of the logarithm function, we see that every $\mathbf{p} \in \mathrm{P}_+$ is associated with a unique $\mathbf{s} \in \mathrm{S}$ where $\mathrm{S} \subset \mathbb{R}^K$ is the set of all feasible logarithmic transmit powers, that is to say,

$$\mathrm{S} := \{\mathbf{s} \in \mathbb{R}^K : \mathbf{s} = \log \mathbf{p}, \mathbf{p} \in \mathrm{P}_+\} . \tag{6.11}$$

Consequently, if $F(\mathbf{p})$ attains its minimum on P_+, for some $\mathbf{p}^*$, then equivalently,

$$F_e(\mathbf{s}) := F(e^{\mathbf{s}}) = \sum_{k\in\mathsf{K}} w_k\, \psi(\mathrm{SIR}_k(e^{\mathbf{s}})) \tag{6.12}$$

attains its minimum on S for $\mathbf{s}^* = \log \mathbf{p}^*$.

Lemma 6.2. *Let* $\mathbf{s}(\mu) := (1-\mu)\hat{\mathbf{s}} + \mu\check{\mathbf{s}}$. *Then,*

$$\mathrm{SIR}_k(e^{\mathbf{s}(\mu)}) \geq \mathrm{SIR}_k(e^{\hat{\mathbf{s}}})^{1-\mu}\mathrm{SIR}_k(e^{\check{\mathbf{s}}})^{\mu}, \quad 1 \leq k \leq K \tag{6.13}$$

for all $\hat{\mathbf{s}}, \check{\mathbf{s}} \in \mathbb{R}^K$ *and* $\mu \in [0,1]$.

Proof. By Hölder's inequality (Theorem A.2),

$$\begin{aligned} I_k(e^{\mathbf{s}(\mu)}) &= \sum_{l\neq k} v_{k,l} e^{s_l(\mu)} + z_k = \sum_{l\neq k} (v_{k,l}e^{\hat{s}_l})^{1-\mu}(v_{k,l}e^{\check{s}_l})^{\mu} + z_k^{1-\mu} z_k^{\mu} \\ &\leq \Big(\sum_{l\neq k} v_{k,l}e^{\hat{s}_l} + z_k\Big)^{1-\mu}\Big(\sum_{l\neq k} v_{k,l}e^{\check{s}_l} + z_k\Big)^{\mu} = I_k(e^{\hat{\mathbf{s}}})^{1-\mu} I_k(e^{\check{\mathbf{s}}})^{\mu} \end{aligned}$$

for all $\mu \in [0,1]$. Thus, considering (6.6) yields

$$\mathrm{SIR}_k(e^{\mathbf{s}(\mu)}) \geq \frac{e^{(1-\mu)\hat{s}_k + \mu\check{s}_k}}{I_k(e^{\hat{\mathbf{s}}})^{1-\mu} I_k(e^{\check{\mathbf{s}}})^{\mu}} = \frac{(e^{\hat{s}_k})^{1-\mu}(e^{\check{s}_k})^{\mu}}{I_k(e^{\hat{\mathbf{s}}})^{1-\mu} I_k(e^{\check{\mathbf{s}}})^{\mu}} = \mathrm{SIR}_k(e^{\hat{\mathbf{s}}})^{1-\mu}\mathrm{SIR}_k(e^{\check{\mathbf{s}}})^{\mu}$$

which completes the proof.

An immediate consequence of Lemma 6.2 is that the logarithmic SIR

$$h_k(\mathbf{s}) := \log(\mathrm{SIR}_k(e^{\mathbf{s}})), \quad 1 \leq k \leq K \tag{6.14}$$

is a concave function of $\mathbf{s} \in \mathrm{S}$. Now we use this result to show that $F_e(\mathbf{s})$ is convex on $\mathbb{R}^K$ [86, 53].

Theorem 6.3. $F_e(\mathbf{s})$ *is convex on* $\mathbb{R}^K$, *i.e., we have*

$$F_e(\mathbf{s}(\mu)) \leq (1-\mu)F_e(\hat{\mathbf{s}}) + \mu F_e(\check{\mathbf{s}}) \tag{6.15}$$

for all $\hat{\mathbf{s}}, \check{\mathbf{s}} \in \mathbb{R}^K$ *and* $\mu \in (0,1)$.

Proof. Let $\hat{\mathbf{s}}, \check{\mathbf{s}} \in \mathbb{R}^K$ with $\hat{\mathbf{s}} \neq \check{\mathbf{s}}$ be arbitrary. For all $\mu \in (0,1)$, we have

$$\begin{aligned}
F_e(\mathbf{s}(\mu)) &= \sum_{k=1}^{K} w_k \psi(\mathrm{SIR}_k(e^{\mathbf{s}(\mu)})) \\
&\overset{(a)}{\leq} \sum_{k=1}^{K} w_k \psi\big(\mathrm{SIR}_k(e^{\hat{\mathbf{s}}})^{1-\mu}\mathrm{SIR}_k(e^{\check{\mathbf{s}}})^{\mu}\big) \\
&= \sum_{k=1}^{K} w_k \psi\big(e^{(1-\mu)h_k(\hat{\mathbf{s}})+\mu h_k(\check{\mathbf{s}})}\big) \\
&= \sum_{k=1}^{K} w_k \psi_e\big((1-\mu)h_k(\hat{\mathbf{s}})+\mu h_k(\check{\mathbf{s}})\big) \\
&\overset{(b)}{\leq} \sum_{k=1}^{K} w_k \Big((1-\mu)\psi_e\big(h_k(\hat{\mathbf{s}})\big)+\mu\psi_e\big(h_k(\check{\mathbf{s}})\big)\Big) \\
&= (1-\mu)F_e(\hat{\mathbf{s}}) + \mu F_e(\check{\mathbf{s}})\,.
\end{aligned} \tag{6.16}$$

While inequality (a) follows from Lemma 6.2 and strict monotonicity of ψ (the function is strictly decreasing), inequality (b) is due to convexity of $\psi_e(x) = \psi(e^x)$.

So, F_e is convex on $\mathrm{S} \subset \mathbb{R}^K$, and therefore, with (obvious) convexity of S, we arrive at an equivalent convex formulation of the power control problem in (6.3).

Corollary 6.4. *Suppose that the ψ-conditions hold, and let $\mathbf{s} = \log \mathbf{p}$ be the logarithmic power vector. Then, the power control problem*

$$\mathbf{s}^* = \arg\min_{\mathbf{s}\in\mathrm{S}} F(e^{\mathbf{s}}) = \arg\min_{\mathbf{s}\in\mathrm{S}} F_e(\mathbf{s}) \tag{6.17}$$

is a convex optimization problem.

These results establish a strong connection to the results in the first part of the book. Indeed, by Theorem B.28, we see that ψ_e is convex on $\mathbb{R}$ (as required by the third ψ-condition) if and only if $\gamma : \mathrm{Q} \to \mathbb{R}_{++}$ with $\gamma(\psi(x)) = x, x > 0$, is log-convex. In other words, if the inverse function of ψ is log-convex, then the power control problem can be transformed into a convex optimization problem. By Sect. 5.3, the log-convexity property implies that the feasible QoS region is a convex set.

6.3 Strong Convexity Conditions

In this section, we strengthen Theorem 6.3 by proving sufficient conditions on strong convexity of $F_e(\mathbf{s})$. For more information about strong convexity, the reader is referred to Sect. B.2.1. The main motivation behind the analysis is to ensure a geometric rate of convergence of the algorithm.

First we are going to show that if $\psi_e : \mathbb{R} \to \mathbb{R}$ is strongly convex on any bounded interval on the real line, then F_e is strongly convex on an arbitrary bounded convex set $\overline{\mathrm{S}} \subset \mathbb{R}^K$. To this end, let $\mathrm{I} \subset \mathbb{R}$ be a bounded interval chosen such that $h_k(\mathbf{s}) \in \mathrm{I}$ for each $k \in \mathsf{K}$ and all $\mathbf{s} \in \overline{\mathrm{S}}$. Since $h_k : \mathbb{R}^K \to \mathbb{R}$ is continuous, it is clear that there exists such an interval. Therefore, by the assumption (see also Definition B.19), there exists some constant $c > 0$ such that $\psi_e\big((1-\mu)h_k(\hat{\mathbf{s}}) + \mu h_k(\check{\mathbf{s}})\big) \leq (1-\mu)\psi_e\big(h_k(\hat{\mathbf{s}})\big) + \mu\psi_e\big(h_k(\check{\mathbf{s}})\big) - 1/2c\mu(1-\mu)\big(h_k(\hat{\mathbf{s}}) - h_k(\check{\mathbf{s}})\big)^2, 1 \leq k \leq K$, for all $\mu \in (0,1)$ and $\hat{\mathbf{s}}, \check{\mathbf{s}} \in \overline{\mathrm{S}}$. Incorporating this into inequality (*b*) in (6.16) yields $F_e(\mathbf{s}(\mu)) \leq (1-\mu)F_e(\hat{\mathbf{s}}) + \mu F_e(\check{\mathbf{s}}) - 1/2c\mu(1-\mu)\|\mathbf{h}(\hat{\mathbf{s}}) - \mathbf{h}(\check{\mathbf{s}})\|_{\mathbf{W}}^2$ for all $\mu \in (0,1)$ and $\hat{\mathbf{s}}, \check{\mathbf{s}} \in \overline{\mathrm{S}}$, where $\mathbf{h}(\mathbf{s}) := (h_1(\mathbf{s}), \ldots, h_K(\mathbf{s}))$ is the vector of the logarithmic SIRs, $\mathbf{W} = \mathrm{diag}(w_1, \ldots, w_K)$ is positive definite, and $\|\mathbf{u}\|_{\mathbf{W}}^2 = \mathbf{u}^T\mathbf{W}\mathbf{u}$. Since $\mathbf{W}$ is positive definite and all norms are equivalent on finite dimensional metric spaces [87], we deduce that there exists a constant $c_1 > 0$ such that

$$F_e(\mathbf{s}(\mu)) \leq (1-\mu)F_e(\hat{\mathbf{s}}) + \mu F_e(\check{\mathbf{s}}) - 1/2c_1\mu(1-\mu)\|\mathbf{h}(\hat{\mathbf{s}}) - \mathbf{h}(\check{\mathbf{s}})\|_2^2$$

for all $\mu \in (0,1)$ and $\hat{\mathbf{s}}, \check{\mathbf{s}} \in \overline{\mathrm{S}}$. Now note that $\mathbf{h} : \mathbb{R}^K \to \mathbb{R}^K$ is a bijection. This immediately follows from the fact that $\mathbb{R}_{++} \to \mathbb{R} : x \to \log(x)$ is bijection and $\mathbf{p}(\boldsymbol{\omega})$ defined by (5.31) is a bijection, provided that $z_k > 0$ for each $k \in \mathsf{K}$. Moreover, $\mathbf{h}(\mathbf{s})$ is Lipschitz continuous on $\overline{\mathrm{S}}$ (Definition B.33) since the Jacobian matrix of $\mathbf{h}(\mathbf{s})$ is bounded in the matrix 2-norm on the bounded set $\overline{\mathrm{S}}$. Therefore, as $\mathbf{h}$ is bijection, it is actually bilipschitz so that there exists a constant $0 < M < +\infty$ with $1/M\|\hat{\mathbf{s}}-\check{\mathbf{s}}\|_2 \leq \|\mathbf{h}(\hat{\mathbf{s}})-\mathbf{h}(\check{\mathbf{s}})\|_2 \leq M\|\hat{\mathbf{s}}-\check{\mathbf{s}}\|_2$ for all $\hat{\mathbf{s}}, \check{\mathbf{s}} \in \overline{\mathrm{S}}$. Combining this with the inequality above implies that there exists a constant $c_2 > 0$ such that $F_e(\mathbf{s}(\mu)) \leq (1-\mu)F_e(\hat{\mathbf{s}}) + \mu F_e(\check{\mathbf{s}}) - 1/2c_2\mu(1-\mu)\|\hat{\mathbf{s}}-\check{\mathbf{s}}\|_2^2$ for all $\mu \in (0,1)$ and $\hat{\mathbf{s}}, \check{\mathbf{s}} \in \overline{\mathrm{S}}$. We summarize these observations in a lemma.

Lemma 6.5. *Let the ψ-conditions be satisfied, let $z_k > 0$, $k \in \mathsf{K}$, and let $\mathbf{V}$ be an arbitrary nonnegative matrix. In addition, suppose that, for any bounded interval $\mathrm{I} \subset \mathbb{R}$, there exists a constant $c > 0$ (dependent on* I*) such that $\psi_e(x) - 1/2cx^2$ is convex on* I. *Then, $F_e(\mathbf{s})$ is strongly convex on any bounded convex subset of $\mathbb{R}^K$.*

Proof. The lemma follows from the discussion above and Observation B.20 saying that any continuous function $f : \mathbb{R}^K \to \mathbb{R}$ is strongly convex (with modulus of strong convexity c) if and only if $f(\mathbf{x}) - 1/2c\|\mathbf{x}\|_2^2$ is convex.

Since ψ is assumed to be twice continuously differentiable, the requirement on strong convexity of ψ_e is equivalent to (Theorem B.22)

$$c \leq \psi_e''(x) = \frac{d^2\psi_e}{dx^2}(x) = e^x\Big(\psi''(e^x)e^x + \psi'(e^x)\Big), \quad x \in \mathrm{I} \subset \mathbb{R}\,. \qquad (6.18)$$

If $\Psi : \mathbb{R}_{++} \to \mathsf{Q}$ is given by (5.24) with $\alpha > 1$, then (6.18) is satisfied by $\psi(x) = -\Psi(x)$. Indeed, taking the second derivative of $\psi_e(x)$ gives $\psi_e''(x) = (\alpha - 1)e^{x(1-\alpha)}, \alpha > 1$, which is positive and bounded away from zero on any bounded interval $\mathsf{I} \subset \mathbb{R}$. The strong convexity condition is also satisfied by $\psi(x) = -\Psi(x)$ with Ψ given by (5.25) since then $\psi_e''(x) = (\alpha - 1)e^x/(1 + e^x)^\alpha, \alpha \geq 2$. In contrast, the requirement is not met when $\Psi(x) = \log(x), x > 0$ in which case the second derivative of $\psi_e(x) = -x$ is identically zero on $\mathbb{R}$.

Note that, in the lemma above, there are no additional limitations on the choice of the channel state matrix $\mathbf{V} \geq 0$. On the extreme, $\mathbf{V}$ may even be the zero matrix, in which case $F_e(\mathbf{s}) = \sum_k w_k \psi(\frac{e^{s_k}}{z_k}) = \sum_{k=1}^K w_k \psi_e(s_k - \log z_k)$. Thus, $\nabla^2 F_e(\mathbf{s}) = \operatorname{diag}\big(w_1 \psi_e''(s_1 - \log z_1), \ldots, w_K \psi_e''(s_K - \log z_K)\big), \mathbf{V} = \mathbf{0}$. Now we see that if (6.18) holds and $\mathbf{V} = \mathbf{0}$, the Hessian of $F(\mathbf{s})$ is positive definite on any bounded subset of $\mathbb{R}^K$. In contrast, if $\psi(x) = -\log(x), x > 0$, we obtain $F_e(\mathbf{s}) = \sum_k w_k \log z_k - \mathbf{w}^T \mathbf{s}$ which is linear in $\mathbf{s} \in \mathbb{R}^K$, and therefore not strongly convex. As $\Psi(x) = \log(x), x > 0$, is of great interest for wireless applications (see Sects. 5.2.3 and 5.3), we prove a sufficient condition under which the strong convexity property of $F_e(\mathbf{s})$ is guaranteed on any bounded convex subset of $\mathbb{R}^K$, provided that the ψ-conditions are satisfied. It turns out that some very mild restrictions on the channel state matrix $\mathbf{V} \geq 0$ are sufficient to reestablish the strong convexity property of Lemma 6.5.

Lemma 6.6. *Let $\mathbf{V} \geq 0$ be a matrix such that, for each $l \in \mathsf{K}$, there exists $k \neq l$ with $v_{k,l} > 0$. Then, $F_e(\mathbf{s})$ is strongly convex on any bounded convex subset of $\mathbb{R}^K$.*

In other words, each column of the matrix $\mathbf{V}$ is required to have at least one positive entry.

Proof. If $\psi(x) = -\log x, x > 0$, then $\psi_e(x) = -x, x \in \mathbb{R}$. So, since $\psi_e(x)$ is convex and strictly decreasing, it is sufficient to consider $\psi_e(x) = -x$. In other words, if the lemma holds for the linear function, then it holds for any function satisfying ψ-conditions.

Suppose that $\overline{\mathsf{S}}$ is any bounded convex subset of $\mathbb{R}^K$. Let $\hat{\mathbf{s}}, \check{\mathbf{s}} \in \overline{\mathsf{S}}$ with $\hat{\mathbf{s}} \neq \check{\mathbf{s}}$ be arbitrary. Note that the lemma has the same setup as Theorem 2.12 and $\mathbb{R} \to \mathbb{R}_{++} : x \to e^x$ is a log-convex function. Hence, proceeding essentially as in the proof of Theorem 2.12 shows that there exists $k_0 \in \mathsf{K}$ such that

$$f_{k_0}(\mu) := I_{k_0}(e^{\mathbf{s}(\mu)}) \qquad \text{and} \qquad g_{k_0}(\mu) := \log f_{k_0}(\mu)$$

are strictly log-convex and strictly convex functions of $\mu \in (0, 1)$, respectively. In fact, $g_{k_0}(\mu)$ is strongly convex. To see this, let $l_0 \in \mathsf{K}$ with $\hat{s}_{l_0} \neq \check{s}_{l_0}$ be arbitrary, and let $k_0 \in \mathsf{K}, k_0 \neq l_0$, be such that $v_{k_0,l_0} > 0$. Note that by assumption, such an index exists. Taking the second derivative of $g_{k_0}(\mu)$ yields

$$g''_{k_0}(\mu) = \frac{\left(\sum_l e^{s_l(\mu)} v_{k_0,l}(\check{s}_l - \hat{s}_l)^2\right) I_{k_0}(e^{\mathbf{s}(\mu)}) - \left(\sum_l e^{s_l(\mu)} v_{k_0,l}(\check{s}_l - \hat{s}_l)\right)^2}{(I_{k_0}(e^{\mathbf{s}(\mu)}))^2}.$$

Now it may be verified that, for any $x_1, \ldots, x_n \in \mathbb{R}$ and nonnegative constants $a_1, \ldots, a_n$, $\left(\sum_{i=1}^n a_i x_i^2\right)\left(\sum_{j=1}^n a_j\right) - \left(\sum_{i=1}^n a_i x_i\right)^2 = \frac{1}{2}\sum_{i,j} a_i a_j (x_i - x_j)^2 \geq 0$. Hence, as $v_{k_0,l_0} > 0$, $z_{k_0} > 0$ and $\hat{\mathbf{s}}, \check{\mathbf{s}}$ are members of a *bounded* set $\overline{\mathsf{S}}$, there exists a constant $c > 0$ such that

$$g''_{k_0}(\mu) \geq z_{k_0} \frac{\sum_l e^{s_l(\mu)} v_{k_0,l}(\check{s}_l - \hat{s}_l)^2}{(I_{k_0}(e^{\mathbf{s}(\mu)}))^2} \geq c(\check{s}_{l_0} - \hat{s}_{l_0})^2$$

for all $\mu \in (0,1)$ and $\hat{\mathbf{s}}, \check{\mathbf{s}} \in \overline{\mathsf{S}}$. From this, it follows that $g_{k_0}(\mu)$ is strongly convex, and hence $h_{k_0}(\mathbf{s}(\mu)) = s_{k_0}(\mu) - g_{k_0}(\mu)$ is strongly concave on $\overline{\mathsf{S}}$. This is equivalent to saying that there exists a constant $c > 0$ such that $h_{k_0}(\mathbf{s}(\mu)) \geq (1-\mu)h_{k_0}(\hat{\mathbf{s}}) + \mu h_{k_0}(\check{\mathbf{s}}) + 1/2c\mu(1-\mu)\|\check{\mathbf{s}} - \hat{\mathbf{s}}\|_2^2$ for all $\mu \in (0,1)$ and $\hat{\mathbf{s}}, \check{\mathbf{s}} \in \overline{\mathsf{S}}$. So, with $\psi_e(x) = -x$, we obtain $\psi_e(h_{k_0}(\mathbf{s}(\mu))) \leq (1-\mu)\psi_e(h_{k_0}(\hat{\mathbf{s}})) + \mu\psi_e(h_{k_0}(\check{\mathbf{s}})) - 1/2c\mu(1-\mu)\|\hat{\mathbf{s}} - \check{\mathbf{s}}\|_2^2$ for all $\mu \in (0,1)$ and $\hat{\mathbf{s}}, \check{\mathbf{s}} \in \overline{\mathsf{S}}$. This implies that for any fixed $\hat{\mathbf{s}}, \check{\mathbf{s}} \in \overline{\mathsf{S}}$, there exists at least one addend in $\sum_k w_k \psi_e(h_k(\mathbf{s}(\mu)))$ for which the inequality above is satisfied, with an appropriately chosen positive constant $c > 0$. From this, strong convexity of $F_e(\mathbf{s})$ on any bounded convex subset of $\mathbb{R}^K$ follows.

Let us summarize both lemmas in a theorem.

Theorem 6.7. *Let the ψ-conditions be satisfied, and let $\mathbf{z} > 0$. Suppose that one of the following holds.*

(i) ψ_e is strongly convex on any bounded interval on the real line.
(ii) Each column of $\mathbf{V}$ has at least one positive entry.

Then, F_e is strongly convex on any bounded convex subset of $\mathbb{R}^K$.

It is important to emphasize that in the setup of Theorem 6.6, the channel state matrix is not necessarily irreducible. To illustrate the result, consider the following matrix

$$\mathbf{V} = \begin{pmatrix} 0 & v_{1,2} & v_{1,3} \\ v_{2,1} & 0 & 0 \\ 0 & 0 & 0 \end{pmatrix} \qquad v_{1,2}, v_{1,3}, v_{2,1} > 0\,.$$

The matrix is reducible and satisfies the condition of Lemma 6.6. It may be verified that with this choice of $\mathbf{V}$, the Hessian of $F_e(\mathbf{s})$ is positive definite on any bounded subset of $\mathbb{R}^K$. The explanation is basically the same as in Sect. 2.3.3: For each l, there is a k such that $v_{k,l} > 0$. This implies that each link is an interferer to some other link. Thus, since the noise term is positive for all $k \in \mathsf{K}$, it follows that for each $l \in \mathsf{K}$, there must exist $k \in \mathsf{K}$ such that $I_k(e^{\mathbf{s}})$ is strictly log-convex along the lth coordinate of $\mathbf{s}$. This in turn implies that

$F_e(\mathbf{s})$ is strongly log-convex on any bounded convex set. However, this is no longer true if we take the transpose of the matrix above

$$\mathbf{V} = \begin{pmatrix} 0 & v_{1,2} & 0 \\ v_{2,1} & 0 & 0 \\ v_{3,1} & 0 & 0 \end{pmatrix} .$$

Now link 3 is exposed to interference from link 1 but it is an interferer to no other link. Consequently, there is no k such that $I_k(e^{\mathbf{s}})$ is strictly log-convex along the third coordinate of $\mathbf{s}$. Choosing $\psi(x) = -\log x, x > 0$ yields $F_e(\mathbf{s})$ that is linear in s_3, and hence the function cannot be strongly convex.

By Theorem B.28, we know that the inverse function of ψ is strictly log-convex on Q if and only if ψ_e is strictly convex on $\mathbb{R}_{++}$, which is true if ψ_e is strongly convex on any bounded interval on the real line. Therefore, Lemma 6.5 corresponds in some sense to Theorem 2.11. In contrast, Lemma 6.6 corresponds to Theorem 2.12, which has the same setup and asserts that, for every $\hat{\mathbf{s}}, \check{\mathbf{s}} \in \mathrm{F}_\gamma$, there exists $k \in \mathsf{K}$ such that $p_k(\boldsymbol{\omega}(\mu))$ is a strictly log-convex function of $\mu \in [0,1]$. The reader should notice a striking analogy between these results. In fact, the proof of Lemma 6.6 is based on the proof of Theorem 2.12. The last theorem in Sect. 2.3.3 (Theorem 2.14) suggests that each addend in $\sum_k w_k \psi_e(h_k(\mathbf{s}))$ would be strongly convex on any bounded convex set if $\mathbf{V}$ was irreducible, regardless of the choice of ψ_e for which the ψ-conditions hold. Indeed, if $\mathbf{V}$ is irreducible, an examination of the proof of Theorem 2.14 reveals that $I_k(e^{\mathbf{s}})$ is strictly log-convex on $\mathbb{R}^K$ for each $k \in \mathsf{K}$. Therefore, proceeding essentially as in the proof of Lemma 6.6 shows that $\psi_e(h_k(\mathbf{s}))$ is strongly log-convex on any bounded convex subset of $\mathbb{R}^K$, provided that $\mathbf{V}$ is an irreducible matrix. This may have a positive effect on the convergence rate of our algorithms.

6.4 Gradient Projection Algorithm

Under the assumption of the ψ-conditions, we consider the following recursive gradient projection algorithm with a *constant step size* $\delta > 0$ (small enough) [83]:

$$\mathbf{s}(n+1) = \Pi_{\mathrm{S}}\big[\mathbf{s}(n) - \delta \nabla F_e(\mathbf{s}(n))\big], \quad \mathbf{s}(0) \in \mathrm{S},\ n \in \mathbb{N}_0 \tag{6.19}$$

where $\Pi_{\mathrm{S}}[\mathbf{x}]$ denotes the projection of $\mathbf{x} \in \mathbb{R}^K$ on S (with respect to the Euclidean norm; see Theorem B.32). The kth partial derivative $\nabla_k F_e(\mathbf{s})$ is equal to

$$\nabla_k F_e(\mathbf{s}) = \frac{\partial F_e}{\partial s_k}(\mathbf{s}) = e^{s_k}\Big(g_k(e^{\mathbf{s}}) - \sum_{l \neq k} v_{l,k}\mathrm{SIR}_l(e^{\mathbf{s}}) g_l(e^{\mathbf{s}})\Big), \quad k \in \mathsf{K}, \tag{6.20}$$

with $\mathbf{s} = \log \mathbf{p}$ and

$$g_k(\mathbf{p}) = \frac{w_k \psi'(\mathrm{SIR}_k(\mathbf{p}))}{I_k(\mathbf{p})}, \quad k \in \mathsf{K}. \tag{6.21}$$

The operation of projecting a vector in $\mathbb{R}^K$ on S can be easily parallelized. We will discuss this in Sect. 6.4.4. In contrast, the problem of parallel computing $\nabla_k F_e(\mathbf{s})$ is anything but trivial. This problem is addressed in Sect. 6.5. In the following section, we show that the sequence $\{\mathbf{s}(n)\}$ generated by (6.19) converges to a stationary point that minimizes $F_e(\mathbf{s})$ over S.

6.4.1 Global Convergence

Although the problem is convex, it is not obvious that the algorithm converges to a stationary point (Definition B.31) of the problem. This is because, in a general case, some step size control is necessary to obtain the convergence. In view of distributed implementation, however, the step size is assumed to be constant. It is well known [83, 88] (see also Sect. B.4) that the gradient projection algorithm with a constant step size converges to a stationary point if each of the following is satisfied:

(i) $F_e(\mathbf{s})$ is bounded below on S,
(ii) $F_e(\mathbf{s})$ is continuously differentiable and the gradient $\nabla F_e(\mathbf{s})$ is Lipschitz continuous on S (see Definition B.33), and
(iii) $0 < \delta < 2/M$ where M is the Lipschitz constant. Note that this condition can be satisfied only if $\nabla F_e(\mathbf{s})$ is Lipschitz continuous.

Whereas the first condition is satisfied by assumption, the Lipschitz continuity condition is not met on S. Indeed, if we let the kth entry of $\hat{\mathbf{s}} = \check{\mathbf{s}} + c \in \mathrm{S}, c \in \mathbb{R}, \hat{\mathbf{s}} \neq \check{\mathbf{s}}$, tend to $-\infty$ while keeping all the other entries constant, it is easy to see from (6.20) and (6.7) that $\|\nabla F_e(\hat{\mathbf{s}}) - \nabla F_e(\check{\mathbf{s}})\|_2$ may grow without bound. [2] However, the problem stems from unboundedness of S and can be evaded by letting the step size depend on the start point. Indeed, it is intuitive to expect that for every given $\mathbf{s}(0) \in \mathrm{S}$, $\nabla F_e(\mathbf{s})$ satisfies the Lipschitz continuity condition on

$$\overline{\mathrm{S}} := \{\mathbf{x} \in \mathrm{S} : F_e(\mathbf{x}) \leq F_e(\mathbf{s}(0)) < +\infty\}. \tag{6.22}$$

Obviously, $\overline{\mathrm{S}}$ is *bounded* and, by convexity of $F_e(\mathbf{s})$, a convex set for every $\mathbf{s}(0) \in \mathrm{S}$. As shown below, the Lipschitz continuity property is a consequence of the fact that the Hessian $\nabla^2 F_e(\mathbf{s})$ exists and is continuous on S.

Lemma 6.8. *Suppose that ψ-conditions hold, $\mathbf{s}(0) \in \mathrm{S}$ is arbitrary, and $\overline{\mathrm{S}}$ is given by (6.22). Then, $\nabla F_e(\mathbf{s})$ is Lipschitz continuous on $\overline{\mathrm{S}}$, that is to say, there exists a constant $M > 0$ such that*

$$\|\nabla F_e(\hat{\mathbf{s}}) - \nabla F_e(\check{\mathbf{s}})\|_2 \leq M \|\hat{\mathbf{s}} - \check{\mathbf{s}}\|_2 \tag{6.23}$$

for all $\hat{\mathbf{s}}, \check{\mathbf{s}} \in \overline{\mathrm{S}}$.

---

[2] This is not the case when $\psi(x) = \log(1/x)$. See the brief discussion at the end of this section.

Proof. Let $\hat{\mathbf{s}}, \check{\mathbf{s}} \in \overline{\mathrm{S}}$ be arbitrary. By the twice continuous differentiability of $\psi_e(x), x \in \mathbb{R}$, each entry of the Hessian matrix $\nabla^2 F_e(\mathbf{s})$ is a continuous function on $\mathbb{R}^K$. This implies that the gradient $\nabla F_e : \mathbb{R}^K \to \mathbb{R}^K$ is Gateaux differentiable (Definition B.10), and hence, by [89, p. 69], one has

$$\|\nabla F_e(\hat{\mathbf{s}}) - \nabla F_e(\check{\mathbf{s}})\|_2 \leq \sup_{0 \leq \mu \leq 1} \|\nabla^2 F_e(\hat{\mathbf{s}} + \mu(\check{\mathbf{s}} - \hat{\mathbf{s}}))\|_2 \|\hat{\mathbf{s}} - \check{\mathbf{s}}\|_2$$

where $\|\mathbf{A}\|_2 = \sqrt{\lambda_{\max}}$ is the induced matrix 2-norm and $\lambda_{\max}$ is the largest eigenvalue of $\mathbf{A}^T\mathbf{A}$ (see (A.6) in Sect. A.2 for the definition of a matrix induced norm). Now, because each entry of $\mathbf{s} \in \overline{\mathrm{S}}$ is bounded, it is obvious that the Hessian $\nabla^2 F_e(\mathbf{s})$ is bounded above over $\overline{\mathrm{S}}$ in the matrix 2-norm. Defining this bound as M yields (6.23).

In fact, $\nabla F_e(\mathbf{s})$ satisfies the Lipschitz continuity condition on every bounded subset of S and, in particular, on the convex set $\overline{\mathrm{S}}$ for any $\mathbf{s}(0) \in \mathrm{S}$. Thus, by Lemma 6.8 and Theorem B.35, if

$$0 < \delta < 2/M, \quad M = \sup_{\mathbf{s} \in \overline{\mathrm{S}}} \|\nabla^2 F_e(\mathbf{s})\|_2 \tag{6.24}$$

the sequence $\{\mathbf{s}(n)\}$ generated by (6.19) will stay within the set $\overline{\mathrm{S}}$ for every $n \in \mathbb{N}_0$. Moreover, the algorithm will decrease the value of $F_e(\mathbf{s})$, unless a stationary point $\mathbf{s}^* \in \mathrm{S}$ has been reached. This point satisfies $(\mathbf{s} - \mathbf{s}^*)^T \nabla F_e(\mathbf{s}^*) \geq 0$ for every $\mathbf{s} \in \mathrm{S}$ (Theorem B.30 and Definition B.31). In fact, due to convexity of $F_e(\mathbf{s})$ shown in Sect. 6.2, we can conclude that $\mathbf{s}^*$ minimizes $F_e(\mathbf{s})$ over S, and therefore $\mathbf{p}^* = e^{\mathbf{s}^*}$ minimizes $F(\mathbf{p})$ over P. Let us summarize these observations in a theorem [53, 55].

Theorem 6.9. *Let the ψ-conditions be satisfied, and let $\{\mathbf{s}(n)\}$ be a sequence generated by (6.19). Then, for sufficiently small $\delta > 0$, $\{\mathbf{s}(n)\}$ converges to a limit point $\mathbf{s}^* \in \mathrm{S}$. Moreover, the limit point minimizes $F_e(\mathbf{s})$ over* S.

Remark 6.10. It is important to emphasize that the choice of δ in (6.19) depends on the start point $\mathbf{s}(0)$. However, this should not pose a significant problem to wireless networks where successful transmission requires some minimum SIR at the output of each linear receiver. This information could be used to predetermine a worst-case step size that would work under any feasible scenario at the possible expense of the convergence rate. In order to ensure some signal-to-interference ratios, nodes may start the iteration process with predefined transmit powers. Furthermore, the effective interference $I_k(\mathbf{p})$ should not exceed some predefined threshold for each $1 \leq k \leq K$. This could be achieved by limiting the number of active links with not too large path attenuations.

The Hessian of $F_e(\mathbf{s})$ can be easily calculated to give

$$\begin{aligned}\nabla^2 F_e(\mathbf{s}) = &\sum_{k=1}^{K} w_k \psi_e''(h_k(\mathbf{s}))\nabla h_k(\mathbf{s})\nabla h_k(\mathbf{s})^T \\ &+ \sum_{k=1}^{K} w_k \psi_e'(h_k(\mathbf{s}))\nabla^2 h_k(\mathbf{s})\,.\end{aligned} \tag{6.25}$$

We see that, by concavity of $h_k(\mathbf{s})$ and strict decreasingness of ψ_e, the second addend is positive semidefinite. The first addend is positive semidefinite as well since $\psi_e(x) = \psi(e^x)$ is convex. As the sum of positive semidefinite matrices is positive semidefinite, this implies that the Hessian matrix is positive semidefinite if the ψ-conditions are satisfied. This is in full agreement with Theorem 6.3, which, however, does not require differentiability.

The Lipschitz constant M in (6.24) is equal to $M = \sup_{\mathbf{s}\in\bar{S}} \lambda_{\max}(\nabla^2 F_e(\mathbf{s}))$. A closed form solution for the supremum is in general an intricate problem. In order to obtain a simpler condition, note that, for any matrix $\mathbf{A}$, we have $\rho(\mathbf{A}) \le \|\mathbf{A}\|_\infty$ and $\rho(\mathbf{A}) \le \|\mathbf{A}^T\|_\infty$ where $\rho(\mathbf{A})$ denotes the spectral radius of $\mathbf{A}$ (Definition A.8) and $\|\mathbf{A}\|_\infty$ is defined by (A.7). Hence, $\rho(\mathbf{A}) \le \min\{\max_i \sum_j |a_{i,j}|, \max_j \sum_i |a_{i,j}|\}$. Now since the Hessian matrix is symmetric, we obtain

$$\lambda_{\max}(\nabla^2 F_e(\mathbf{s})) \le \kappa(\mathbf{s}) := \max_i \sum_j |(\nabla^2 F_e(\mathbf{s}))_{i,j}|\,.$$

Therefore, choosing $0 < \delta < 2/M$ with $M = \sup_{\mathbf{s}\in\bar{S}} \kappa(\mathbf{s})$ ensures the convergence of the algorithm. When compared with (6.24), the Lipschitz constant here is significantly easier to estimate due to a simple relationship between $\kappa(\mathbf{s})$ and the entries of the Hessian matrix. In a special case, when $\Psi(x) = -\psi(x) = \log(x), x > 0$, an examination of (6.25) reveals that $\nabla^2 F_e(\mathbf{s}) = -\sum_k w_k \nabla^2 h_k(\mathbf{s})$ where

$$\left(\nabla^2 h_k(\mathbf{s})\right)_{i,j} = \begin{cases} \frac{(e^{s_i} v_{k,i})^2 - e^{s_i} v_{k,i} I_k(e^{\mathbf{s}})}{I_k(e^{\mathbf{s}})^2} & i = j \neq k \\ \frac{e^{s_i} v_{k,i} e^{s_j} v_{k,j}}{I_k(e^{\mathbf{s}})^2} & \text{elsewhere}\,. \end{cases}$$

Since $I_k(e^{\mathbf{s}}) \ge z_k > 0$ for all $\mathbf{s} \in S$, the Hessian matrix is bounded above in the matrix 2-norm on S. Therefore, in this special case, the Lipschitz continuity condition is satisfied on S. This in turn implies that there is a step size δ that works for any start point.

6.4.2 Rate of Convergence

The rate (or speed) of converge says how fast the method converges to the optimal solution, and therefore a fast convergence rate is highly desired. However, due to the dynamic nature of wireless networks as well as strict limitations on wireless resources, it is clear that only a relatively small number of iterations can be carried out in wireless networks. For this reason, the algorithms are usually required to have a faster initial convergence.

As far as the rate of convergence is concerned, much more powerful algorithms than that in (6.19) can be devised to solve the power control problem [88]. These algorithms, however, usually require extensive coordination between nodes in a network. For instance, the choice of the step size can be optimized in every iteration to improve the rate of convergence. On the other hand, such a step size control may require the exchange of a significant amount of information between nodes, and hence waste scarce wireless resources. Newton-like methods may provide a super-linear rate of convergence but require the inverse of the Hessian matrix which is prohibitive for wireless network applications.

In case of convex problems (each local optimum is a global one), rate of convergence is evaluated in terms of an error function $e : \mathbb{R}^K \to \mathbb{R}$ satisfying $e(\mathbf{x}) \geq 0$ for all $\mathbf{x} \in \mathbb{R}^K$ and $e(\mathbf{x}) = 0$ if and only if $\mathbf{x} = \mathbf{x}^*$ where $\mathbf{x}^*$ is a global optimum [88]. A typical choice of an error function that is assumed in what follows is the Euclidean distance

$$e(\mathbf{x}) = \|\mathbf{x} - \mathbf{x}^*\|_2 \,. \tag{6.26}$$

As for the power control algorithm presented above, everything we can guarantee is a geometric (or linear) rate of convergence defined as follows [88].

Definition 6.11. *A sequence of real-valued vectors* $\{\mathbf{x}(n)\}$ *is said to converge geometrically to* $\mathbf{x}^*$ *if there exist constants* $a > 0$ *and* $\beta \in (0,1)$ *such that*

$$e(\mathbf{x}(n)) \leq a\beta^n \,. \tag{6.27}$$

Provided that δ is sufficiently small, each iteration update of the power control algorithm in (6.19) stays within the bounded convex set $\overline{\mathrm{S}}$ defined by (6.22). At the same time, Theorem 6.7 asserts that $F_e(\mathbf{s})$ is strongly convex on any bounded subset of $\mathbb{R}^K$. Thus, considering Theorem B.36 yields the following corollary.

Corollary 6.12. *Suppose that one of the following holds:*

(i) $\psi_e : \mathbb{R}_{++} \to \mathrm{Q}$ *is strongly convex on any bounded interval in* $\mathbb{R}$,
(ii) each column of $\mathbf{V}$ *has at least one positive entry.*

Then, provided that δ *is chosen positive and small enough, the sequence* $\{\mathbf{s}(n)\}$ *generated by (6.19) converges to* $\mathbf{s}^*$ *geometrically.*

Geometric convergence is obtained if $\limsup_{k\to\infty} \frac{e(\mathbf{x}(n+1))}{e(\mathbf{x}(n))} \leq \beta$ for some $\beta \in (0,1)$. Thus, asymptotically ($n \to \infty$), the error is reduced by a factor of at least $\beta \in (0,1)$ at each iteration. So, a geometric convergence rate is a fairly satisfactory rate of convergence, provided the factor β is not too close to unity. Among others, this factor is influenced by the step size δ. For this reason, it may be beneficial to determine an appropriate step size at the beginning of each frame interval. As already mentioned before, the step size cannot be too large since otherwise divergence will occur. On the other hand, a small step size ensures convergence but the rate of convergence may be very slow. To speed up the rate of convergence, an appropriate scaling can be performed.

6.4.3 Diagonal Scaling

The rate of convergence of gradient methods depends on the condition number of $\nabla^2 F_e(\mathbf{s}(n))$, which is defined as the ratio of the largest eigenvalue of the Hessian to its smallest one [88]. If the Hessian is positive definite on $\overline{\mathrm{S}}$ (a geometric convergence), the condition number is finite but it can still be relatively large causing gradient methods to converge very slow. In such cases, the problem can often be alleviated by appropriately scaling the update direction. The scaled power control algorithm with a constant step size takes the form

$$\mathbf{s}(n+1) = \Pi_{\mathrm{S}}^n \left[\mathbf{s}(n) - \delta\, \mathbf{D}(n)\nabla F_e(\mathbf{s}(n))\right], \qquad \mathbf{s}(0) \in \mathrm{S} \tag{6.28}$$

where $\mathbf{D}(n)$ is a symmetric positive definite matrix for every n. The projection in (6.28) is performed with respect to a different norm given by $\|\mathbf{x}\|_{\mathbf{D}(n)} = \sqrt{\mathbf{x}^T\mathbf{D}(n)\mathbf{x}}$. Thus, for any fixed $\mathbf{x} \in \mathbb{R}^K$, $\Pi_{\mathrm{S}}^n[\mathbf{x}]$ is a unique vector that minimizes $\|\mathbf{y} - \mathbf{x}\|_{\mathbf{D}(n)}$ over all $\mathbf{y} \in \mathrm{S}$.

Ideally, $\mathbf{D}(n)$ should be the inverse of the Hessian matrix of $F_e(\mathbf{s}(n))$ but this would require extensive global coordination and centralized computation. Thus, a reasonable choice of $\mathbf{D}(n)$ is a matrix for which all the diagonal entries of $\mathbf{D}(n)^{\frac{1}{2}}\nabla^2 F_e(\mathbf{s}(n))\mathbf{D}(n)^{\frac{1}{2}}$ are approximately equal to unity. This may be achieved by a diagonal matrix $\mathbf{D}(n)$ whose kth diagonal element $d_k(n)$ is given by

$$d_k(n) = \left(\frac{\partial^2 F_e}{\partial s_k^2}(\mathbf{s}(n))\right)^{-1}$$

where the Hessian matrix of $F_e(\mathbf{s})$ is given by (6.25).

6.4.4 Projection on a Closed Convex Set

In general, gradient projection algorithms are not amenable to distributed implementation as the computation of the projection (6.19) may involve all components of the update vector. Fortunately, the geometric structure of S makes a parallel implementation possible.

First of all, Theorem B.32 asserts that the projection exists and is unique since S is a closed convex set (Definitions B.1 and B.15). By Theorem B.32, given an arbitrary $n \in \mathbb{N}_0$, the projection $\Pi_{\mathrm{S}}[\mathbf{u}(n)]$ of the update vector $\mathbf{u}(n) = \mathbf{s}(n) - \delta\nabla F_e(\mathbf{s}(n))$ on S with respect to the Euclidean norm is equal to

$$\Pi_{\mathrm{S}}[\mathbf{u}(n)] = \arg\min_{\mathbf{x}\in\mathrm{S}} \|\mathbf{u}(n) - \mathbf{x}\|_2^2 .$$

In what follows, assume that $n \in \mathbb{N}_0$ is arbitrary but fixed, and let $\mathbf{u} = \mathbf{u}(n)$. From (4.6) and $\mathbf{s} = \log(\mathbf{p})$, we see that S is the N-fold Cartesian product $\mathrm{S} = \mathrm{S}_1 \times \cdots \times \mathrm{S}_N$ where

$$\mathrm{S}_m = \left\{ \mathbf{x} \in \mathbb{R}^{|\mathsf{K}(m)|} : \sum_{k=1}^{|\mathsf{K}(m)|} e^{x_k} \leq P_m \right\} .$$

Here P_m is the individual power constraint on node $m \in \mathsf{N}$. Each of these sets, say set S_m, is a closed subset of $\mathbb{R}^{|\mathsf{K}(m)|}$ with $\mathbb{R}^K = \mathbb{R}^{|\mathsf{K}(1)|} \times \cdots \times \mathbb{R}^{|\mathsf{K}(N)|}$. Therefore, it follows that (see also [88]) the projection of $\mathbf{u}$ on S can be accomplished by projecting $\mathbf{u}^{(m)}$ on $\mathrm{S}_m \subset \mathbb{R}^{|\mathsf{K}(m)|}$ where $\mathbf{u}^{(m)} \in \mathbb{R}^{|\mathsf{K}(m)|}$ is a subvector of $\mathbf{u}$ such that $\mathbf{u}^{(m)} = (u_k)_{k\in\mathsf{K}(m)}$.

Obviously, the projection of $\mathbf{u}^{(m)}$ on S_m can be carried out at node $m \in \mathsf{N}$ without any coordination with other nodes. In other words, each node, say node m, must solve the following problem

$$\Pi_{\mathrm{S}_m}[\mathbf{u}^{(m)}] = \arg\min_{\mathbf{x}\in\mathrm{S}_m} \|\mathbf{u}^{(m)} - \mathbf{x}\|_2^2$$

which is a standard quadratic optimization problem over a closed convex set $\mathrm{S}_m \subset \mathbb{R}^{|\mathsf{K}(m)|}$. Obviously, if $\mathbf{u}^{(m)} \in \mathrm{S}_m$, then $\Pi_{\mathrm{S}_m}[\mathbf{u}^{(m)}] = \mathbf{u}^{(m)}$. Finally, note that in the special case when there are individual power constraints on *each link* $P_1, \ldots, P_K$, then the projection of $\mathbf{u}$ on S is obtained by projecting the kth component of $\mathbf{u}$ on $(-\infty, \log P_k]$ (the projection on a box). So, in this case, the projection is a straightforward operation.

6.5 Distributed Implementation

An essential advantage of the algorithm is its amenability to efficient implementation in distributed networks. In particular, there is no need for step size control or complex operations such as matrix inversion. The projection operation can be performed without any coordination between nodes. Actually, the major problem is to parallelize the computation of $\nabla F_e(\mathbf{s})$ in such a way that each node, say node n, can calculate $\nabla_k F_e(\mathbf{s}) = \frac{\partial F_e}{\partial s_k}(\mathbf{s})$ for all $k \in \mathsf{K}(n)$, without resorting to extensive internode communication. The parallelization can be seen as separating the algorithm into K local algorithms operating concurrently at different transmitter–receiver pairs.

6.5.1 Local and Global Parts of the Gradient Vector

Consider the nth iteration in (6.19) and assume that $\mathbf{p} = \mathbf{p}(n) = e^{\mathbf{s}(n)}$ is the nth power vector. Using $\mathbf{s} = \log \mathbf{p}$, it is easy to see that $\nabla F_e(\mathbf{s})$ defined by

(6.20) is a version of $\nabla F(\mathbf{p})$ scaled with the positive definite diagonal matrix $\mathbf{P} = \text{diag}(p_1, \dots, p_K) = \text{diag}(e^{s_1}, \dots, e^{s_K})$:

$$\nabla F_e(\mathbf{s}) = \mathbf{P}\nabla F(\mathbf{p}), \quad \mathbf{s} = \log \mathbf{p}. \tag{6.29}$$

Thus, $\nabla_k F_e(\mathbf{s})$ can be easily obtained from $\nabla_k F(\mathbf{p})$ by multiplying it with $p_k = e^{s_k}$. In what follows, we focus on $\nabla F(\mathbf{p})$. Considering (6.20) reveals that we can rewrite the gradient vector as follows

$$\nabla F(\mathbf{p}) = \underbrace{(\mathbf{I} + \mathbf{\Gamma}(\mathbf{p}))\mathbf{g}(\mathbf{p})}_{\boldsymbol{\eta}(\mathbf{p})<0} - \underbrace{(\mathbf{I} + \mathbf{V}^T)\mathbf{\Gamma}(\mathbf{p})\mathbf{g}(\mathbf{p})}_{\boldsymbol{\theta}(\mathbf{p})<0} = \boldsymbol{\eta}(\mathbf{p}) - \boldsymbol{\theta}(\mathbf{p}). \tag{6.30}$$

Here and hereafter, $\boldsymbol{\eta}(\mathbf{p}) = (\eta_1(\mathbf{p}), \dots, \eta_K(\mathbf{p}))$, $\boldsymbol{\theta}(\mathbf{p}) = (\theta_1(\mathbf{p}), \dots, \theta_K(\mathbf{p}))$, $\mathbf{g}(\mathbf{p}) := (g_1(\mathbf{p}), \dots, g_K(\mathbf{p}))$ with $g_k(\mathbf{p})$ defined by (6.21) and

$$\mathbf{\Gamma}(\mathbf{p}) := \text{diag}\big(\text{SIR}_1(\mathbf{p}), \dots, \text{SIR}_K(\mathbf{p})\big).$$

So the problem of computing $\nabla_k F(\mathbf{p})$ at the kth logical transmitter is equivalent to the computation of both $\eta_k(\mathbf{p})$ and $\theta_k(\mathbf{p})$ at this transmitter. Let us first focus on $\eta_k(\mathbf{p})$. The problem of computing $\boldsymbol{\theta}(\mathbf{p})$ is deferred to the next section.

It follows from (6.21) that

$$\eta_k(\mathbf{p}) = \big(1 + \text{SIR}_k(\mathbf{p})\big)g_k(\mathbf{p}) = \big(1 + \text{SIR}_k(\mathbf{p})\big)\frac{w_k\psi'(\text{SIR}_k(\mathbf{p}))}{I_k(\mathbf{p})}. \tag{6.31}$$

Hence, $\eta_k(\mathbf{p})$ can be easily calculated at a node where link k originates, provided that the signal-to-interference ratio $\text{SIR}_k(\mathbf{p})$ is known at this node. In fact, knowledge of a good estimate of the signal-to-interference ratio in every iteration and for each logical transmitter–receiver pair is crucial for the algorithm to be implemented. In order to obtain an estimate of SIR, each transmitter may send a training sequence[3] that is known to the respective receiver. It is important to emphasize that all logical transmitters must be synchronized so that the transmission can take place simultaneously on all links. Using some standard estimation method (see for instance [90] for further information and references), each logical receiver estimates the signal-to-interference ratio on its link and sends the estimate back to the corresponding transmitter (node) using a reliable low-rate feedback channel.

In this book, we do not address the problem of estimating SIR. Instead, it is assumed that a good estimate of SIR is available at the corresponding transmitter and receiver side. Based on this information, each logical transmitter, say the transmitter on link k, is able to calculate the estimate of $g_k(\mathbf{p}) < 0$ and $\eta_k(\mathbf{p})$. We also assume that the kth receiver is able to determine the interference $I_k(\mathbf{p})$ based on the SIR estimate.

[3] By a training sequence, we mean a deterministic sequence of symbols generated by a pseudorandom number generator. We assume that the elements of this sequence "approximate" zero-mean independent and identically distributed random variables.

6.5.2 Adjoint Network

In contrast to $\eta_k(\mathbf{p})$, the problem of computing

$$\theta_k(\mathbf{p}) = \mathrm{SIR}_k(\mathbf{p})g_l(\mathbf{p}) + \sum_{l \neq k} v_{l,k}\mathrm{SIR}_k(\mathbf{p})g_l(\mathbf{p}) \tag{6.32}$$

is significantly more tricky. Interestingly, $\theta_k(\mathbf{p})$ can be estimated at each link (node) by a scheme that relies on the concept of an *adjoint* network defined as follows [54, 53, 55].

Definition 6.13 (Adjoint Network). *Consider an arbitrary wireless network with K logical links and the channel state matrix $\mathbf{V}$. Let us call it the primal network. Then, a network with K logical links and the channel state matrix $\mathbf{U} \in \mathbb{R}_+^{K\times K}$ is said to be adjoint to the primal network if $\mathbf{U} = \mathbf{V}^T$.*

Note that for any given primal network, an adjoint network is not unique in general. The definition above merely states that the channel state matrix of an adjoint network is a transpose matrix of the channel matrix of the primal network. In a special case, if $\mathbf{V} = \mathbf{V}^T$, then any network is adjoint to itself. In what follows, assume that a primal network with the channel state matrix $\mathbf{V}$ is given.

The reason for introducing the definition becomes more clear if we have a look at $\boldsymbol{\theta}(\mathbf{p})$ in (6.30) or (6.32). We see that $\boldsymbol{\theta}(\mathbf{p})$ results from the multiplication of the vector $\boldsymbol{\Gamma}(\mathbf{p})\mathbf{g}(\mathbf{p})$ with $(\mathbf{I} + \mathbf{V}^T)$. This suggests that the entries of the vector $\boldsymbol{\theta}(\mathbf{p})$ may be made available to some nodes in the network by transmitting appropriately scaled pilot symbols over an adjoint network. Obviously, the following two conditions should be satisfied:

(i) the kth logical transmitter in an adjoint network has an access to the kth coordinate of the vector $\boldsymbol{\Gamma}(\mathbf{p})\mathbf{g}(\mathbf{p})$, and
(ii) the kth coordinate of $(\mathbf{I} + \mathbf{U})\boldsymbol{\Gamma}(\mathbf{p})\mathbf{g}(\mathbf{p})$ corresponds to the kth logical transmitter in the primal network where $\mathbf{U}$ is the channel state matrix of an adjoint network.

In order to satisfy both conditions, we consider a so-called reversed network defined as follows.

Definition 6.14 (Reversed Network). *We call a network reversed if the roles of transmitters and receivers on each logical link in a primal network are reversed. In a reversed network, logical link $k \in \mathsf{K}$ is a link between the kth logical receiver in the primal network and the kth logical transmitter.*

By the reversed roles we mean that, in each transmitter–receiver pair, say the pair on logical link k, the kth transmitter becomes the kth receiver and vice versa. The corresponding link in a reversed network is labeled by k. A nice feature of a reversed network is that the first condition can be easily met. In fact, the kth logical transmitter in a reversed network knows $\mathrm{SIR}_k(\mathbf{p})$ and

$I_k(\mathbf{p})$ since it is the kth receiver in the primal network (see the previous section). Consequently, since $\psi'(x)$ is common for all links, the kth transmitter in the primal network only needs to inform the kth receiver about its weight w_k. However, this must be done only once before starting the iteration process. The second condition is automatically satisfied by a reversed network.

Unfortunately, in general, a reversed network is not adjoint to the primal network. To see this, let us write $\mathbf{V}$ as $\mathbf{V} = \mathbf{DG}$, where

$$\mathbf{D} = \mathrm{diag}\big(1/V_1, \ldots, 1/V_K\big)$$

is a diagonal matrix of the inverse path gains (see Sect. 4.3.1) and

$$\mathbf{G} = (V_{k,l})_{1 \leq k,l \leq K} \in \mathbb{R}_+^{K \times K}, \quad \mathrm{trace}(\mathbf{G}) = 0$$

incorporates the path gains (coupling factors) between different links. Now since the roles of logical transmitters and receivers are reversed on each link in a reversed network, its channel state matrix $\mathbf{U}$ is equal to

$$\mathbf{U} = \mathbf{DG}^T.$$

Obviously, we have $\mathbf{U} \neq \mathbf{V}^T$, unless $\mathbf{D}$ is a scaled identity $\alpha\mathbf{I}$ for some $\alpha > 0$.

Unfortunately, $\mathbf{D}$ is not a scaled identity in general, which is simply due to different channel realizations on different logical links. In contrast, if the wireless channel is an additive white Gaussian channel (AWGN), then $\mathbf{D}$ is a scaled identity, provided that all receivers are normalized appropriately. For instance, consider a network based on code division multiple access (CDMA) from Sect. 4.3.4. It follows that if $h_{k,k} = 1$ for each $k \in \mathsf{K}$ (AWGN channel), then $V_k = |\langle \mathbf{c}_k, \mathbf{s}_k \rangle|^2$ where $\mathbf{c}_k \in \mathbb{R}^J$ and $\mathbf{s}_k \in \mathbb{R}^J$ are the logical receiver on link k and the corresponding spreading sequence, respectively. Thus, normalizing all receivers such that $|\langle \mathbf{c}_k, \mathbf{s}_k \rangle|^2 = 1/\alpha$ yields $\mathbf{D} = \alpha\mathbf{I}$. Without loss of generality, we can assume that $|\langle \mathbf{c}_k, \mathbf{s}_k \rangle|^2 = 1$.

Remark 6.15. In what follows, it is assumed that X_k is an information-bearing symbol transmitted over the wireless channel. In particular, we do not care about the physical-layer realization. So, depending on the transmission technique, X_k may be either a symbol spread by some spreading sequence (CDMA) or by some beam-forming vector.

Using the above definitions, the channel state matrix of an adjoint network is $\mathbf{V} = \mathbf{G}^T\mathbf{D}$. Comparing this with $\mathbf{DG}^T$ of a reversed network, we see that instead of multiplying $\mathbf{G}^T$ by $\mathbf{D}$ on the right, the matrix is multiplied by $\mathbf{D}$ on the left. A straightforward examination shows that the right-hand side multiplication is achieved if each logical transmitter in a reversed network inverts its "own" wireless channel such that the resulting path attenuation between each transmitter–receiver pair is equal to 1. The effect of this on the channel state matrix $\mathbf{U} = \mathbf{DG}^T$ of the reversed network is that $\mathbf{D} = \mathbf{I}$ (due to the channel inversions) and $(\mathbf{G}^T)_{k,l} = V_{l,k}/V_l$ (due to the effect of the channel inversions on other links). Therefore, in this case, we obtain $\mathbf{U} = \mathbf{V}^T$. We summarize these observations in a theorem.

Theorem 6.16. *Assume a flat fading wireless channel, with the channel coefficient of link* $k \in \mathsf{K}$ *being equal to* $h_{k,k} \in \mathbb{C}, h_{k,k} \neq 0$. *Then, a reversed network is adjoint to a given primal network if each logical transmitter in the reversed network inverts its channel by multiplying transmit symbols by* $1/h_{k,k}$.

We point out that these ideas can be extended to frequency-selective channels, provided that all channels are invertible. In such a case, each symbol should be convoluted with the inverse impulse response of the corresponding wireless channel. Given some primal network, in all that follows, the adjoint network refers to the reversed network in which each transmitter performs the channel inversion. To illustrate the theorem, we neglect the Gaussian noise (see a remark on the noisy case in Section 6.5.3) and assume that $V_{k,l} = |h_{k,l}|^2$ and $V_k = |h_{k,k}|^2$ with $|h_{k,k}| > 0$ where $h_{k,l} \in \mathbb{C}$ is a given channel coefficient. In other words, $h_{k,l}, 1 \leq k, l \leq K$, are realizations of the wireless channel at the beginning of some frame interval. Now if all logical transmitters in the reversed network concurrently transmit sequences of zero-mean random symbols X_k multiplied by $1/h_{k,k}$, then the resulting network may be easily seen to have the channel state matrix $\mathbf{V}^T$.

The concept of an adjoint network is illustrated in Fig. 6.1 under the assumption of *noiseless links.* In this example, the signal-to-interference ra-

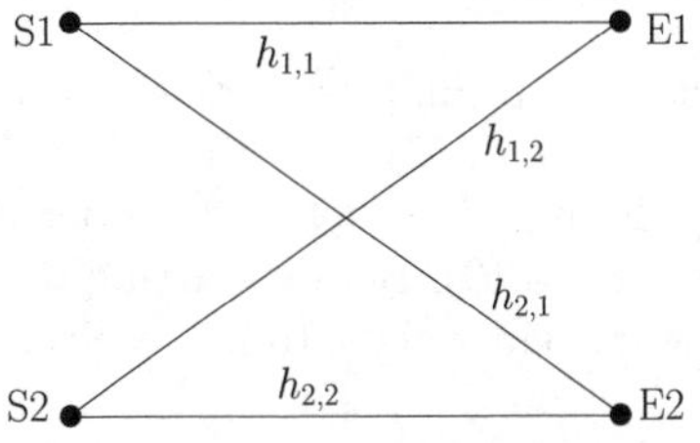

Fig. 6.1. In the primal network, the received signal samples at E1 and E2 are $y_1 = h_{1,1}X_1 + h_{1,2}X_2$ and $y_2 = h_{2,2}X_2 + h_{2,1}X_1$, respectively, where X_1, X_2 are zero-mean independent information-bearing symbols with $E[|X_1|^2] = p_1, E[|X_2|^2] = p_2$. In the adjoint network, E1 and E2 transmits $X_1/h_{1,1}$ and $X_2/h_{2,2}$, respectively, so that the received signal samples are $\tilde{y}_1 = X_1 + h_{2,1}/h_{2,2}X_2$ and $\tilde{y}_2 = X_2 + h_{1,2}/h_{1,1}X_1$.

tios in the primal network at E1 and E2 are $\text{SIR}_1(\mathbf{p}) = p_1/(v_{1,2}p_2)$ with $v_{1,2} = |h_{1,2}|^2/|h_{1,1}|^2$ and $\text{SIR}_2(\mathbf{p}) = p_2/(v_{2,1}p_1)$ with $v_{2,1} = |h_{2,1}|^2/|h_{2,2}|^2$, respectively. So, the channel state matrix is

$$\mathbf{V} = \begin{pmatrix} 0 & \frac{|h_{1,2}|^2}{|h_{1,1}|^2} \\ \frac{|h_{2,1}|^2}{|h_{2,2}|^2} & 0 \end{pmatrix}.$$

In the adjoint network, we have $\text{SIR}_1(\mathbf{p}) = p_1/(v_{2,1}p_2)$ (at node S1) and $\text{SIR}_2(\mathbf{p}) = p_2/(v_{1,2}p_1)$. The channel state matrix for the adjoint network is therefore given by

$$\begin{pmatrix} 0 & \frac{|h_{2,1}|^2}{|h_{2,2}|^2} \\ \frac{|h_{1,2}|^2}{|h_{1,1}|^2} & 0 \end{pmatrix} = \mathbf{V}^T .$$

This procedure straightforwardly extends to networks with an arbitrary number of links, provided that the network and signal model introduced in Chapt. 4 holds.

Finally, we point out that the channel inversion in the adjoint network may cause some problems. Indeed, if $|h_{k,k}|$ is small, then the transmit power on link k in the adjoint network can be unacceptably high, thereby violating some power constraints. A simple but effective solution to this problem is to define a certain threshold and prevent those links from transmission for which the path attenuation falls below this threshold. In addition, each logical transmitter in the adjoint network, say the transmitter on link k, can scale its training sequence by $\alpha/h_{k,k}$ for some $0 < \alpha \leq 1$ common to all links. The effect of such a scaling can be easily corrected at the receiver, provided that all transmitters use the same scaling factor. Obviously, a good choice of α depends on the realization of the wireless channel, and therefore such an approach requires some coordination between nodes.

6.5.3 Distributed Handshake Protocol

The basic idea is to use the primal network and the adjoint network alternately to obtain an estimate of $\nabla_k F$ at the kth transmitter in the primal network. To illustrate the principle, let us consider the nth iteration of the power control algorithm in (6.19). Before starting the iteration process, each transmitter reports the current weight to its receiver.

Primal Network: Assume that the "local part" $\eta_k(\mathbf{p}(n))$ of the gradient vector has already been estimated using the procedure described in Sect. 6.5.1. Let $\hat{\eta}_k(\mathbf{p}(n))$ be the estimate such that

$$\hat{\eta}_k(\mathbf{p}(n)) \approx \eta_k(\mathbf{p}(n)) = g_k(\mathbf{p}(n)) + \mathrm{SIR}_k(\mathbf{p}(n))g_k(\mathbf{p}(n)) .$$

By the procedure of Sect. 6.5.1, both $\mathrm{SIR}_k(\mathbf{p}(n))$ and $I_k(\mathbf{p}(n))$ are known to the kth receiver. Based on this information as well as on the knowledge of w_k, the kth logical receiver computes $g_k(\mathbf{p}(n)) < 0$ given by (6.21).

Adjoint Network: All logical transmitters concurrently send sequences of zero-mean independent symbols X_k (not necessarily known to the receivers) with $E[|X_k|^2] = |\mathrm{SIR}_k(\mathbf{p}(n))g_k(\mathbf{p}(n))|$ for each $k \in \mathsf{K}$. Note that this involves the channel inversion as specified in Theorem 6.16 so that the actual transmit powers are higher than $|\mathrm{SIR}_k(\mathbf{p}(n))g_k(\mathbf{p}(n))|$. Each logical receiver, say receiver k, estimates the received power by averaging over all symbol intervals and multiplies the result by -1 (since $g_k(\mathbf{p}(n)) < 0$) to obtain

$$\hat{\theta}_k(\mathbf{p}(n)) \approx -\Big(\big|\mathrm{SIR}_k(\mathbf{p}(n))g_k(\mathbf{p}(n))\big| + \sum_{l\neq k} v_{l,k}\big|\mathrm{SIR}_k(\mathbf{p}(n))g_k(\mathbf{p}(n))\big|\Big).$$

If the Gaussian noise is not negligible (when compared with the multiple access interference), then the noise variance must be estimated (if not known) and subtracted from the estimated received power.

Now since the kth receiver in the adjoint network is the kth transmitter in the primal network, the latter one computes

$$\nabla_k \hat{F}(\mathbf{p}(n)) = \hat{\eta}_k(\mathbf{p}(n)) - \hat{\theta}_k(\mathbf{p}(n)), \quad 1 \le k \le K$$

which is "close" to $\nabla_k F(\mathbf{p}(n))$, provided that the estimates are accurate enough.

The following list summarizes the whole procedure for the nth iteration. In the following description, "transmitter" and "receiver" refer to logical transmitters and logical receivers in the primal network. We assume that the function ψ is known at all nodes and that the weight w_k is known at both sides of link $k \in \mathsf{K}$.

---

1. Concurrent transmission of training sequences at powers $(p_1(n), \ldots, p_K(n))$.
2. Receiver side estimation of the signal-to-interference ratios and interferences. Based on these estimations, each receiver calculates $g_k(\mathbf{p}(n)), k \in \mathsf{K}$.
3. All receivers feed the signal-to-interference ratios back to the corresponding transmitters using a per-link control channel. Transmitter-side computation of $g_k(\mathbf{p}(n))$, and then $\eta_k(\mathbf{p}(n))$ for each $k \in \mathsf{K}$.
4. Concurrent transmission of sequences of zero-mean independent symbols X_k with
$$E[|X_k|^2] = |\mathrm{SIR}_k(\mathbf{p}(n)) \cdot g_k(\mathbf{p}(n))|, \ k \in \mathsf{K}$$
over the adjoint network. Note that the transmission over the adjoint network involves channel inversion (Theorem 6.16).
5. Transmitter side estimation of the received power and subtraction of noise variances from the estimates to obtain $\theta_k(\mathbf{p}(n))$. Since $\eta_k(\mathbf{p}(n))$ and $\theta_k(\mathbf{p}(n))$ are known at transmitter k, the transmitter computes
$$\begin{aligned}\nabla_k \hat{F}(\mathbf{p}(n)) &= \eta_k(\mathbf{p}(n)) - \theta_k(\mathbf{p}(n)) \\ &= g_k(\mathbf{p}(n)) - (\mathbf{V}^T\mathbf{\Gamma}(\mathbf{p}(n))\mathbf{g}(\mathbf{p}(n)))_k\end{aligned}$$
where we assumed that all the variables have been estimated perfectly.
6. Update of transmit powers according to (6.19) with $\mathbf{s}(n) = \log \mathbf{p}(n)$
7. $n \to n+1$.

6.5.4 Noisy Measurements

In the previous section, we assumed that all unknown variables such as the received powers or the signal-to-interference ratios can be estimated with accuracy allowing for the treatment of the algorithm within the framework of *deterministic* optimization theory. However, due to estimation errors and other distorting factors such as quantization noise, this assumption is not adequate for many real world wireless networks. Even if all the estimators are consistent or strongly consistent,[4] larger estimation inaccuracies in steps 2 and 5 of the above scheme may appear simply by virtue of strongly limited estimation time. Also the neglect of the background noise in the computation steps or an erroneous assumption about the noise variance in the adjoint network may result in *biased* estimates. Indeed, since information conveyed over the adjoint network is contained in the average received power, the gradient estimates will converge (say in probability if the estimator is consistent) to the true value plus the noise variance when the background noise is neglected and independent of the estimator and the received signal. Clearly, in such a case, the algorithm will not converge to the optimum even under the assumption of perfect estimation.

In the case of such uncertainties, the proposed algorithm has to be analyzed in a more general context of stochastic approximation theory. The topic is too broad to be considered here in more detail. Moreover, it requires mathematical tools and concepts that are different from those used so far in this book. For these reasons, we only mention some basic ideas from this theory. Reference [55] provides a preliminary analysis of the above scheme in this context. A comprehensive reference for stochastic approximation algorithms is [92]. The reader is also referred to [93, 94, 95, 83].

For simplicity, the uncertainty of the estimation of SIRs in step 2) in the above procedure is neglected. So, we focus on the problem of estimating the gradient components in step 5) which are assumed to be random variables of the form

$$\Delta_k(n) = \nabla_k F_e(\mathbf{s}(n)) + M_k(n),\ k \in \mathsf{K},\ n \in \mathbb{N}_0 .$$

In general, the estimation noise processes $\{M_k(n)\}, k \in \mathsf{K}$, are dependent on the estimator type, the estimation time, and the receiver noise process. In the literature, the following assumptions are often made to simplify the analysis.

- A1 The receiver noise processes are martingale-differences uncorrelated with transmit symbols [96, 92] and have variances $\sigma_k^2 < \infty$. In particular, this includes additive white Gaussian noise.
- A2 The estimation noise is zero-mean and *exogenous*, in that its future evolution, given the past history of the iterates and the receiver noise, depends only on the noise.

[4] An estimator is said to be consistent if the estimate converges in probability to the quantity being estimated as the estimation time grows. It is said to be strongly consistent if the estimate converges to the true value almost surely [91]

Note that while A1 is not restrictive, A2 is not necessarily fulfilled by the scheme. For instance, if an erroneous assumption about the receiver noise variance is made, the estimates of $\nabla F_e(\mathbf{s}(n))$ may be biased. Also, A2 can be violated as the evolution of the estimation noise may depend on the iterate. This is simply because the transmit power in the adjoint network (and thus also the estimation accuracy in case of a limited estimation time) depends on $\mathbf{s}(n)$. This dependency can be reduced by extending the estimation time of each $\Delta_k(n), k \in \mathsf{K}, n \in \mathbb{N}_0$.

In what follows, we assume that the two assumptions are satisfied. So, for each $k \in \mathsf{K}$, we can write

$$M_k(n+1) = \Delta_k(n+1) - E\big[\Delta_k(n+1)|\mathbf{s}(0), \Delta_k(m), m \leq n\big]$$

where we have $E[\Delta_k(n+1)|\mathbf{s}(0), \Delta_k(m), m \leq n] = \nabla_k F_e(\mathbf{s}(n))$ and $\Delta_k(n) = \sum_{m=0}^{n-1} M_k(m)$. In words, the estimation noise process $\{M_k(n)\}, k \in \mathsf{K}$, is a martingale-difference independent of transmit symbols and with finite variance. Moreover, $\{\Delta_k(n)\}, k \in \mathsf{K}$, is a martingale sequence. Recall that $X(n) = Y(n+1) - Y(n)$ is said to be a martingale difference if the process $\{Y(n)\}$ (a sequence of random variables) is martingale, that is to say, if $E[Y(n+1)|Y(i), i \leq n] = Y(n)$ with probability one for all n [96, 92]. Thus, the expectation of $X(n)$ conditioned on the past is zero. Moreover, since the variance is finite, the martingale differences are uncorrelated in that for $m \neq n$, we have $E[Y(n+1) - Y(n)][Y(m+1) - Y(m)] = 0$.

It can be inferred from the landmark paper [93] that taking many observations of the noise corrupted gradient samples in each iteration step and then averaging them to obtain a good estimate of the gradient vector is in general inefficient. Instead, they proposed considering a diminishing step size $\delta(n)$. So, in case of noisy measurements, the power control algorithm (6.19) usually takes the form

$$\mathbf{s}(n+1) = \Pi_{\mathsf{S}}\big[\mathbf{s}(n) - \delta(n)\boldsymbol{\Delta}(n)\big], \quad \mathbf{s}(0) \in \mathsf{S} \tag{6.33}$$

where $\boldsymbol{\Delta}(n) = (\Delta_1(n), \ldots, \Delta_K(n))$ is the vector of noisy measurements of the gradient in the nth iteration and $\{\delta(n)\}$ is a non-increasing sequence of positive real numbers satisfying $\sum_{n=0}^{\infty} \delta(n) = \infty, \lim_{n\to\infty} \delta(n) = 0$ and $\epsilon \leq \delta(n) \leq M - \epsilon$ for all $n \in \mathbb{N}_0$ and some $\epsilon > 0$, where M denotes a Lipschitz constant (see Section 6.4.1). The choice of the step size sequence $\{\delta(n)\}$ is central to the effectiveness of the algorithm (6.33). A typical choice of the step size satisfies $\sum_{n=1}^{\infty} \delta^2(n) < \infty$. However, this condition can often be weakened considerably [92, 94]. A standard example of a step size that satisfy the above conditions is $\delta(n) = C/(n+1)$ for some sufficiently small $C > 0$.

Under the assumptions A1-2, some standard results can be exploited to deduce convergence of the power control algorithm (6.33) to a stationary point of the problem. Usually, two types of convergence are of interest: convergence in distribution (or weak convergence) or almost sure convergence

[92, 96]. Convergence in the mean square sense was considered in [83]. For instance, with A1-2, it follows from [92] that the algorithm converges weakly to a stationary point if $\delta(n) = cn^{-\mu}$ for some sufficiently small $c > 0$ and $\mu \in (0, 1]$. Almost sure convergence of the algorithm follows from [94, 92] when the noise process is subject to some additional constraints. In [92], the reader can find further useful results for correlated and/or non-exogenous (state dependent) noise processes. Finally, in [95], methods for averaging the iterates (in parallel to the stochastic recursion (6.33)) are presented to improve the convergence rate of the algorithm. In [55], it is shown that the averaging scheme of [95] significantly decreases the variance of the convergence curve of the algorithm (6.33) for reasonable SIR values.

Part III

Appendices

A

Some Concepts and Results from Matrix Analysis

The appendix provides some (very) basic concepts and results from linear algebra that are vital to understanding the theory presented in this book. This is also a good opportunity to introduce the notation used throughout the book. Proofs are provided only for the most important results such as the Perron–Frobenius theorem. For other proofs and a detailed treatment of this material, the reader is referred to any linear algebra book and [5, 3, 7, 2, 12, 4].

A.1 Vectors and Vector Norms

Vectors and matrices can be defined over an arbitrary field $\mathbb{K}$. It could be $\mathbb{R}$, the field of real numbers, or $\mathbb{C}$, the field of complex numbers. These are the most common choices. Unless something otherwise stated, all matrices in this section are over $\mathbb{K} = \mathbb{R}$. Elements of $\mathbb{K}$ are called scalars.

The set of all n-tuples over $\mathbb{R}$ with two algebraic operations called vector addition and scalar multiplication form an n-dimensional vector space denoted by $\mathbb{R}^n$. Elements of $\mathbb{R}^n$ are referred to as vectors and are written as $\mathbf{u} = (u_1, \ldots, u_n)$, which, in connection with vectors or matrices, should be regarded as a column vector. In the book, we use the following notation: $\mathbf{0} = (0, \ldots, 0)$ is the zero vector, $\mathbf{1} = (1, \ldots, 1)$ is the vector of ones, and $\mathbf{e}_i = (0, \ldots, 0, 1, 0, \ldots, 0)$ is a unit vector with 1 in the ith position and zeros elsewhere.

Throughout the book, we use partial ordering on $\mathbb{R}^n$ defined as follows: For any $\mathbf{u}, \mathbf{v} \in \mathbb{R}^n$, there holds

$$\begin{aligned} \mathbf{u} \leq \mathbf{v} &\Leftrightarrow \forall_{1 \leq i \leq n}\, u_i \leq v_i \quad \mathbf{u} < \mathbf{v} \Leftrightarrow \forall_{1 \leq i \leq n}\, u_i < v_i \\ \mathbf{u} = \mathbf{v} &\Leftrightarrow \forall_{1 \leq i \leq n}\, u_i = v_i\,. \end{aligned}$$

When $\forall_{1 \leq i \leq n}\, u_i \geq c$ for some constant c, we write $\mathbf{u} \geq c$ with equality if and only if $u_i = c$ for each $1 \leq i \leq n$. So we have $\mathbf{v} \geq \mathbf{u}$ if and only if $\mathbf{s} = \mathbf{v} - \mathbf{u} \geq 0$ and $\mathbf{v} > \mathbf{u}$ if and only if $\mathbf{s} = \mathbf{v} - \mathbf{u} > 0$ where $\mathbf{s} \in \mathbb{R}^n$. Moreover, if $\mathbf{u} \leq \mathbf{v}$ and $\mathbf{v} \leq \mathbf{s}$ for any $\mathbf{u}, \mathbf{v}, \mathbf{s} \in \mathbb{R}^n$, then $\mathbf{u} \leq \mathbf{s}$.

We say that $\mathbb{R}^n$ is a normed vector space if there is a map $\|\cdot\| : \mathbb{R}^n \to \mathbb{R}$ (called a norm on $\mathbb{R}^n$) satisfying the following properties:

$$\begin{array}{ll} \forall_{\mathbf{u}\in\mathbb{R}^n} \|\mathbf{u}\| \geq 0 & \forall_{\lambda\in\mathbb{R},\mathbf{u}\in\mathbb{R}^n} \|\lambda\mathbf{u}\| = |\lambda|\cdot\|\mathbf{u}\| \\ \|\mathbf{u}\| = 0 \Leftrightarrow \mathbf{u} = 0 & \forall_{\mathbf{u},\mathbf{v}\in\mathbb{R}^n} \|\mathbf{u}+\mathbf{v}\| \leq \|\mathbf{u}\| + \|\mathbf{v}\|\,. \end{array} \tag{A.1}$$

All the norms used in this book are l^p-norms and the maximum norm: For any $p \geq 1$, the l^p-norm and the maximum norm of $\mathbf{u} \in \mathbb{R}^n$ are defined to be

$$\|\mathbf{u}\|_p = \left(\sum_{i=1}^n |u_i|^p\right)^{\frac{1}{p}} \quad \text{and} \quad \|\mathbf{u}\|_\infty = \max(|u_1|, \ldots, |u_n|)$$

respectively. A vector space equipped with a l^p-norm is called the l^p space. Minkowski's inequality establishes that the l^p spaces are normed vector spaces.

Theorem A.1 (Minkowski's Inequality). *For $1 \leq p \leq \infty$,*

$$\|\mathbf{u}+\mathbf{v}\|_p \leq \|\mathbf{u}\|_p + \|\mathbf{v}\|_p\,. \tag{A.2}$$

The l^p spaces are also metric spaces with the distance $d(\mathbf{u},\mathbf{v})$ from $\mathbf{u} \in \mathbb{R}^n$ to $\mathbf{v} \in \mathbb{R}^n$ given by $d(\mathbf{u},\mathbf{v}) = \|\mathbf{u}-\mathbf{v}\|$ for some l^p-norm on $\mathbb{R}^n$. The vector space $\mathbb{R}^n$ becomes an inner product space if it is equipped with the inner product defined as

$$\langle\mathbf{u},\mathbf{v}\rangle = \sum_{i=1}^n u_i\, v_i, \quad \mathbf{u},\mathbf{v} \in \mathbb{R}^n\,.$$

Hölder's inequality provides a relationship between the inner product of two vectors and their norms. The following theorem presents Hölder's inequality for nonnegative vectors. Hölder's inequalities for general complex vectors (along with the equality conditions) can be found in [97, pp. 50-54].

Theorem A.2 (Hölder's Inequality). *Let $\mathbf{u} \in \mathbb{R}^n_+$ and $\mathbf{v} \in \mathbb{R}^n_+$ be arbitrary. Then, there holds*

$$\langle\mathbf{u},\mathbf{v}\rangle \leq \|\mathbf{u}\|_p\|\mathbf{v}\|_q \tag{A.3}$$

where $p, q > 1$ are chosen so that $1/p + 1/q = 1$. Equality holds if and only if $v_k = c\,u_k^{p-1}, 1 \leq k \leq K$, for some constant $c > 0$.

When $p = 2$, $\mathbb{R}^n$ is called Euclidean space with the Euclidean distance $d(\mathbf{u},\mathbf{v}) = \|\mathbf{u}-\mathbf{v}\|_2$ and the Euclidean norm $\|\mathbf{u}\|_2 = \sqrt{\langle\mathbf{u},\mathbf{u}\rangle}$. The Euclidean space $\mathbb{R}^n$ is a Hilbert space since it is complete with respect to its norm induced by the inner product. When $p = q = 2$, the general form of the inequality (A.3) is better known as the Cauchy–Schwarz inequality:

$$|\langle\mathbf{u},\mathbf{v}\rangle| \leq \|\mathbf{u}\|_2\|\mathbf{v}\|_2 \tag{A.4}$$

for all $\mathbf{u},\mathbf{v} \in \mathbb{R}^n$, with equality if and only if $\mathbf{u} = c\mathbf{v}$ for some constant $c \neq 0$.

A.2 Matrices and Matrix Norms

If $n, m \geq 1$, a matrix $\mathbf{X}$ of size $n \times m$ with entries in $\mathbb{K}$ is an array with n rows and m columns:

$$\mathbf{X} = \begin{pmatrix} x_{1,1} & \cdots & x_{1,m} \\ \vdots & \ddots & \vdots \\ x_{n,1} & \cdots & x_{n,m} \end{pmatrix}.$$

We write $\mathbf{X} = (x_{i,j})_{1 \leq i \leq n, 1 \leq j \leq m}$ or simply $\mathbf{X} = (x_{i,j})$ if the matrix size is clear. The entries of $\mathbf{X}$ are also denoted by $(\mathbf{X})_{i,j}$. If $m = n$, the matrix $\mathbf{X}$ is said to be square. A $n \times n$ diagonal matrix $\mathbf{X}$ denoted by $\mathbf{X} = \mathrm{diag}(\mathbf{x}) = \mathrm{diag}(x_1, \ldots, x_n)$ is a matrix with diagonal entries $x_1, \ldots, x_n$ and all off-diagonal entries being equal to zero. In particular, $\mathbf{I} = \mathrm{diag}(\mathbf{1}) = \mathrm{diag}(1, \ldots, 1)$ is referred as the identity matrix or, simply, the identity.

The set of all $n \times m$ matrices over $\mathbb{R}$ form a vector space denoted by $\mathbb{R}^{n \times m}$. As in the case of vectors, we define a partial ordering on $\mathbb{R}^{n \times m}$ as follows: For any $\mathbf{A} \in \mathbb{R}^{n \times m}$ and $\mathbf{B} \in \mathbb{R}^{n \times m}$,

$$\begin{aligned} \mathbf{A} \leq \mathbf{B} &\Leftrightarrow \forall_{i,j}\, a_{i,j} \leq b_{i,j} \quad \mathbf{A} < \mathbf{B} \Leftrightarrow \forall_{i,j}\, a_{i,j} < b_{i,j} \\ \mathbf{A} = \mathbf{B} &\Leftrightarrow \forall_{i,j}\, a_{i,j} = b_{i,j}\,. \end{aligned}$$

Again, if $\forall_{i,j} a_{i,j} \geq c$ for some constant c, we write $\mathbf{A} \geq c$. For any two matrices $\mathbf{A}, \mathbf{B} \in \mathbb{R}^{n \times m}$, their Hadamard product $\mathbf{A} \circ \mathbf{B}$ is the entry-wise product of $\mathbf{A}$ and $\mathbf{B}$. When considered in connection with matrices, vectors are to be regarded as column vectors.

Given a matrix $\mathbf{X} \in \mathbb{R}^{n \times m}$, a matrix norm of $\mathbf{X}$ is denoted by $\|\mathbf{X}\|$. General matrix norms satisfy the conditions in (A.1), with the vector $\mathbf{u}$ replaced by some matrix. Additionally, if $\mathbf{AB}$ exists, we have

$$\|\mathbf{AB}\| \leq \|\mathbf{A}\| \|\mathbf{B}\|\,.$$

The simplest notion of a matrix norm of $\mathbf{X} \in \mathbb{R}^{n \times m}$ is the Frobenius norm given by

$$\|\mathbf{X}\|_F^2 = \sum_{i,j} |x_{i,j}|^2 = \mathrm{trace}(\mathbf{X}^T \mathbf{X}) \tag{A.5}$$

where $\mathrm{trace}(\mathbf{X}) = \sum_i x_{i,i}$ is the trace of a matrix $\mathbf{X}$. Other widely used matrix norms are induced matrix norms: For any $\mathbf{X} \in \mathbb{R}^{n \times m}$, define

$$\|\mathbf{X}\| = \max_{\mathbf{u} \in \mathbb{R}^m, \|\mathbf{u}\| = 1} \|\mathbf{X}\mathbf{u}\| \tag{A.6}$$

where $\|\cdot\|$ denotes any vector norm satisfying (A.1). If the l^2-norm is used, then $\|\mathbf{X}\|_2$ is the matrix 2-norm equal to $\sqrt{\lambda_{\max}}$ where $\lambda_{\max}$ is the largest eigenvalue of $\mathbf{X}^T\mathbf{X}$ (see below) and $\mathbf{X}^T$ is the transpose of $\mathbf{X}$ defined as follows.

Definition A.3 (Matrix Transpose). *The transpose matrix of* $\mathbf{X} = (x_{i,j}) \in \mathbb{R}^{n\times m}$ *is defined as a matrix* $\mathbf{X}^T \in \mathbb{R}^{m\times n}$ *such that* $(\mathbf{X}^T)_{i,j} = x_{j,i}$.

Furthermore, we have

$$\begin{aligned} \|\mathbf{X}\|_1 &= \max_{\|\mathbf{u}\|_1=1} \|\mathbf{X}\mathbf{u}\|_1 = \max_j \sum_{i=1}^{n} |x_{i,j}| \\ \|\mathbf{X}\|_\infty &= \max_{\|\mathbf{u}\|_\infty=1} \|\mathbf{X}\mathbf{u}\|_\infty = \max_i \sum_{j=1}^{m} |x_{i,j}| \,. \end{aligned} \tag{A.7}$$

Finally, we point out that every matrix $\mathbf{X} \in \mathbb{R}^{n\times m}$ can be considered as a linear map from $\mathbb{R}^m$ into $\mathbb{R}^n$. Recall that a map $f : \mathbb{R}^m \to \mathbb{R}^n$ is said to be linear if $f(\mathbf{u}+\mathbf{v}) = f(\mathbf{u}) + f(\mathbf{v})$ and $f(a\mathbf{u}) = a\, f(\mathbf{u})$ for every $\mathbf{u}, \mathbf{v} \in \mathbb{R}^m$ and $a \in \mathbb{R}$. We write $\mathbb{R}^m \to \mathbb{R}^n : \mathbf{u} \to \mathbf{X}\mathbf{u} = \mathbf{v}$. The image of $\mathbf{u}$ under $\mathbf{X}$ is the vector $\mathbf{v}$. The image of $\mathbb{R}^m$ under $\mathbf{X}$ is called the range of $\mathbf{X}$. The kernel of $\mathbf{X}$ is the set $\ker(\mathbf{X})$ of those $\mathbf{u} \in \mathbb{R}^m$ for which $\mathbf{X}\mathbf{u} = 0$.

A.3 Square Matrices and Eigenvalues

From now on we focus on square matrices of size $n \times n$ over $\mathbb{R}$.

Definition A.4 (Eigenvalues and Eigenvectors). *For an arbitrary* $n\times n$ *matrix* $\mathbf{X}$, *scalars* $\lambda \in \mathbb{C}$ *and* n*-vectors* $\mathbf{p} \neq 0$ *(with* $\mathbf{p} \in \mathbb{C}^n$*) satisfying* $\lambda\mathbf{p} = \mathbf{X}\mathbf{p}$ *(over* $\mathbb{C}$*) are called eigenvalues and right eigenvectors of* $\mathbf{X}$, *respectively. The pair* $(\lambda, \mathbf{p})$ *is an eigenpair of* $\mathbf{X}$ *or, equivalently,* $\mathbf{p}$ *is an eigenvector of* $\mathbf{X}$ *associated with an eigenvalue* λ. *A left eigenvector* $\mathbf{q} \neq 0$ *of* $\mathbf{X}$ *associated with* λ *is a* n*-vector satisfying* $\lambda\mathbf{q} = \mathbf{X}^T\mathbf{q}$.

It is important to emphasize that even if $\mathbf{X}$ is a real-valued matrix, its eigenvalues are in general complex numbers. It is easy to see that eigenvalues of $\mathbf{X}$ are exactly those scalars $\lambda \in \mathbb{C}$ for which both of the following hold:

(i) the matrix $\mathbf{A} = \lambda\mathbf{I} - \mathbf{X}$ is singular, that is, $\mathbf{A}$ is not invertible. This means that there is no matrix $\mathbf{A}^{-1} \in \mathbb{R}^{n\times n}$ such that $\mathbf{A}\mathbf{A}^{-1} = \mathbf{A}^{-1}\mathbf{A} = \mathbf{I}$.

(ii) $p(\lambda) = \det(\lambda\mathbf{I} - \mathbf{X}) = 0$ in $\mathbb{C}$ where $\det(\mathbf{A})$ denotes determinant of $\mathbf{A}$.

Recall that

$$\det(\mathbf{A}) = \sum_{\sigma} \epsilon(\sigma) \prod_{j=1}^{n} a_{1,\sigma_j}$$

where the sum is taken over all permutations $\sigma = (\sigma_1, \ldots, \sigma_n)$ of $(1, 2, \ldots, n)$ and $\epsilon(\sigma) = \pm 1$ is equal to 1 if σ is the product of an even number of transpositions, and -1 otherwise.

The polynomial $p(\lambda)$ is called the characteristic polynomial of $\mathbf{X}$. Hence the second statement says that the eigenvalues of $\mathbf{X}$ are the roots of its characteristic polynomial in the field of complex numbers. As the order of the polynomial is n, we see that $\mathbf{X} \in \mathbb{R}^{n \times n}$ has altogether n eigenvalues. Some of these eigenvalues, however, can be repeated.

Definition A.5 (Algebraic and Geometric Multiplicity). *The multiplicity of λ as a root of the characteristic polynomial is called algebraic multiplicity of the eigenvalue λ. The geometric multiplicity of λ is the dimension of* $\ker(\lambda\mathbf{I} - \mathbf{X})$ *in* $\mathbb{R}^n$.

The geometric multiplicity is smaller than or equal to the algebraic multiplicity. Eigenvalues with algebraic multiplicity 1 are called simple.

The following fundamental result is invoked throughout the book whenever we say that the eigenvalues of a matrix are continuous functions of its entries [5].

Theorem A.6. *Let $\mathbf{X} \in \mathbb{R}^{n \times n}$ be arbitrary, and let us fix some norm on $\mathbb{R}^n$. Suppose that $\lambda \in \mathbb{C}$ is any eigenvalue of $\mathbf{X}$, with multiplicity μ, and d is the distance from λ to the other eigenvalues of $\mathbf{X}$. Let $B_\rho(\lambda)$ be an open disk of radius ρ with $0 < \rho < d$ centered at λ. Then, there exists $\epsilon > 0$ such that if $\mathbf{A} \in \mathbb{R}^n$ and $\|\mathbf{A}\| < \epsilon$, the sum of the algebraic multiplicities of the eigenvalues of $\mathbf{X} + \mathbf{A}$ in $B_\rho(\lambda)$ is equal to μ.*

For a definition of an open disk, the reader is referred to Sect. B.1.

A.3.1 Spectral Radius and Neumann Series

First let us define the matrix spectrum.

Definition A.7 (Matrix Spectrum). *The set of* distinct *eigenvalues of $\mathbf{X}$ is referred to as the spectrum of $\mathbf{X}$ and is denoted by $\sigma(\mathbf{X})$.*

Since the roots of a polynomial with real coefficients occur in conjugate pairs, $\lambda \in \sigma(\mathbf{X})$ implies that $\overline{\lambda} \in \sigma(\mathbf{X})$ where $\overline{x}$ denotes the conjugate complex. Furthermore, we have $\sigma(\mathbf{X}) = \sigma(\mathbf{X}^T)$.

Definition A.8 (Spectral Radius). *For any square matrix $\mathbf{X} \in \mathbb{R}^{n \times n}$, we define $\rho : \mathbb{R}^{n \times n} \to \mathbb{R}$ as*

$$\rho(\mathbf{X}) = \max\{|\lambda| : \lambda \in \sigma(\mathbf{X})\}. \tag{A.8}$$

The real number $\rho(\mathbf{X})$ is called the spectral radius of $\mathbf{X}$.

Thus, in order to obtain an upper bound on the magnitudes of all eigenvalues of $\mathbf{X}$, it is sufficient to bound above the spectral radius. A crude upper bound is given by the following observation

Observation A.9. *For any matrix norm and $\mathbf{X} \in \mathbb{R}^{n \times n}$, there holds*

$$\rho(\mathbf{X}) \le \|\mathbf{X}\| \tag{A.9}$$

Proof. Suppose that $(\lambda, \mathbf{p})$ with $\mathbf{p} \neq 0$ is any eigenpair of $\mathbf{X}$. Let $\|\cdot\|$ be any matrix norm, and let $\mathbf{P} \in \mathbb{R}^{n\times n}$ be a matrix all of whose columns are equal to $\mathbf{p}$. Then, $|\lambda|\|\mathbf{P}\| = \|\lambda\mathbf{P}\| = \|\mathbf{X}\mathbf{P}\| \leq \|\mathbf{X}\|\|\mathbf{P}\|$. Hence, $|\lambda| \leq \rho(\mathbf{X}) \leq \|\mathbf{X}\|$ for any matrix norm.

In particular, the observation implies that $\rho(\mathbf{X}) = \rho(\mathbf{X}^k)^{1/k} \leq \|\mathbf{X}^k\|^{1/k}$ for any $k \geq 1$. This and the equivalence of the norms in finite dimensional spaces are key ingredients in proving the following result.

Theorem A.10. *Let $\mathbf{X} \in \mathbb{R}^{n\times n}$ be arbitrary. Then,*

$$\rho(\mathbf{X}) = \lim_{k\to\infty} \|\mathbf{X}^k\|^{1/k} \tag{A.10}$$

for every matrix norm $\|\cdot\|$.

Given any $\mathbf{X} \in \mathbb{R}^{n\times n}$, the convergence of the Neumann series $\sum_{k=0}^{\infty} \mathbf{X}^k$ is fundamental for some of the concepts introduced in this book. The following result provides a necessary and sufficient condition for the Neumann series to converge

Theorem A.11. *Let $\mathbf{X} \in \mathbb{R}^{n\times n}$ be arbitrary. Then, the following statements are equivalent.*

(i) $\sum_{k=0}^{\infty} \mathbf{X}^k$ *converges.*
(ii) $\rho(\mathbf{X}) < 1$.
(iii) $\lim_{k\to\infty} \mathbf{X}^k = 0$.

In these cases, $(\mathbf{I} - \mathbf{X})^{-1}$ exists, and $(\mathbf{I} - \mathbf{X})^{-1} = \sum_{k=0}^{\infty} \mathbf{X}^k$.

A.3.2 Orthogonal, Symmetric and Positive Semidefinite Matrices

A matrix $\mathbf{X} \in \mathbb{R}^{n\times n}$ is said to be normal if $\mathbf{X}$ commutes with $\mathbf{X}^T$. In other words, if $\mathbf{X}$ is normal, then $\mathbf{X}\mathbf{X}^T = \mathbf{X}^T\mathbf{X}$. Important subsets of normal matrices are the sets of orthogonal and symmetric matrices.

Definition A.12 (Orthogonal and Symmetric Matrices). *We say that $\mathbf{X} \in \mathbb{R}^{n\times n}$ is an orthogonal matrix if $\mathbf{X}^T\mathbf{X} = \mathbf{X}\mathbf{X}^T = \mathbf{I}$. It is said to be symmetric if $\mathbf{X}^T = \mathbf{X}$.*

If $\mathbf{X}$ is orthogonal and $\mathbf{p} \neq 0$ is an eigenvector of $\mathbf{X}$ associated with $\lambda \in \sigma(\mathbf{X})$, then $|\lambda|^2\|\mathbf{p}\|_2^2 = (\lambda\mathbf{p})^T(\lambda\mathbf{p}) = \mathbf{p}^T\mathbf{X}^T\mathbf{X}\mathbf{p} = \|\mathbf{p}\|_2^2$. Hence, the eigenvalues of any orthogonal matrix are (in general) complex numbers of modulus one. For symmetric matrices, we have the following standard results.

Theorem A.13. *The eigenvalues of symmetric matrices are real.*

Theorem A.14. *Symmetric matrices are orthogonally similar to a diagonal matrix. In other words, given a symmetric matrix $\mathbf{X}$, there exists an orthogonal matrix $\mathbf{U}$ such that $\mathbf{U}\mathbf{X}\mathbf{U}^T$ is diagonal.*

Theorem A.15. *Let* $\hat{\mathbf{X}}, \check{\mathbf{X}}$ *be arbitrary symmetric matrices, and let* $\mathbf{X}(\mu) = (1-\mu)\hat{\mathbf{X}} + \mu\check{\mathbf{X}}$ *for* $\mu \in [0,1]$. *Then,*

$$\rho(\mathbf{X}(\mu)) \le (1-\mu)\rho(\hat{\mathbf{X}}) + \mu\rho(\check{\mathbf{X}}) \tag{A.11}$$

for all $\mu \in (0,1)$.

In words, the theorem says that the spectral radius is convex on the set of symmetric matrices (for the definition of convexity, the reader is referred to Appendix B). This immediately becomes obvious when one considers that

$$\rho(\mathbf{X}) = \sup\{\mathbf{u}^T\mathbf{X}\mathbf{u} : \mathbf{u}^T\mathbf{u} = 1, \mathbf{X} = \mathbf{X}^T\}.$$

Another fundamental notion in matrix analysis is that of positive semidefiniteness.

Definition A.16 (Positive Semidefinite Matrix). *We say that a symmetric matrix* $\mathbf{X} \in \mathbb{R}^{n\times n}$ *is positive semidefinite if* $\mathbf{u}^T\mathbf{X}\mathbf{u} \ge 0$ *for all* $\mathbf{u} \in \mathbb{R}^n$. *If there is strict inequality for all* $\mathbf{u} \in \mathbb{R}^n$, *then* $\mathbf{X}$ *is said to be positive definite.*

Clearly, every symmetric positive semidefinite matrix is orthogonally similar to a diagonal matrix and all its eigenvalues are real. In addition, however, all eigenvalues are nonnegative.

Theorem A.17. *A symmetric matrix* $\mathbf{X} \in \mathbb{R}^{n\times n}$ *is positive semidefinite if and only if* $\sigma(\mathbf{X})$ *is a subset of the nonnegative reals. It is positive definite if and only if* $\sigma(\mathbf{X})$ *is a subset of the positive reals.*

A.4 Perron–Frobenius Theory

In this section, we focus on vectors and $n \times n$ matrices defined over $\mathbb{R}_+ \subset \mathbb{R}$ and $\mathbb{R}_{++} \subset \mathbb{R}_+$, the sets of nonnegative and positive reals, respectively.

Definition A.18. *Any square matrix* $\mathbf{X} \in \mathbb{R}^{n\times n}$ *whose entries are nonnegative (positive) reals is called a nonnegative (positive) matrix. The set of all* $n \times n$ *nonnegative (positive) matrices is denoted by* N_n (P_n).

The following implications can be easily verified.

$$\forall_{k\ge1} \quad \mathbf{X} \ge 0 \Rightarrow \mathbf{X}^k \ge 0 \quad \text{and} \quad \mathbf{X} > 0 \Rightarrow \mathbf{X}^k > 0 \tag{A.12}$$

$$\mathbf{X} > 0, \mathbf{u} \ge 0, \mathbf{u} \ne \mathbf{0} \Rightarrow \mathbf{X}\mathbf{u} > 0 \tag{A.13}$$

$$\mathbf{X} \ge 0, \mathbf{u} > 0, \mathbf{X}\mathbf{u} = 0 \Rightarrow \mathbf{X} = 0. \tag{A.14}$$

If $|\mathbf{X}| = (|x_{i,j}|)$, then we further have $|\mathbf{X}^m| \le |\mathbf{X}|^m$ for all $m \in \mathbb{N}$ and if $0 \le \mathbf{A} \le \mathbf{B}$, then $0 \le \mathbf{A}^m \le \mathbf{B}^m$. Moreover, if $|\mathbf{A}| \le |\mathbf{B}|$, then $\|\mathbf{A}\|_F \le \|\mathbf{B}\|_F$

where $\|\cdot\|_F$ is the Frobenius norm defined by (A.5). Finally, it is clear that $\|\mathbf{X}\|_F = \||\mathbf{X}|\|_F$. Combining these observations yields

$$\|\mathbf{A}^m\|_F^{1/m} \leq \||\mathbf{A}|^m\|_F^{1/m} \leq \|\mathbf{B}^m\|_F^{1/m}$$

for all $m \in \mathbb{N}$ and $\mathbf{A}, \mathbf{B} \in \mathbb{R}^{n\times n}$ such that $|\mathbf{A}| \leq \mathbf{B}$. Now if we let $m \to \infty$ and apply Theorem A.10, we obtain the following important result.

Theorem A.19. *Let* $\mathbf{A}, \mathbf{B} \in \mathbb{R}^{n\times n}$. *If* $|\mathbf{A}| \leq \mathbf{B}$, *then* $\rho(\mathbf{A}) \leq \rho(|\mathbf{A}|) \leq \rho(\mathbf{B})$.

Therefore, if $\mathbf{X} \geq 0$, then the spectral radius $\rho(X)$ is monotonic with respect to the matrix entries.

An important set of nonnegative matrices is the set of stochastic matrices.

Definition A.20 (Stochastic Matrix). $\mathbf{X} \in \mathrm{N}_n$ *is said to be stochastic if*

$$\forall_{1\leq i\leq n} \sum_{j=1}^{n} x_{i,j} = 1\,.$$

If both $\mathbf{X}$ *and* $\mathbf{X}^T$ *are stochastic, then* $\mathbf{X}$ *is said to be doubly stochastic.*

It is easy to see that for any stochastic matrix $\mathbf{X}$, we have $\mathbf{X1} = \mathbf{1}$. Furthermore, since $\|\mathbf{Xu}\|_\infty \leq \|\mathbf{X}\|_\infty \|\mathbf{u}\|_\infty = \|\mathbf{u}\|_\infty$, there holds $\rho(\mathbf{X}) = 1$ in the case of any stochastic matrix $\mathbf{X}$.

The Perron–Frobenius theory addresses the problem to what extent the nonnegativity (positivity) property of a matrix is inherited by its eigenvalues and eigenvectors [2, 3, 4, 12]. Intuitively, it can be expected that the spectral properties (in terms of positivity) somehow depend on the number and position of positive entries of $\mathbf{X}$. For instance, consider $\mathbf{X} = \left(\begin{smallmatrix} 0 & 1 \\ 0 & 0 \end{smallmatrix}\right)$ in which case there is neither a positive eigenvalue ($\sigma(\mathbf{X}) = \{0\}$) nor a positive eigenvector. In contrast, $\mathbf{X} = \left(\begin{smallmatrix} 0 & 1 \\ 1 & 0 \end{smallmatrix}\right)$ has significantly stronger properties than the previous example, although the matrices differ from each other only in one position. Indeed, the spectrum is now $\sigma(\mathbf{X}) = \{+1, -1\}$ so that the matrix has a *simple positive* eigenvalue $\lambda_p = \rho(\mathbf{X}) = 1$. Moreover, $(1/\sqrt{2}, 1/\sqrt{2})$ is a positive right eigenvector associated with λ_p. It should be emphasized that these additional properties (a simple positive eigenvalue that equals the spectral radius and an associated positive eigenvector) are not exclusively due to the increased number of positive entries in $\mathbf{X}$. In fact, what really matters is a combination of the number of positive entries and their right positions. This becomes clear after considering the matrix $\mathbf{X} = \left(\begin{smallmatrix} 1 & 1 \\ 0 & 0 \end{smallmatrix}\right)$. Despite having the same number of positive entries as the previous example, it is not possible to associate a positive eigenvector with $\lambda_p = \rho(\mathbf{X}) = 1$. An important difference between these two matrices is captured by the notion of reducibility.

A.4.1 Perron–Frobenius Theorem for Irreducible Matrices

Definition A.21 (Reducible and irreducible matrices). *A nonnegative matrix* $\mathbf{X} \in \mathrm{N}_n$ *is said to be* reducible *if there exists a permutation matrix* $\mathbf{P}$ *such that*

$$\mathbf{P}^T\mathbf{X}\mathbf{P} = \begin{pmatrix} \mathbf{A} & \mathbf{0} \\ \mathbf{B} & \mathbf{C} \end{pmatrix}$$

where $\mathbf{A}$ *and* $\mathbf{C}$ *are both square matrices. Otherwise,* $\mathbf{X}$ *is said to be irreducible. The set of all* $n \times n$ *nonnegative irreducible matrices is denoted by* $\mathrm{X}_n \subset \mathrm{N}_n$.

The following result used later in the proof of Theorem A.36 is sometimes stated as the definition of irreducible matrices [2].

Lemma A.22. *We have* $\mathbf{X} \in \mathrm{X}_n$ *if and only if, for each* (i, j) *with* $1 \le i, j \le n$, *there exists* $k \ge 0$ *such that* $x_{i,j}^{(k)} := (\mathbf{X}^k)_{i,j} > 0$.

Irreducibility of $\mathbf{X}$ is equivalent to strong connectivity of the graph $\mathrm{G}(\mathbf{X})$ of $\mathbf{X} \in \mathrm{N}_n$ which is defined to be the directed graph of $\mathsf{N} = \{1, \ldots, n\}$ nodes in which there is a directed edge leading from node $j \in \mathsf{N}$ to $i \in \mathsf{N}$ if and only if $x_{i,j} > 0$. Now we say that $\mathrm{G}(\mathbf{X})$ is strongly connected if for each pair of nodes (i, j) there is a sequence of directed edges leading from i to j. The following connection between irreducibility and strong connectivity is well known [12].

Observation A.23. $\mathbf{X} \in \mathrm{X}_n$ *if and only if* $\mathrm{G}(\mathbf{X})$ *is strongly connected.*

This observation is for instance useful to verify the irreducibility property of relatively small matrices. So it may be easily seen that $\mathbf{X} = \left(\begin{smallmatrix} 0 & 1 \\ 1 & 0 \end{smallmatrix}\right)$ is irreducible, whereas the other two examples above are not.

The following lemma (which is directly connected to the previous one) shows that irreducible matrices can be easily converted into positive ones. The proof can be found in many textbooks (see references at the beginning of this appendix).

Lemma A.24. *If* $\mathbf{X} \in \mathrm{X}_n$, *then* $(\mathbf{I} + \mathbf{X})^{n-1} > 0$.

Now we use this result to prove the Perron–Frobenius theorem which is of central importance for the theory presented in this book.

Theorem A.25 (Perron–Frobenius Theorem). *Let* $\mathbf{X} \in \mathrm{X}_n$. *Then, there exists an eigenvalue* $\lambda_p \in \sigma(\mathbf{X})$ *such that*

(i) $\lambda_p = \rho(\mathbf{X}) > 0$, *and hence* $\lambda_p \ge |\lambda|$ *for any eigenvalue* $\lambda \neq \lambda_p$,
(ii) strictly positive left and right eigenvectors can be associated with λ_p,
(iii) the eigenvectors associated with λ_p *are unique up to constant multiples,*
(iv) λ_p *is a simple root of the characteristic equation of* $\mathbf{X}$.

Proof. Let $\mathbf{s} \in \mathbb{R}_+^n$ with $\mathbf{s} \neq \mathbf{0}$, and let

$$\lambda_p(\mathbf{s}) := \min_i \frac{(\mathbf{X}\mathbf{s})_i}{s_i}$$

where the ratio is assumed to be $+\infty$ if $s_i = 0$. It is clear that $0 \le \lambda_p(\mathbf{s}) < +\infty$ for all $\mathbf{s} \ge 0, \mathbf{s} \neq \mathbf{0}$. Moreover, it follows that $\lambda_p(\mathbf{s})s_i \le (\mathbf{X}\mathbf{s})_i$ for all $1 \le i \le n$.

Therefore, $\lambda_p(\mathbf{s})\mathbf{s} \le \mathbf{X}\mathbf{s}$, from which we obtain $\lambda_p(\mathbf{s}) \le \mathbf{1}^T\mathbf{X}\mathbf{s}/\mathbf{1}^T\mathbf{s} \le \|\mathbf{X}\|_1$. In words, $\lambda_p(\mathbf{s})$ is uniformly bounded above for all $\mathbf{s} \ge 0$ with $\mathbf{s} \ne \mathbf{0}$. Since $\lambda_p(\mathbf{1}) > 0$, this implies that

$$\lambda_p = \sup_{\substack{\mathbf{s}\in\mathbb{R}_+^n \\ \mathbf{s}\ne\mathbf{0}}} \min_i \frac{(\mathbf{X}\mathbf{s})_i}{s_i} \tag{A.15}$$

satisfies $0 < \lambda_p(\mathbf{1}) \le \lambda_p \le \|\mathbf{X}\|_1 < +\infty$. Furthermore, as $\lambda_p(\alpha\mathbf{s}) = \lambda_p(\mathbf{s})$ for all $\alpha > 0$, one has

$$\lambda_p = \sup_{\mathbf{s}\in\mathrm{C}} \min_i \frac{(\mathbf{X}\mathbf{s})_i}{s_i}$$

where $\mathrm{C} = \{\mathbf{s} \in \mathbb{R}_+^n : \mathbf{s}^T\mathbf{s} = 1\} \subset \mathbb{R}^n$. Hence, λ_p is attained for some vector $\mathbf{p} \in \mathbb{R}_+^n$. In other words, there must exist $\mathbf{p} \in \mathbb{R}_+^n$ with $\mathbf{p} \ne \mathbf{0}$ such that $\lambda_p = \min_i(\mathbf{X}\mathbf{p})_i/p_i$. From this it follows that $\forall_{1\le i\le n}(\mathbf{X}\mathbf{p})_i \ge \lambda_p p_i$ with equality for some i. Thus, $\mathbf{u} = \mathbf{X}\mathbf{p} - \lambda_p\mathbf{p} \in \mathbb{R}_+^n$. Suppose that $\mathbf{u} \ge 0, \mathbf{u} \ne \mathbf{0}$. Then, by Lemma A.24 and (A.13),

$$(\mathbf{I}+\mathbf{X})^{n-1}\mathbf{u} = (\mathbf{I}+\mathbf{X})^{n-1}(\mathbf{X}\mathbf{p} - \lambda_p\mathbf{p}) > 0 \Rightarrow \mathbf{X}\mathbf{y} - \lambda_p\mathbf{y} > 0$$

where $\mathbf{y} = (\mathbf{I}+\mathbf{X})^{n-1}\mathbf{p}$. So the last inequality yields

$$\forall_{1\le i\le n}\ \lambda_p < \frac{(\mathbf{X}\mathbf{y})_i}{y_i}\,.$$

But this contradicts (A.15), and therefore $\mathbf{u} = 0$ or, equivalently, $\mathbf{X}\mathbf{p} = \lambda_p\mathbf{p}$. This implies that λ_p is a real-valued eigenvalue of $\mathbf{X}$. Now we show that $\lambda_p \ge |\lambda|$ for all $\lambda \in \sigma(\mathbf{X})$. Let $\lambda\mathbf{s} = \mathbf{X}\mathbf{s}, \mathbf{s} \ne \mathbf{0}$. Taking the modulus of both sides yields $|\lambda||\mathbf{s}| \le \mathbf{X}|\mathbf{s}|$ where $|\mathbf{s}| = (|s_1|, \ldots, |s_n|)$. Therefore, by (A.15),

$$|\lambda| \le \min_{1\le i\le n} \frac{(\mathbf{X}|\mathbf{s}|)_i}{|s_i|} \le \lambda_p\,.$$

This completes the proof of (i).

(ii) Multiplying $\mathbf{X}\mathbf{p} = \lambda_p\mathbf{p}$ by $(\mathbf{I}+\mathbf{X})^{n-1} > 0$ gives $(\mathbf{I}+\mathbf{X})^{n-1}\mathbf{X}\mathbf{p} = \mathbf{X}(\mathbf{I}+\mathbf{X})^{n-1}\mathbf{p} = \mathbf{X}\mathbf{y} = \lambda_p\mathbf{y}$. Therefore, by (A.13), a right eigenvector of $\mathbf{X}$ associated with λ_p is positive. Obviously, $\mathbf{X} \in \mathrm{X}_n$ if and only if $\mathbf{X}^T \in \mathrm{X}_n$ so that the proof of (i) can be applied to $\mathbf{X}^T$. In particular, since $\sigma(\mathbf{X}) = \sigma(\mathbf{X}^T)$, there must exist $\mathbf{q} \in \mathbb{R}_+^n, \mathbf{q} \ne \mathbf{0}$, such that $\lambda_p\mathbf{q} = \mathbf{X}^T\mathbf{q}$ with $\lambda_p \ge |\lambda|$ for all $\lambda \in \sigma(\mathbf{X}^T)$. Finally, proceeding as above shows that it is possible to associate with λ_p a positive left eigenvector.

(iii) Assume that the claim is not true, and let $\mathbf{s} > 0$ and $\mathbf{u} > 0$ be two linearly independent right eigenvectors of $\mathbf{X}$ associated with λ_p. Then, there must exist constants α and β such that $\mathbf{p} = \alpha\mathbf{s} + \beta\mathbf{u} \ge 0$ has at least one zero coordinate and $\mathbf{p} \ne \mathbf{0}$ is a right eigenvector of $\mathbf{X}$ associated with λ_p. However, this contradicts $\mathbf{p} > 0$, so $\mathbf{p}$ must be unique up to constant multiples. The same reasoning applies to $\mathbf{q}$.

(iv) Let $\mathbf{A}(\lambda) \in \mathbb{R}^{n\times n}$ with $\lambda \in \mathbb{R}$ be the adjugate matrix of the characteristic matrix $(\lambda\mathbf{I} - \mathbf{X})$. So we have $\mathbf{A}(\lambda) = (a_{i,j}(\lambda)) = p(\lambda)(\lambda\mathbf{I} - \mathbf{X})^{-1}$ where $p(\lambda)$ is the characteristic polynomial of the matrix $\mathbf{X}$. Furthermore, the derivative of $p(\lambda)$ is $p'(\lambda) = \sum_{j=1}^{n} a_{j,j}(\lambda)$. We are going to show that $p'(\lambda_p) \neq 0$. Due to (ii) and (iii), λ_p has a unique (up to constant multiples) positive right eigenvector $\mathbf{p} > 0$. Therefore, $\mathbf{A}(\lambda_p) \neq \mathbf{0}$ and each column of $\mathbf{A}(\lambda_p)$ is either the zero vector or a vector of all whose elements are nonzero and have the same sign. Considering the transpose $\mathbf{X}^T$ shows that the same applies to the rows of $\mathbf{A}(\lambda_p)$. Thus, since $\mathbf{A}(\lambda_p) \neq \mathbf{0}$, it follows that all entries of $\mathbf{A}(\lambda_p)$ are non-zero and have the same sign as, say $s \neq 0$. This implies that $s \cdot p'(\lambda_p) = s\sum_{j=1}^{n} a_{j,j}(\lambda_p) > 0$. So $p'(\lambda_p) \neq 0$, and therefore λ_p is a simple root of $p(\lambda_p) = 0$ or, equivalently, a simple eigenvalue of $\mathbf{X}$.

In fact, in the above proof, $p(\lambda)$ is increasing for all $\lambda \leq \lambda_p$ since λ_p is the largest root of $p(\lambda) = 0$ over $\mathbb{R}$. Hence, we have $p'(\lambda_p) > 0$.

Definition A.26. *$\lambda_p = \rho(\mathbf{X}) > 0$ of Theorem A.25 is called the Perron root of $\mathbf{X} \in \mathrm{X}_K$. The unique vector $\mathbf{p}$ defined by*

$$\mathbf{X}\mathbf{p} = \lambda_p\mathbf{p} \qquad \mathbf{p} \in \mathbb{R}^n_{++} \qquad \|\mathbf{p}\|_1 = 1 \tag{A.16}$$

is referred to as the right Perron eigenvector of $\mathbf{X}$. If $\mathbf{X}$ is replaced by $\mathbf{X}^T$ in (A.16), then $\mathbf{p}$ is called the left Perron eigenvector of $\mathbf{X}$.

The proof of (i) in Theorem A.25 gives rise to the so-called Collatz–Wielandt formula for the Perron root of irreducible matrices.

Theorem A.27 (Collatz–Wielandt Formula). *For every $\mathbf{X} \in \mathrm{X}_n$, there holds*

$$\lambda_p = \rho(\mathbf{X}) = \max_{\mathbf{s}\in\mathbb{R}^n_{++}} \min_{1\leq i\leq n} \frac{(\mathbf{X}\mathbf{s})_i}{s_i} = \min_{\mathbf{s}\in\mathbb{R}^n_{++}} \max_{1\leq i\leq n} \frac{(\mathbf{X}\mathbf{s})_i}{s_i}\,. \tag{A.17}$$

The maximum is attained if and only if $\mathbf{s}$ is a positive right eigenvector of $\mathbf{X}$ associated with $\rho(\mathbf{X})$.

The proof of the min-max characterization in (A.17) proceeds along similar lines as the proof of the max-min part in Theorem A.25.

A.4.2 Perron–Frobenius Theorem for Primitive Matrices

A set of primitive matrices constitutes an important subset of irreducible matrices.

Definition A.28 (Primitive Matrices). *A nonnegative matrix $\mathbf{X} \in \mathrm{N}_n$ is said to be primitive if there exists $k \geq 1$ such that $\mathbf{X}^k > 0$.*

Comparing this definition with Lemma A.22 and Lemma A.24 reveals that every primitive matrix is irreducible. The converse however does not need to hold. Furthermore, a primitive matrix does not need to be positive, although every positive matrix is primitive. For example, $\mathbf{X} = \left(\begin{smallmatrix} 1 & 1 \\ 1 & 0 \end{smallmatrix}\right)$ is primitive since $\mathbf{X}^2 = \left(\begin{smallmatrix} 2 & 1 \\ 1 & 1 \end{smallmatrix}\right)$. For primitive matrices, there is a somewhat stronger version of the Perron–Frobenius theorem. Since the primitivity is not necessary for the results presented in the book, we omit the proof.

Theorem A.29 (Perron Theorem). *Let $\mathbf{X} \in \mathrm{N}_n$ be primitive. Then, there exists an eigenvalue $\lambda_p \in \sigma(\mathbf{X})$ such that*

(i) $\lambda_p = \rho(\mathbf{X}) > 0$ and $\lambda_p > |\lambda|$ for any eigenvalue $\lambda \neq \lambda_p$,
(ii) positive left and right eigenvectors can be associated with λ_p,
(iii) the eigenvectors associated with λ_p are unique up to constant multiples,
(iv) λ_p is a simple root of the characteristic equation of $\mathbf{X}$.

As any primitive matrix is irreducible, (ii)–(iv) follow from Theorem A.25. Moreover, we have $\lambda_p = \rho(\mathbf{X}) > 0$. So, the only difference from irreducible matrices is that $\lambda_p > |\lambda|$ for any $\lambda \in \sigma(\mathbf{X})$ with $\lambda \neq \lambda_p$. In other words, the Perron root of any primitive matrix is the only eigenvalue on the boundary of a disk centered at zero and with radius $\rho(\mathbf{X})$, and therefore all other eigenvalues are interior points of the disk.

A.4.3 Some Remarks on Reducible Matrices

The following result, which is sometimes referred to as the weak form of the Perron–Frobenius theorem [5], shows that some of the spectral properties of irreducible matrices carry over to general nonnegative matrices.

Theorem A.30 (Weak Form of the Perron–Frobenius Theorem). *If $\mathbf{X} \in \mathrm{N}_n$, then $\lambda_p = \rho(\mathbf{X})$ is an eigenvalue of $\mathbf{X}$ associated with a nonnegative eigenvector $\mathbf{p} \neq \mathbf{0}$.*

Note that except for the nonnegativity property, there are no additional constraints on $\mathbf{X}$.

Proof. It is clear that any nonnegative matrix $\mathbf{X}$ can be written as a limit value of a sequence of positive matrices $\{\mathbf{X}^{(k)}\}$: $\mathbf{X} = \lim_{k\to\infty} \mathbf{X}^{(k)}$ with $\mathbf{X}^{(k)} > 0, k = 1, 2, \ldots$. Let $\lambda_p^{(k)} = \rho(\mathbf{X}^{(k)}) \in \sigma(\mathbf{X}^{(k)})$ for every $\mathbf{X}^{(k)} > 0$. By the limit above, $\lambda_p = \lim_{k\to\infty} \lambda_p^{(k)}$ exists and λ_p is a real-valued eigenvalue of $\mathbf{X}$. Furthermore, since $\lambda_p^{(k)} > |\lambda^{(k)}| \geq 0$ for every $\lambda^{(k)} \in \sigma(\mathbf{X}^{(k)})$ and $k \geq 1$, it follows from the limit and the continuity of the eigenvalues (Theorem A.6) that $\lambda_p \geq |\lambda| \geq 0, \quad \lambda \in \sigma(\mathbf{X})$. Therefore, $\lambda_p = \rho(\mathbf{X}) \in \sigma(\mathbf{X})$. The same limit implies that an associated eigenvector $\mathbf{p}$ is nonnegative. Moreover, it is possible to choose a subsequence of eigenvectors of $\mathbf{X}^{(k)}$ associated with $\lambda_p^{(k)}$ such that its limit $\mathbf{p} \neq \mathbf{0}$. Thus, $\lambda_p \mathbf{p} = \mathbf{X}\mathbf{p}$ with $\mathbf{p} \neq \mathbf{0}$.

The eigenvalue $\lambda_p = \rho(\mathbf{X})$ can be expressed in terms of the Perron roots of block-diagonal irreducible submatrices of $\mathbf{X}$. Indeed, if $\mathbf{X}$ is reducible, it follows from Definition A.21 that there exists a permutation matrix $\mathbf{P}$ such that

$$\mathbf{P}^T\mathbf{X}\mathbf{P} = \begin{pmatrix} \mathbf{X}^{(1)} & \mathbf{0} & \cdots & \mathbf{0} \\ \mathbf{X}^{(2,1)} & \mathbf{X}^{(2)} & \ldots & \mathbf{0} \\ \cdots & \cdots & \cdots & \cdots \\ \mathbf{X}^{(s,1)} & \mathbf{X}^{(s,2)} & \ldots & \mathbf{X}^{(s)} \end{pmatrix} \tag{A.18}$$

where $\mathbf{X}^{(1)}, \ldots, \mathbf{X}^{(s)}$ are irreducible (square) matrices. From this, we see that the spectral radius $\rho(\mathbf{X})$ is given by [4]

$$\rho(\mathbf{X}) = \max\{\rho(\mathbf{X}^{(i)}) : i = 1, 2, \ldots, s\} \tag{A.19}$$

where $\mathbf{X}^{(i)}, 1 \leq i \leq s$, is the ith irreducible block of $\mathbf{X}$ and $\rho(\mathbf{X}^{(i)})$ its Perron root. The following definition is needed to formulate a necessary and sufficient condition for the existence of a positive right eigenvector of $\mathbf{X}$.

Definition A.31 (Maximal and Isolated Diagonal Blocks). *In the matrix (A.18), we say that the ith diagonal block* $\mathbf{X}^{(i)}$ *is*

(i) maximal if $\rho(\mathbf{X}^{(i)}) = \rho(\mathbf{X})$, *and*
(ii) isolated if $\mathbf{X}^{(i,j)} = \mathbf{0}$ *for each* $1 \leq j < i$.

Now we are in a position to state the following result [4].

Theorem A.32. *Suppose that* $\mathsf{I} \subseteq \{1, \ldots, s\}$ *and* $\mathsf{M} \subseteq \{1, \ldots, s\}$ *are the sets of indices corresponding to isolated and maximal diagonal blocks, respectively. Then, there exists a vector* $\mathbf{p} > 0$ *with* $\rho(\mathbf{X})\mathbf{p} = \mathbf{X}\mathbf{p}$ *if and only if* $\mathsf{I} = \mathsf{M}$.

In words, the theorem says that there exists a positive right eigenvector of $\mathbf{X}$ associated with $\rho(\mathbf{X})$ if and only if every isolated block is maximal and there are no other maximal diagonal blocks. However, it is important to emphasize that this is not sufficient for the matrix $\mathbf{X}$ to have positive left *and* right eigenvectors associated with $\rho(\mathbf{X})$. In order to completely identify the set of nonnegative matrices for which there exist positive left and right eigenvectors associated with $\rho(\mathbf{X})$, we consider the following definition.

Definition A.33 (Block Irreducibility). *We say that* $\mathbf{X} \in \mathrm{N}_n$ *is block-irreducible if there exists a permutation matrix* $\mathbf{P}$ *such that*

$$\mathbf{P}^T\mathbf{X}\mathbf{P} = \begin{pmatrix} \mathbf{X}^{(1)} & \mathbf{0} & \cdots & \mathbf{0} \\ \mathbf{0} & \mathbf{X}^{(2)} & \ddots & \vdots \\ \vdots & \vdots & \cdots & 0 \\ \mathbf{0} & \mathbf{0} & \cdots & \mathbf{X}^{(s)} \end{pmatrix} \tag{A.20}$$

where all $\mathbf{X}^{(1)}, \ldots, \mathbf{X}^{(s)}$ *are square nonnegative irreducible matrices.*

In other words, $\mathbf{X}$ is block-irreducible if every diagonal block in the matrix (A.18) is isolated. In the case of block-irreducible matrices, we can further strengthen the weak form of the Perron–Frobenius theorem [4].

Theorem A.34. *Let $\mathbf{X} \in \mathrm{N}_n$ be arbitrary, and let $\lambda_p = \rho(\mathbf{X})$. The eigenvalue λ_p can be associated with positive left and right eigenvectors if and only if $\mathbf{X}$ is block-irreducible and every diagonal block is maximal.*

A.4.4 The Existence of a Positive Solution p to $(\alpha\mathbf{I} - \mathbf{X})\mathbf{p} = \mathbf{b}$

Chapter 2 deals with a positive solution to a system of linear equations with nonnegative coefficients. This section provides two well-known results on the existence of such a solution [2].

Theorem A.35. *Let $\mathbf{X} \in \mathrm{N}_n$ be arbitrary, and let $\alpha > 0$ be any scalar. A necessary and sufficient condition for a solution $\mathbf{p} \geq 0, \mathbf{p} \neq \mathbf{0}$, to*

$$(\alpha\mathbf{I} - \mathbf{X})\mathbf{p} = \mathbf{b} \tag{A.21}$$

to exist for any $\mathbf{b} > 0$ is that $\alpha > r = \rho(\mathbf{X})$. In this case, there is only one solution $\mathbf{p}$, which is strictly positive and given by $\mathbf{p} = (\alpha\mathbf{I} - \mathbf{X})^{-1}\mathbf{b}$.

We emphasize that in the setup of the theorem, $\mathbf{X} \neq \mathbf{0}$ is an arbitrary nonnegative matrix. Instead, the vector $\mathbf{b}$ is required to be positive.

Proof. Assume that $\mathbf{p} \geq 0$ exists. Since $\mathbf{b} > 0$, it follows from $\alpha\,\mathbf{p} = \mathbf{X}\mathbf{p} + \mathbf{b}$ that $\mathbf{X}\mathbf{p} < \alpha\mathbf{p}$. Clearly, as $\mathbf{X}\mathbf{p} \geq 0$, this can hold only if $\alpha > 0$ and $\mathbf{p} > 0$. Now let $\mathbf{q} \geq 0, \mathbf{q} \neq \mathbf{0}$, be a left eigenvector of $\mathbf{X}$ associated with r. By Theorem A.30, such an eigenvector exists. Hence, $\mathbf{q}^T\mathbf{X}\mathbf{p} = r\mathbf{q}^T\mathbf{p} < \alpha\mathbf{q}^T\mathbf{p}$, from which it follows that $r < \alpha$ since $\mathbf{p} > 0$ and $\mathbf{q} \neq \mathbf{0}$, and therefore $\mathbf{q}^T\mathbf{p} > 0$.

Now assume that $\alpha > r = \rho(\mathbf{X})$. By Theorem A.11, the following Neumann series converges $(\alpha\mathbf{I} - \mathbf{X})^{-1} = \alpha^{-1}(\mathbf{I} - \alpha^{-1}\mathbf{X})^{-1} = \alpha^{-1}\sum_{l=0}^{\infty}(\alpha^{-1}\mathbf{X})^l$. By nonnegativity of $\mathbf{X}$, there holds $(\alpha\mathbf{I} - \mathbf{X})^{-1} \geq 0$. Furthermore, since $(\alpha^{-1}\mathbf{X})^0 = \mathbf{I}$, each row of $(\alpha\mathbf{I} - \mathbf{X})^{-1}$ has at least one positive entry. Hence, since $\mathbf{b} > 0$, we must have $(\alpha\mathbf{I} - \mathbf{X})^{-1}\mathbf{b} > 0$ for any $\mathbf{b} > 0$. Putting $\mathbf{p} = (\alpha\mathbf{I} - \mathbf{X})^{-1}\mathbf{b} > 0$ yields the sought vector.

A combination of positivity of $\mathbf{b}$ and nonnegativity of $\mathbf{X}$ guarantees the existence of a *positive* solution $\mathbf{p}$ to (A.21), provided that $\rho(\mathbf{X}) < \alpha$. It is clear that if $\mathbf{b}$ is an arbitrary nonnegative vector (not necessarily positive) and $\rho(\mathbf{X}) < \alpha$, then the solution $\mathbf{p} \geq 0$ with $\mathbf{p} = (\alpha\mathbf{I} - \mathbf{X})^{-1}\mathbf{b}$ exists but does not need to be positive. If $\mathbf{b}$ is an arbitrary nonnegative vector, the following result shows that the positivity of $\mathbf{p}$ is recovered when $\mathbf{X} \in \mathrm{N}_n$ is irreducible.

Theorem A.36. *Let $\alpha > 0$ be any scalar, and let $\mathbf{X} \in \mathrm{N}_n$ be irreducible. A necessary and sufficient condition for a solution $\mathbf{p} \geq 0, \mathbf{p} \neq \mathbf{0}$ to*

$$(\alpha\mathbf{I} - \mathbf{X})\mathbf{p} = \mathbf{b}$$

to exist for any $\mathbf{b} \geq 0, \mathbf{b} \neq \mathbf{0}$, *is that* $\alpha > r = \rho(\mathbf{X})$. *In this case, there is only one solution* $\mathbf{p}$, *which is strictly positive and given by* $\mathbf{p} = (\alpha\mathbf{I} - \mathbf{X})^{-1}\mathbf{b}$.

First we prove the following lemma.

Lemma A.37. *Let* $\mathbf{X} \geq 0$ *be irreducible,* $\alpha > 0$, *and* $\mathbf{y} \geq 0, \mathbf{y} \neq \mathbf{0}$, *a vector satisfying*

$$\mathbf{X}\mathbf{y} \leq \alpha\,\mathbf{y}\,. \tag{A.22}$$

Then $\mathbf{y} > 0$ *and* $\alpha \geq r = \rho(\mathbf{X})$ *where* r *is the Perron root of* $\mathbf{X}$. *Moreover,* $\alpha = r$ *if and only if* $\mathbf{X}\mathbf{y} = \alpha\,\mathbf{y}$.

Proof. First suppose that $\mathbf{y}$ is not positive and $y_i = 0$ for some fixed $1 \leq i \leq n$. By (A.22), $\mathbf{X}^k\mathbf{y} \leq \alpha^k\mathbf{y}$, and therefore $\sum_{j=1}^{n} x_{i,j}^{(k)} y_j \leq \alpha^k\, y_i$. As $\mathbf{X} \geq 0$ is irreducible, we know from Lemma A.22 that, for each $1 \leq j \leq n$, there is a natural number k such that $x_{i,j}^{(k)} > 0$. Thus, since $y_j > 0$ for some j, there holds $x_{i,j}^{(k)} y_j > 0$ for some $k \geq 1$ and j. This in turn implies $y_i > 0$, thereby contradicting the assumption $y_i = 0$. As a consequence, we must have $\mathbf{y} > 0$. Now letting $\mathbf{q}$ be any positive left eigenvector of $\mathbf{X}$ yields $\alpha\,\mathbf{q}^T\mathbf{y} \geq \mathbf{q}^T\mathbf{X}\mathbf{y} = r\mathbf{q}^T\mathbf{y}$. From this, we have $\alpha \geq r$ since $\mathbf{q}^T\mathbf{y} > 0$.

Now suppose that $\mathbf{X}\mathbf{y} \leq r\mathbf{y}$ with strict inequality in at least one place. Then, $\mathbf{q}^T\mathbf{X}\mathbf{y} = r\mathbf{q}^T\mathbf{y} < r\mathbf{q}^T\mathbf{y}$. So $r < r$, which does not make sense, and therefore we must have $\mathbf{X}\mathbf{y} = \alpha\mathbf{y}$ if $\alpha = r$. Proceeding in a similar way shows that $\alpha = r$ implies $\mathbf{X}\mathbf{y} = \alpha\,\mathbf{y}$.

Now we are in a position to prove Theorem A.36.

Proof. First assume that $\mathbf{p} \geq 0, \mathbf{p} \neq \mathbf{0}$, exists. Since $\mathbf{b} \geq 0, \mathbf{b} \neq \mathbf{0}$, and $\mathbf{b} + \mathbf{X}\mathbf{p} = \alpha\mathbf{p}$, we have $\mathbf{X}\mathbf{p} \leq \alpha\mathbf{p}$ with at least one inequality. Of course, this can be satisfied only if $\alpha > 0$. Moreover, by Lemma A.37, we have $\alpha > r = \rho(\mathbf{X}) > 0$.

Conversely, if $\alpha > r = \rho(\mathbf{X})$, then the Neumann series converges $(\alpha\mathbf{I} - \mathbf{X})^{-1} = \alpha^{-1}\sum_{l=0}^{\infty}(\alpha^{-1}\mathbf{X})^l$. Furthermore, as $\mathbf{X}$ is irreducible, Lemma A.22 implies that $(\alpha\mathbf{I} - \mathbf{X})^{-1} > 0$. Therefore, by (A.13), $(\alpha\mathbf{I} - \mathbf{X})^{-1}\mathbf{b} > 0$ for any $\mathbf{b} \geq 0$ with $\mathbf{b} \neq \mathbf{0}$. Defining $\mathbf{p} = (\mathbf{I} - \mathbf{X})^{-1}\mathbf{b} > 0$ completes the proof.

Finally we point out a well-known connection to M-matrices.

Definition A.38. *A real nonsingular matrix is said to be an M-matrix if all off-diagonal entries are non-positive and its inverse is a nonnegative matrix.*

Theorem A.39. *Let* $\mathbf{A} \in \mathbb{R}^{n\times n}$ *be any nonsingular matrix with non-positive off-diagonal entries. The following statements are equivalent.*

(i) $\mathbf{A}$ *is an M-matrix.*

(ii) There is a matrix $\mathbf{X} \geq 0$ *and a real number* $\alpha > \rho(\mathbf{X})$ *such that* $\mathbf{A} = \alpha\mathbf{I} - \mathbf{X}$.
(iii) All principal minors of $\mathbf{A}$ *are positive.*
(iv) $Re(\lambda) > 0$ *for all* $\lambda \in \sigma(\mathbf{A})$.

In the setup of Theorems A.35 and A.36, there exists a positive solution $\mathbf{p}$ to $(\alpha\mathbf{I} - \mathbf{X})\mathbf{p} = \mathbf{b}$ if and only if $\rho(\mathbf{X}) < \alpha$. On the other hand, by (ii) in the above theorem, $\alpha\mathbf{I} - \mathbf{X}$ with $\mathbf{X} \geq 0$ is an M-matrix if and only if $\rho(\mathbf{X}) < \alpha$. So we can conclude that a positive solution $\mathbf{p}$ exists if and only if $\alpha\mathbf{I} - \mathbf{X}$ is an M-matrix.

B

Some Concepts and Results from Convex Analysis

In this chapter, we collect definitions, notational conventions and several results from convex analysis that may be helpful in better understanding the material covered in this manuscript. Proofs are provided only for selected results concerning the notion of log-convexity and the convergence of gradient projection algorithms. For other proofs, the reader is referred to any standard analysis book (e.g., [98]) and [83, 88, 11].

B.1 Sets and Functions

If A is a set, then $x \in \mathrm{A}$ means that x is a member (or an element) of A. If x is not a member of A, then we write $x \notin \mathrm{A}$. The set with no elements is called empty and is denoted by $\emptyset$. For any two sets A and B, $\mathrm{A} \subset \mathrm{B}$ means that every member of A is a member of B. In such a case, A is said to be a subset of B. If, in addition, there is an element of B which is not in A, then A is said to be a proper subset of B. If $\mathrm{A} \subset \mathrm{B}$ and $\mathrm{B} \subset \mathrm{A}$, we write $\mathrm{A} = \mathrm{B}$. Otherwise, we have $\mathrm{A} \neq \mathrm{B}$. In the book, we use $\mathbb{N}$ and $\mathbb{N}_0$ to denote the set of natural numbers and the set of nonnegative integers.

Now suppose $\mathbb{R}^n$ is a metric space. Let $\|\cdot\| : \mathbb{R}^n \to \mathbb{R}$ be a norm defined on $\mathbb{R}^n$ and $d(\mathbf{p}, \mathbf{q}) = \|\mathbf{p} - \mathbf{q}\|$ the distance from $\mathbf{p} \in \mathbb{R}^n$ to $\mathbf{q} \in \mathbb{R}^n$ (Appendix A.1). Let A be some subset of $\mathbb{R}^n$, which is also a metric space equipped with the same metric. In the definition below, elements of $\mathbb{R}^n$ are referred to as points. These points are vectors if $\mathbb{R}^n$ is a vector space. Note that $\mathbb{R}^n$ could be replaced by an arbitrary metric space.

Definition B.1. *All points and sets mentioned below are understood to be elements and subsets of* $\mathbb{R}^n$.

(a) An open ball $\mathrm{B}_r(\mathbf{p})$ *of radius* $r > 0$ *centered at point* $\mathbf{p}$ *is defined to be*

$$\mathrm{B}_r(\mathbf{p}) := \{\mathbf{q} \in \mathbb{R}^n : d(\mathbf{p}, \mathbf{q}) < r\}.$$

$\mathrm{B}_r(\mathbf{p})$ *is an open interval if* $n = 1$ *and an open disk if* $n = 2$.

(b) *A point* $\mathbf{p}$ *is a limit point of the set* A *if, for all* $r > 0$, $B_r(\mathbf{p})$ *contains a point* $\mathbf{q} \neq \mathbf{p}$ *with* $\mathbf{q} \in$ A.
(c) A *is closed if every limit point of* A *is a point of* A.
(d) *A point* $\mathbf{p}$ *is an interior point of* A *if there is* $r > 0$ *such that* $B_r(\mathbf{p}) \subset$ A.
(e) A *is open if every point of* A *is an interior point of* A.
(f) *The complement of* A *(denoted by* A^c*) in* $\mathbb{R}^n$ *is the set of all points* $\mathbf{p} \in \mathbb{R}^n$ *such that* $\mathbf{p} \notin$ A.
(g) A *is bounded if there is a real number* M *and a point* $\mathbf{q} \in \mathbb{R}^n$ *such that* $d(\mathbf{p}, \mathbf{q}) < M$ *for all* $\mathbf{p} \in$ A. *Otherwise, it is said to be unbounded.*

According to these definitions, $\mathbb{R}^n$ is a closed, open and unbounded set. Throughout the manuscript, for any $a < b$ with $a, b \in \mathbb{R}$, $[a, b]$ is called a closed interval on the real line, $[a, b)$ and $(a, b]$ are half-open intervals, and (a, b) is an open interval (also referred to as a segment). Now let us introduce the notion of compactness.

Definition B.2 (Open Cover and Compact Set). *An open cover of* A *in* $\mathbb{R}^n$ *is a collection* $\{\mathrm{G}_\alpha\}$ *of open subsets of* $\mathbb{R}^n$ *such that* $\mathrm{A} \subset \cup_\alpha \mathrm{G}_\alpha$. *A subset* A *of* $\mathbb{R}^n$ *is said to be compact if every open cover of* A *contains a finite subcover. In other words, if* $\{\mathrm{G}_\alpha\}$ *is an open cover of* A, *then there are finitely many indices* $\alpha_1, \ldots, \alpha_m$ *such that* $\mathrm{A} \subset \mathrm{G}_{\alpha_1} \cup \cdots \cup \mathrm{G}_{\alpha_m}$.

The Heine–Borel theorem stated below is implicitly invoked in the book when the existence of maxima or/and minima should be guaranteed (see Theorem B.8).

Theorem B.3. *For a subset* A *of the Euclidean space* $\mathbb{R}^n$, *the following are equivalent.*

(i) A *is closed and bounded.*
(ii) A *is compact.*

The Euclidean space is defined in Appendix A. It should be emphasized that (i) and (ii) cease to be equivalent in general metric spaces.

For any function (map) from the set A into the set B, we write $f : \mathrm{A} \to \mathrm{B}$ or $\mathrm{A} \to \mathrm{B} : x \to f(x)$. The set A is called the domain of f, and the elements $f(x)$ of B are called values of f. The set of all values of f is called the range of f denoted by $f(\mathrm{A}) \subset \mathrm{B}$. If $f(\mathrm{A}) = \mathrm{B}$, we say that f maps A *onto* B. If $f(x_1) \neq f(x_2)$ whenever $x_1 \neq x_2, x_1 \in \mathrm{A}, x_2 \in \mathrm{A}$, then f is said to be a one-to-one mapping (function) from A into B.

Definition B.4 (Bijective Function). *We say that* f *is* bijective *if* $f : \mathrm{A} \to \mathrm{B}$ *is a one-to-one map from* A *onto* B *(one-to-one and onto).*

The notion of bijectivity (specialized to real-valued functions) is used (explicitly or implicitly) at many points.

Theorem B.5. *Let* $\mathrm{A}, \mathrm{B} \subset \mathbb{R}$. *A function* $f : \mathrm{A} \to \mathrm{B}$ *is bijective if and only if there is a function* $g : \mathrm{B} \to \mathrm{A}$ *such that* $g(f(x)) = x, x \in \mathrm{A}$, *and* $f(g(x)) = x, x \in \mathrm{B}$ *(identity function).*

In what follows, the domain of f is a subset of a metric space $\mathbb{R}^n$ denoted by D. The range of f is the set of all reals $\mathbb{R}$, in which case f is said to be a real-valued function. The metric on $\mathbb{R}$ is simply $d(p,q) = |p-q|$ for all $p, q \in \mathbb{R}$, whereas $\mathbb{R}^n$ is a metric space equipped with the metric $d : \mathbb{R}^n \to \mathbb{R}_+$ induced by some norm (see above).

Definition B.6 (Function Limit). *Suppose that $f : \mathrm{D} \to \mathbb{R}$, and $\mathbf{p}$ is a limit point of* D. *We write $f(\mathbf{x}) \to q$ as $\mathbf{x} \to \mathbf{p}$, or, equivalently, $\lim_{x\to p} f(x) = q$ if there is a point $q \in \mathbb{R}$ with the following property: For every $\epsilon > 0$, there exists $\delta > 0$ such that $|f(\mathbf{x}) - q| < \epsilon$ for all $\mathbf{x} \in \mathrm{D}$ with $d(\mathbf{x}, \mathbf{p}) < \delta$.*

Note that in the definition above, $\mathbf{p}$ does not need to be a member of D. So f does not need to be defined at $\mathbf{p}$.

Definition B.7 (Continuous Function). *Suppose that $f : \mathrm{D} \to \mathbb{R}$, and $\mathbf{p} \in \mathrm{D}$. Then, f is said to be continuous at $\mathbf{p}$ if for every $\epsilon > 0$, there exists $\delta > 0$ such that $|f(\mathbf{x}) - f(\mathbf{p})| < \epsilon$ for all points $\mathbf{x} \in \mathrm{D}$ for which $d(\mathbf{x}, \mathbf{p}) < \delta$. If f is continuous for all points in* D, *then f is said to be continuous on* D, *or simply continuous.*

Theorem B.8. *Suppose that $f : \mathrm{D} \to \mathbb{R}$ is continuous and* D *is a compact subset of $\mathbb{R}^n$. Let*

$$M = \sup_{\mathbf{x}\in \mathrm{D}} f(\mathbf{x}) \quad \textit{and} \quad m = \inf_{\mathbf{x}\in \mathrm{D}} f(\mathbf{x}) \tag{B.1}$$

be the least upper bound of all $f(\mathbf{x})$ with $\mathbf{x}$ ranging over D *and the greatest lower bound of this set of numbers, respectively. Then, there exist points $\mathbf{p}, \mathbf{q} \in \mathrm{D}$ such that $f(\mathbf{p}) = M$ and $f(\mathbf{q}) = m$.*

In words, the theorem asserts that a continuous function f defined on a compact set $\mathrm{D} \subset \mathbb{R}^n$ attains its minimum and maximum on this set. We point out that the theorem can be somewhat modified to include upper semicontinuous and lower semicontinuous functions. If $\mathbb{R}^n$ is Euclidean space, then, by Theorem B.3, D in Theorem B.8 is a closed and bounded subset of $\mathbb{R}^n$.

Definition B.9 (Partial Derivatives and Gradient). *Let $\mathrm{D} \subset \mathbb{R}^n$ be an open set, let $f : \mathrm{D} \to \mathbb{R}$ be given, and let $\mathbf{e}_i$ be the ith unit vector (all components are zero except for the ith component which is 1). For any $\mathbf{x} \in \mathrm{D}$, we define*

$$\nabla_i f(\mathbf{x}) = \frac{\partial f}{\partial x_i}(\mathbf{x}) = \lim_{h\to 0} \frac{f(\mathbf{x} + h\mathbf{e}_i) - f(\mathbf{x})}{h} \tag{B.2}$$

provided that the limit exists. We call $\nabla_i f(\mathbf{x})$ the ith partial derivative of f at point $\mathbf{x} \in \mathrm{D}$. Assuming that the partial derivative exists for each $1 \le i \le n$, the gradient of f at $\mathbf{x}$ is defined to be

$$\nabla f(\mathbf{x}) = \left(\frac{\partial f}{\partial x_1}(\mathbf{x}), \ldots, \frac{\partial f}{\partial x_n}(\mathbf{x})\right). \tag{B.3}$$

Now let fix $\mathbf{x} \in \mathrm{D}$ where D is some open nonempty subset of $\mathbb{R}^n$, and let $\mathbf{u} \in \mathbb{R}^n$ be a vector of unit norm ($\|\mathbf{u}\| = 1$). Define

$$D_{\mathbf{u}} f(\mathbf{x}) = \lim_{t \to 0} \frac{f(\mathbf{x} + t\mathbf{u}) - f(\mathbf{x})}{t} \tag{B.4}$$

provided that the limit exists. We call $D_{\mathbf{u}} f(\mathbf{x})$ *the directional derivative* of f at $\mathbf{x}$, in the direction of the unit vector $\mathbf{u}$.

Definition B.10 (Gateaux Differentiability). *We say that f is (Gateaux) differentiable at $\mathbf{x} \in \mathrm{D}$ if the gradient exists and satisfies $\nabla f(\mathbf{x})^T \mathbf{u} = D_{\mathbf{u}} f(\mathbf{x})$. The function f is called differentiable over $\mathrm{D} \subset \mathbb{R}^n$ if it is differentiable at every $\mathbf{x} \in \mathrm{D}$.*

If f is differentiable over an open set D and the gradient $\nabla f(\mathbf{x})$ is a continuous function of $\mathbf{x}$, then f is said to be *continuously differentiable*. Continuously differentiable functions are Frechet differentiable, which implies Gateaux differentiability [98].

Remark B.11. Throughout the book, all functions are assumed to be continuously differentiable over some open set. In this case, Gateaux differentiability is equivalent to Frechet differentiability, and therefore we make no distinction between these two concepts.

Now assume that each of the partial derivatives of a function $f : \mathrm{D} \to \mathbb{R}$ is a continuously differentiable function at $\mathbf{x} \in \mathrm{D}$. We use the notation

$$h_{i,j}(\mathbf{x}) = \frac{\partial^2 f}{\partial x_i \partial x_j}(\mathbf{x}) = \frac{\partial}{\partial x_i} \frac{\partial f}{\partial x_j}(\mathbf{x})$$

to indicate the ith partial derivative of $\frac{\partial f}{\partial x_j}$ at a point $\mathbf{x} \in \mathrm{D}$.

Definition B.12 (Hessian Matrix). *The matrix $\mathbf{H}(\mathbf{x}) = (h_{i,j}(\mathbf{x})) \in \mathbb{R}^{n \times n}$ is called the Hessian matrix (or simply the Hessian) of $f : \mathrm{D} \to \mathbb{R}$ at $\mathbf{x} \in \mathrm{D} \subset \mathbb{R}^n$.*

When $\mathrm{D} \subset \mathbb{R}$, the first derivative (if it exists) of $f : \mathrm{D} \to \mathbb{R}$ at $x \in \mathrm{D} \subset \mathbb{R}$ is denoted by $f'(x) = \frac{df}{dx}(x)$. If f' is itself differentiable, we denote the derivative of f' at x by $f''(x) = \frac{d^2 f}{dx^2}(x)$ and call f'' the second derivative of f. When $f''(x)$ exists for all $x \in \mathrm{D}$, f is said to be twice differentiable. If, in addition, the second derivative is a continuous function, we say that f is twice continuously differentiable. Note that for $f''(x)$ to exist at x, $f'(y)$ must exist in some open ball centered at x (we consider functions over an open interval) and $f'(x)$ must be differentiable at x.

A fundamental result of the analysis is the mean value theorem.

Theorem B.13 (Mean Value Theorem). *Suppose that $f : \mathrm{D} \to \mathbb{R}$ is continuously differentiable over an open interval $\mathrm{D} \subset \mathbb{R}$. Then, for every $a, b \in \mathrm{D}$, there exists some $\xi \in [a, b]$ such that*

$$f(b) - f(a) = f'(\xi)(b - a). \tag{B.5}$$

here is also the corresponding mean value theorem for Gateaux differentiable functions $f : \mathrm{D} \to \mathbb{R}$ with $\mathrm{D} \subset \mathbb{R}^n$ [89]. This however is in general not true for mappings from $\mathrm{D} \subset \mathbb{R}^n$ into $\mathbb{R}^m$ with $m > 1$.

Finally, we define monotonic functions.

Definition B.14 (Monotonic Functions). *Let* $\mathrm{D} \subset \mathbb{R}$ *be an open interval (segment). Then* f *is said to be monotonically increasing (decreasing) on* D *if* $x < y$ *for any* $x, y \in \mathrm{D}$ *implies* $f(x) \leq f(y)$ $(f(x) \geq f(y))$*. If the last inequality is strict inequality, then we say that* f *is strictly increasing (decreasing).*

If $f : \mathrm{D} \to \mathbb{R}$ is differentiable, then f is monotonically increasing (decreasing) on D if and only if its first derivative is nonnegative (nonpositive) on this set.

B.2 Convex Sets and Functions

Definition B.15 (Convex Set). *We say that* $\mathrm{D} \subset \mathbb{R}^n$ *is a convex set if*

$$(1 - \mu)\hat{\mathbf{x}} + \mu\check{\mathbf{x}} \in \mathrm{D} \tag{B.6}$$

for all $\mu \in (0, 1)$ *and* $\hat{\mathbf{x}}, \check{\mathbf{x}} \in \mathrm{D}$.

Throughout this section, unless otherwise stated, it is assumed that $\mathrm{D} \subset \mathbb{R}^n$ is a nonempty *convex* set. Moreover, we use $\mathbf{x}(\mu)$ to denote the convex combination of some $\hat{\mathbf{x}} \in \mathbb{R}^n$ and $\check{\mathbf{x}} \in \mathbb{R}^n$, that is $\mathbf{x}(\mu) = (1 - \mu)\hat{\mathbf{x}} + \mu\check{\mathbf{x}}$.

Definition B.16 (Convex Function). *A function* $f : \mathrm{D} \to \mathbb{R}$ *is said to be convex if* $\mathrm{D} \subset \mathbb{R}^n$ *is a convex set and*

$$f(\mathbf{x}(\mu)) \leq (1 - \mu) f(\hat{\mathbf{x}}) + \mu f(\check{\mathbf{x}}) \tag{B.7}$$

for all $\mu \in (0, 1)$ *and* $\hat{\mathbf{x}}, \check{\mathbf{x}} \in \mathrm{D}$*. The function* f *is said to be strictly convex if there is a strict inequality in (B.7) for all* $\mu \in (0, 1)$*. We say that a function* f *is concave if* $-f$ *is convex.*

Now we provide necessary and sufficient conditions for differentiable and twice continuously differentiable functions to be convex.

Theorem B.17 (First-order Condition). *Let* $f : \mathrm{D} \to \mathbb{R}$ *be differentiable (over* $\mathrm{D} \subset \mathbb{R}^n$*). Then,* f *is convex if and only if*

$$\forall_{\mathbf{x},\mathbf{z} \in \mathrm{D}}\ f(\mathbf{z}) \geq f(\mathbf{x}) + (\mathbf{z} - \mathbf{x})^T \nabla f(\mathbf{x})\,. \tag{B.8}$$

The function f *is strictly convex whenever there is strict inequality for all* $\mathbf{z} \neq \mathbf{x}$.

Theorem B.18 (Second-order Condition). *Let* D *be a convex set with a nonempty interior. Suppose that* $f : \mathrm{D} \to \mathbb{R}$ *is a twice continuously differentiable function. Then,* f *is convex if and only if* $\nabla^2 f(\mathbf{x})$ *is positive semidefinite for all* $\mathbf{x} \in \mathrm{D}$.

In the above theorem, if $\nabla^2 f(\mathbf{x})$ is positive definite for all $\mathbf{x} \in \mathrm{D}$, then f is strictly convex. The converse, however, does not need to hold. A sufficient condition for the Hessian to be positive definite is that f is strongly convex.

B.2.1 Strong Convexity

Strongly convex functions are defined as follows.

Definition B.19. *A function* $f : \mathrm{D} \to \mathbb{R}$ *is said to be strongly convex (with modulus of strong convexity* c*) if there exists* $c > 0$ *such that*

$$f(\mathbf{x}(\mu)) \leq (1-\mu)f(\hat{\mathbf{x}}) + \mu f(\check{\mathbf{x}}) - \frac{1}{2}c\mu(1-\mu)\|\hat{\mathbf{x}} - \check{\mathbf{x}}\|_2^2 \tag{B.9}$$

for all $\hat{\mathbf{x}}, \check{\mathbf{x}} \in \mathrm{D}$ *and* $\mu \in (0,1)$.

Observation B.20. *A function* $f : \mathrm{D} \to \mathbb{R}$ *is strongly convex with modulus of strong convexity* c *if and only if* $g(\mathbf{x}) = f(\mathbf{x}) - 1/2c\|\mathbf{x}\|_2^2$ *is convex.*

Proof. Let $\hat{\mathbf{x}}, \check{\mathbf{x}} \in \mathrm{D}$ be arbitrary. Suppose that f is strongly convex. Then, by the definition, we have

$$\begin{aligned}
f((1-\mu)\hat{\mathbf{x}} + \mu\check{\mathbf{x}}) &\leq (1-\mu)f(\hat{\mathbf{x}}) + \mu f(\check{\mathbf{x}}) - \frac{1}{2}c\mu(1-\mu)\|\hat{\mathbf{x}} - \check{\mathbf{x}}\|_2^2 \\
&= (1-\mu)f(\hat{\mathbf{x}}) + \mu f(\check{\mathbf{x}}) - \frac{1}{2}c\mu(1-\mu)\big(\|\hat{\mathbf{x}}\|_2^2 - 2\langle\hat{\mathbf{x}}, \check{\mathbf{x}}\rangle + \|\check{\mathbf{x}}\|_2^2\big) \\
&= (1-\mu)f(\hat{\mathbf{x}}) + \mu f(\check{\mathbf{x}}) + \frac{1}{2}c(1-\mu)^2\|\hat{\mathbf{x}}\|_2^2 - \frac{1}{2}c(1-\mu)\|\hat{\mathbf{x}}\|_2^2 \\
&\quad + \frac{1}{2}c\mu^2\|\check{\mathbf{x}}\|_2^2 - \frac{1}{2}c\mu\|\check{\mathbf{x}}\|_2^2 + c\mu(1-\mu)\langle\hat{\mathbf{x}}, \check{\mathbf{x}}\rangle \\
&= (1-\mu)f(\hat{\mathbf{x}}) + \mu f(\check{\mathbf{x}}) + \frac{1}{2}c\|(1-\mu)\hat{\mathbf{x}} + \mu\check{\mathbf{x}}\|_2^2 \\
&\quad - \frac{1}{2}c(1-\mu)\|\hat{\mathbf{x}}\|_2^2 - \frac{1}{2}c\mu\|\check{\mathbf{x}}\|_2^2
\end{aligned}$$

for all $\mu \in (0,1)$. Hence, $f((1-\mu)\hat{\mathbf{x}} + \mu\check{\mathbf{x}}) - \frac{1}{2}c\|(1-\mu)\hat{\mathbf{x}} + \mu\check{\mathbf{x}}\|_2^2 \leq (1-\mu)f(\hat{\mathbf{x}}) + \mu f(\check{\mathbf{x}}) - \frac{1}{2}c(1-\mu)\|\hat{\mathbf{x}}\|_2^2 - \frac{1}{2}c\mu\|\check{\mathbf{x}}\|_2^2$ which is just convexity of $g(x) = f(x) - 1/2c\|\mathbf{x}\|_2^2$. Assuming convexity of g and proceeding in reverse order proves the converse.

By the observation, it is clear that any strongly convex function is strictly convex. However, as already mentioned above, the converse does not hold in general. A standard example of a strictly convex function that is not strongly convex is $\mathbb{R} \rightarrow \mathbb{R}_+ : x \rightarrow x^4$. Another example is the function $\mathbb{R}_{++} \rightarrow \mathbb{R}_{++} : x \rightarrow 1/x$, which is strictly convex on $\mathbb{R}_{++}$ but not strongly convex. Yet it is strongly convex on any closed bounded interval on $\mathbb{R}_{++}$. Intuitively, strong convexity is equivalent to assuming that the curvature of f is positive and bounded away from zero at every point. When $f : \mathrm{D} \rightarrow \mathbb{R}$ is continuously differentiable, then the following can be shown to hold [88].

Theorem B.21. *Suppose that $f : \mathrm{D} \rightarrow \mathbb{R}$ is continuously differentiable. Then, f is strongly convex (with modulus of strong convexity c) if and only if there exists a constant $c > 0$ such that*

$$\left(\nabla f(\mathbf{x}) - \nabla f(\mathbf{y})\right)^T (\mathbf{x} - \mathbf{y}) \geq c\|\mathbf{x} - \mathbf{y}\|_2^2 \tag{B.10}$$

for all $\mathbf{x}, \mathbf{y} \in \mathrm{D}$.

In words, the theorem above says that strong convexity of f is equivalent to strong monotonicity of the gradient ∇f. When $f : \mathrm{D} \rightarrow \mathbb{R}$ is twice continuously differentiable, then we have the following result [88].

Theorem B.22. *Let $\mathrm{D} \subset \mathbb{R}^n$ be a convex set with nonempty interior. Then, $f : \mathrm{D} \rightarrow \mathbb{R}$ is strongly convex (with modulus of strong convexity c) if and only if*

$$\nabla^2 f(\mathbf{x}) - c\mathbf{I} \tag{B.11}$$

is positive semidefinite for all $\mathbf{x} \in \mathrm{D}$.

B.3 Log-Convex Functions

Definition B.23 (Log-convex function). *A function $f : \mathrm{D} \rightarrow \mathbb{R}_{++}$ is said to be log-convex if $\mathrm{D} \subset \mathbb{R}^n$ is a convex set and $\log f$ is convex, i.e., if we have*

$$\log f\big((1-\mu)\hat{\mathbf{x}} + \mu\check{\mathbf{x}}\big) \leq (1-\mu)\log f(\hat{\mathbf{x}}) + \mu \log f(\check{\mathbf{x}}) \tag{B.12}$$

for all $\mu \in (0,1)$ and $\hat{\mathbf{x}}, \check{\mathbf{x}} \in \mathrm{D}$. If there is strict inequality in (B.12) for all $\mu \in (0,1)$, we say that f is strictly log-convex.

Similarly, we say that f is log-concave if $\log f$ is concave. Note that f is log-concave if and only if $1/f$ is log-convex. The list below presents some examples of log-convex and log-concave functions [11].

(i) $f(x) = e^{cx}, x \in \mathbb{R}$, is both log-convex and log-concave for any real c.
(ii) $f(x) = x^c$ on $\mathbb{R}_{++}$ is log-convex for $c \leq 0$ and log-concave for $c \geq 0$.
(iii) $f(x) = e^x/(1-e^x), x < 0$, is log-convex.
(iv) The Gamma function $f(x) = \int_0^\infty u^{x-1}e^{-u}du$ is log-convex for $x \geq 1$.

In all that follows, we exclusively focus on *log-convex* functions. For more information about log-concavity, the reader is referred to [11].

Remark B.24. As the logarithm is not defined at zero, any log-convex function f is by definition positive. However, it is often convenient to allow f to take on the value zero, in which case one takes $\log f(x) = -\infty$ [11]. A nonnegative function f is said to be log-convex if the extended-value function $\log f$ is convex.

Theorem B.25. *Let* $\mathrm{D} \subset \mathbb{R}^n$ *be a convex nonempty set. A positive function* $f : \mathrm{D} \to \mathbb{R}_{++}$ *is log-convex on* D *if and only if*

$$f(\mathbf{x}(\mu)) \leq f(\hat{\mathbf{x}})^{1-\mu} f(\check{\mathbf{x}})^{\mu} \tag{B.13}$$

for all $\mu \in (0,1)$ *and* $\hat{\mathbf{x}}, \check{\mathbf{x}} \in \mathrm{D}$.

Proof. Let $\hat{\mathbf{x}}, \check{\mathbf{x}} \in \mathrm{D}$ and $\mu \in (0,1)$ be arbitrary and note that due to convexity of D, $\mathbf{x}(\mu) \in \mathrm{D}$. Writing the right-hand side of (B.12) as $\log(f(\hat{\mathbf{x}})^{1-\mu} f(\check{\mathbf{x}})^{\mu})$ and considering the monotonicity of the logarithm yields (B.13). Conversely, taking the logarithm on both sides of (B.13) and rearranging gives (B.12).

The next result relates log-convexity to convexity.

Theorem B.26. *Let* $f : \mathrm{D} \to \mathbb{R}_{++}$ *be any log-convex function. Then,*

(i) f *is convex.*
(ii) f *is strictly convex on* $\mathrm{D} \subset \mathbb{R}$ *if* f *is strictly monotonic.*

Proof. Let $\hat{\mathbf{x}}, \check{\mathbf{x}} \in \mathrm{D}$ be arbitrary, and let f be log-convex. Since[1]

$$a^{1-\mu} b^{\mu} \leq (1-\mu)a + \mu b \tag{B.14}$$

for any positive constants a, b and $\mu \in (0,1)$, it follows from (B.13) that

$$f(\mathbf{x}(\mu)) \leq f(\hat{\mathbf{x}})^{1-\mu} f(\check{\mathbf{x}})^{\mu} \leq (1-\mu) f(\hat{\mathbf{x}}) + \mu f(\check{\mathbf{x}})$$

for $\mu \in (0,1)$. Hence, f is convex. To prove (ii), suppose that the strict convexity assertion is false. Then, there exist $\hat{x}, \check{x} \in \mathrm{D} \subset \mathbb{R}$ with $\hat{x} \neq \check{x}$ and $\mu \in (0,1)$ such that $f(x(\mu)) = f(\hat{x})^{1-\mu} f(\check{x})^{\mu} = (1-\mu) f(\hat{x}) + \mu f(\check{x})$. Since equality holds in (B.14) if and only if $a = b$, this implies that $f(\hat{x}) = f(\check{x}) = c$ for some positive c. Hence, by strict monotonicity, $\hat{x} = \check{x}$, which contradicts $\hat{x} \neq \check{x}$.

Note that the converse to Theorem B.26 does not hold since log-convexity is stronger than convexity. For instance, $\mathbb{R} \to \mathbb{R}_{++} : x \to e^x - 1$ is convex but not log-convex.

In the case of twice continuously differentiable functions, we have the following result.

---

[1] In words, this inequality says that the arithmetic mean bounds the geometric mean [84].

Theorem B.27. *Let* $\mathrm{D} \subset \mathbb{R}^n$ *be an open convex set and suppose that* $f : \mathrm{D} \to \mathbb{R}_{++}$ *is twice continuously differentiable. Then* f *is log-convex on* D *if and only if*

$$\nabla f(\mathbf{x})\nabla f(\mathbf{x})^T \leq f(\mathbf{x})\nabla^2 f(\mathbf{x}) \tag{B.15}$$

for all $x \in \mathrm{D}$.

Proof. Let $g(\mathbf{x}) = \log f(\mathbf{x}), \mathbf{x} \in \mathrm{D}$. Since f is positive, $g : \mathrm{D} \to \mathbb{R}$ is well defined and twice continuously differentiable on D. Thus, the theorem immediately follows by utilizing Theorem B.18.

B.3.1 Inverse Functions of Monotonic Log-Convex Functions

Now assume that $f : \mathrm{D} \to \mathbb{R}_{++}$ is a continuous bijection where $\mathrm{D} \subset \mathbb{R}$ is any open interval on the real line and $g : \mathbb{R}_{++} \to \mathrm{D}$ is the inverse function so that (Theorem B.5)

$$f(g(x)) \equiv x, \quad x > 0 \qquad \Leftrightarrow \qquad g(f(x)) \equiv x, x \in \mathrm{D}\,. \tag{B.16}$$

Thus, f is a strict monotone (either increasing or decreasing) function. Moreover, f is strictly increasing (decreasing) if and only if g is strictly increasing (decreasing).

Theorem B.28. *Let* f *and* g *be as defined above. Define* $g_e(x) = g(e^x)$ *for all* $x \in \mathbb{R}$. *Then,* f *is log-convex if and only if*

(i) $g_e : \mathbb{R} \to \mathrm{D}$ *is convex when* f *is strictly decreasing.*
(ii) $g_e : \mathbb{R} \to \mathrm{D}$ *is concave when* f *is strictly increasing.*

Moreover, f *is strictly log-convex if and only if* g_e *is, respectively, strictly convex or strictly concave.*

Proof. Let $\hat{x}, \check{x} \in \mathrm{D}$ be arbitrary, and let f be log-convex. Then, by Theorem B.25, we have $f((1-\mu)\hat{x} + \mu\check{x}) \leq f(\hat{x})^{1-\mu} f(\check{x})^{\mu}$ for all $\mu \in (0,1)$. Combining this with (B.16) yields

$$\begin{aligned} (1-\mu)\hat{x} + \mu\check{x} &\leq g\big(f(\hat{x})^{1-\mu} f(\check{x})^{\mu}\big) \quad f \text{ strictly increasing} \\ (1-\mu)\hat{x} + \mu\check{x} &\geq g\big(f(\hat{x})^{1-\mu} f(\check{x})^{\mu}\big) \quad f \text{ strictly decreasing}\,. \end{aligned}$$

Define $\hat{z} = \log f(\hat{x}) \in \mathbb{R}$ and $\check{z} = \log f(\check{x}) \in \mathbb{R}$. Hence, $\hat{x} = g(e^{\hat{z}}), \check{x} = g(e^{\check{z}})$ and $f(\hat{x}) = e^{\hat{z}}, f(\check{x}) = e^{\check{z}}$ from which it follows that, for all $\hat{z}, \check{z} \in \mathbb{R}$ and $\mu \in (0,1)$, one has

$$\begin{aligned} (1-\mu)g(e^{\hat{z}}) + \mu g(e^{\check{z}}) &\leq g\big(e^{(1-\mu)\hat{z}+\mu\check{z}}\big) \quad f \text{ strictly increasing} \\ (1-\mu)g(e^{\hat{z}}) + \mu g(e^{\check{z}}) &\geq g\big(e^{(1-\mu)\hat{z}+\mu\check{z}}\big) \quad f \text{ strictly decreasing}\,. \end{aligned}$$

This proves one direction of the theorem. Reversing the order of the reasoning proves the converse. The proof for the strictly convex case is identical except that strict inequalities are used and strict monotonicity is utilized.

In the case of twice continuously differentiable functions, we have the following relationship between f and g.

Theorem B.29. *Suppose that* $f : \mathrm{D} \to \mathbb{R}_{++}$ *and* $g : \mathbb{R}_{++} \to \mathrm{D}$ *are twice continuously differentiable. Then,* f *is log-convex if and only if*

$$\begin{aligned} 0 &\le g'(x) + xg''(x) \quad f \textit{ strictly decreasing} \\ 0 &\ge g'(x) + xg''(x) \quad f \textit{ strictly increasing} \end{aligned} \tag{B.17}$$

for all $x > 0$.

Proof. By Theorem B.28, f is log-convex on D if and only if $g_e(x) = g(e^x)$ is either convex on $\mathbb{R}$ or concave on $\mathbb{R}$ depending on whether f is strictly decreasing or strictly increasing. Taking the second derivative of $g_e(x)$ yields $g_e''(x) = e^x(g''(e^x)e^x + g'(e^x))$ for all $x \in \mathbb{R}$. Thus, by Theorem B.18, f is log-convex if and only if

$$\begin{aligned} 0 &\le g'(e^x) + e^x g''(e^x) \quad f \text{ strictly decreasing} \\ 0 &\ge g'(e^x) + e^x g''(e^x) \quad f \text{ strictly increasing} \end{aligned}$$

for all $x \in \mathbb{R}$. Since $\mathbb{R} \to \mathbb{R}_{++} : x \to e^x$ is bijective and $e^x > 0$ for all $x \in \mathbb{R}$, this is equivalent to (B.17).

B.4 Convergence of Gradient Projection Algorithms

Here we present some standard results that are utilized in Chapt. 6 to prove global convergence of the power control algorithm. For a thorough treatment of the convex optimization theory, the reader is referred to [83, 88, 11].

Suppose that $f : \mathbb{R}^n \to \mathbb{R}$ attains its minimum over $\mathrm{D} \subset \mathbb{R}^n$. Throughout this section, it is assumed that f is continuously differentiable and $\mathrm{D} \subset \mathbb{R}^n$ is a *nonempty, closed, and convex set.* The first result proves necessary and sufficient conditions for a vector $\mathbf{x} \in \mathrm{D}$ to be optimal.

Theorem B.30. *Let* $f : \mathrm{D} \to \mathbb{R}$ *be given.*

(i) Suppose that $\mathbf{x} \in \mathrm{D}$ *is a local minimum of* f *over* D. *Then,*

$$\forall_{\mathbf{z} \in \mathrm{D}} \; \nabla f(\mathbf{x})^T(\mathbf{z} - \mathbf{x}) \ge 0 \tag{B.18}$$

(ii) If f *is convex, then (B.18) is also sufficient for* $\mathbf{x} \in \mathrm{D}$ *to minimize* f *over* D.

Proof. Let $\mathbf{x} \in \mathrm{D}$ be a local minimum of f over D. Suppose that $\nabla f(\mathbf{x})^T(\mathbf{z} - \mathbf{x}) < 0$ for some $\mathbf{z} \in \mathrm{D}$. By the mean value theorem, for every $\xi \in [0, 1]$, there exists $s \in [0, 1]$ such that $f(\mathbf{x}+\xi(\mathbf{z}-\mathbf{x})) = f(\mathbf{x})+\xi \nabla f\big(\mathbf{x}+s\xi(\mathbf{z}-\mathbf{x})\big)^T(\mathbf{z}-\mathbf{x})$. Since ∇f is continuous and $\nabla f(\mathbf{x})^T(\mathbf{z} - \mathbf{x}) < 0$ (by assumption), we must have $\nabla f(\mathbf{x} + s\xi(\mathbf{z} - \mathbf{x}))^T(\mathbf{z} - \mathbf{x}) < 0$ for sufficiently small $\xi > 0$. Therefore, $f(\mathbf{x} + \xi(\mathbf{z} - \mathbf{x})) < f(\mathbf{x})$ where $\mathbf{x} + \xi(\mathbf{z} - \mathbf{x})$ is feasible for all $\xi \in [0, 1]$ since D is convex. This , however, contradicts the local optimality of $\mathbf{x}$, and hence proves (i).

(ii) Considering Theorem B.17 shows that if f is convex, then $f(\mathbf{z}) \geq f(\mathbf{x}) + \nabla f(\mathbf{x})^T(\mathbf{z} - \mathbf{x})$ for all $\mathbf{z} \in \mathrm{D}$. Thus, by (B.18), it follows that $f(\mathbf{z}) \geq f(\mathbf{x})$ for all $\mathbf{z} \in \mathrm{D}$.

Definition B.31 (Stationary Point). *Any vector (point) satisfying (B.18) is referred to as a stationary point of f.*

Note that if $\mathbf{x}$ in (B.18) is an interior point of D or $\mathrm{D} = \mathbb{R}^n$, then the optimality condition reduces to $\nabla f(\mathbf{x}) = 0$. An important component of the gradient projection algorithm is the projection on a closed convex subset of $\mathbb{R}^n$. We prove that the projection is well defined and unique.

Theorem B.32 (Projection Theorem). *For all $\mathbf{y} \in \mathbb{R}^n$, a vector $\Pi_{\mathrm{D}}[\mathbf{y}] \in \mathrm{D}$ is said to be the projection of $\mathbf{y}$ on $\mathrm{D} \subset \mathbb{R}^n$ if*

$$\Pi_{\mathrm{D}}[\mathbf{y}] = \arg\min_{\mathbf{x} \in \mathrm{D}} \|\mathbf{y} - \mathbf{x}\|_2^2 \,. \tag{B.19}$$

The minimum exists and is unique. Moreover, given some $\mathbf{y} \in \mathbb{R}^n$, $\Pi_{\mathrm{D}}[\mathbf{y}]$ is the unique solution to (B.19) if and only if

$$\forall_{\mathbf{x} \in \mathrm{D}} \left(\mathbf{y} - \Pi_{\mathrm{D}}[\mathbf{y}]\right)^T \left(\mathbf{x} - \Pi_{\mathrm{D}}[\mathbf{y}]\right) \leq 0 \,. \tag{B.20}$$

Proof. By assumption, D is closed but not necessarily bounded. However, the problem in (B.19) is equivalent to minimizing the same metric over all $\mathbf{x} \in \mathrm{D}$ such that $\|\mathbf{y} - \mathbf{x}\|_2 \leq \|\mathbf{y} - \mathbf{z}\|_2$ for some arbitrary $\mathbf{z} \in \mathrm{D}$. Since this is a compact set and the norm is a continuous function of the vector elements, it follows that the minimum exists. Moreover, the minimum is unique since $\|\mathbf{y} - \mathbf{x}\|_2^2$ is a strictly convex function of $\mathbf{x} \in \mathrm{D}$. The proof of (B.20) proceeds along the same lines as the proof of (B.18).

Now let us introduce the notion of Lipschitz continuity.

Definition B.33 (Lipschitz Continuity Condition). *Let $\mathbb{R}^n$ be Euclidean space, and let $\mathrm{D} \subset \mathbb{R}^n$. A function $f : \mathrm{D} \to \mathbb{R}^n$ is said to satisfy the Lipschitz continuity condition (or is called Lipschitz continuous) if there exists a constant $M > 0$ such that*

$$\|f(\mathbf{x}) - f(\mathbf{y})\|_2 \leq M\|\mathbf{x} - \mathbf{y}\|_2 \tag{B.21}$$

for all $\mathbf{x}, \mathbf{y} \in \mathrm{D}$.

We point out that this definition can be extended to functions between arbitrary metric spaces. The following lemma is also known as the descent lemma and is a key ingredient in proving the convergence of gradient methods to a stationary point

Lemma B.34. *If* $f : \mathrm{D} \to \mathbb{R}$ *is continuously differentiable and its gradient is Lipschitz continuous with some Lipschitz constant* $M > 0$*, then*

$$f(\mathbf{x}+\mathbf{y}) \le f(\mathbf{x}) + \mathbf{y}^T \nabla f(\mathbf{x}) + \frac{M}{2}\|\mathbf{y}\|_2^2 \tag{B.22}$$

for all $\mathbf{x}, \mathbf{y} \in \mathrm{D}$.

Proof. We have

$$\begin{aligned} f(\mathbf{x}+\mathbf{y}) - f(\mathbf{x}) &= \int_0^1 \mathbf{y}^T \nabla f(\mathbf{x}+\mu\mathbf{y}) d\mu \\ &= \int_0^1 \mathbf{y}^T \big(\nabla f(\mathbf{x}) + \nabla f(\mathbf{x}+\mu\mathbf{y}) - \nabla f(\mathbf{x})\big) d\mu \\ &\le \mathbf{y}^T \nabla f(\mathbf{x}) + \|\mathbf{y}\|_2 \int_0^1 \|\nabla f(\mathbf{x}+\mu\mathbf{y}) - \nabla f(\mathbf{x})\|_2 \, d\mu \\ &\le \mathbf{y}^T \nabla f(\mathbf{x}) + \|\mathbf{y}\|_2^2 M \int_0^1 \mu \, d\mu = \mathbf{y}^T \nabla f(\mathbf{x}) + \|\mathbf{y}\|_2^2 \frac{M}{2} \end{aligned}$$

for all $\mathbf{x}, \mathbf{y} \in \mathrm{D}$.

We use this lemma to prove the convergence of the gradient projection algorithm:

$$\mathbf{x}(n+1) = T(\mathbf{x}(n)), \quad \mathbf{x}(0) \in \mathrm{D} \tag{B.23}$$

where $T : \mathbb{R}^n \to \mathrm{D}$ is defined to be

$$T(\mathbf{x}) := \Pi_{\mathrm{D}}\big[\mathbf{x} - \delta \nabla f(\mathbf{x})\big] \tag{B.24}$$

and δ is a positive constant (sufficiently small).

Theorem B.35. *Let* $f : \mathrm{D} \to \mathbb{R}$ *be continuously differentiable and bounded below on a nonempty, closed and convex set* $\mathrm{D} \subset \mathbb{R}^n$. *Suppose that* $\nabla f : \mathrm{D} \to \mathbb{R}^n$ *is Lipschitz continuous with the Lipschitz constant* $M > 0$. *Let* $0 < \delta < 2/M$ *and* $\mathbf{x} \in \mathrm{D}$ *be arbitrary. Then,*

(i) $F(T(\mathbf{x})) \le F(\mathbf{x}) - (1/\delta - M/2)\|T(\mathbf{x}) - \mathbf{x}\|_2^2$.
(ii) $T(\mathbf{x}) = \mathbf{x}$ *if and only if* $\mathbf{x}$ *is a stationary point. Moreover, if* f *is convex, we have* $T(\mathbf{x}) = \mathbf{x}$ *if and only if* $\mathbf{x}$ *minimizes* f *over* D.

Proof. Note that $T(\mathbf{x})$ is the projection of $\mathbf{x} - \delta\nabla f(\mathbf{x})$ on D. Hence, it follows from (B.20) that $\forall_{\mathbf{z}\in\mathrm{D}}\ (\mathbf{z} - T(\mathbf{x}))^T(\mathbf{x} - \delta\nabla f(\mathbf{x}) - T(\mathbf{x})) \le 0$. Particularizing this to $\mathbf{z} = \mathbf{x} \in \mathrm{D}$ yields $(T(\mathbf{x}) - \mathbf{x})^T \nabla f(\mathbf{x}) \le -1/\delta\|T(\mathbf{x}) - \mathbf{x}\|_2^2$. On the other hand, considering Lemma B.34 gives $(T(\mathbf{x}) - \mathbf{x})^T \nabla f(\mathbf{x}) \ge f(T(\mathbf{x})) - f(\mathbf{x}) - M/2\|T(\mathbf{x}) - \mathbf{x}\|_2^2$. Therefore, $f(T(\mathbf{x})) \le f(\mathbf{x}) - (1/\delta - M/2)\|T(\mathbf{x}) - \mathbf{x}\|_2^2$ which proves (i).

(ii) By (B.24), $T(\mathbf{x})$ is the projection of $\mathbf{x} - \delta\nabla f(\mathbf{x})$ on D. Therefore, if $T(\mathbf{x}) = \mathbf{x}$, (B.20) implies that $\delta\nabla f(\mathbf{x})^T(\mathbf{z} - \mathbf{x}) \ge 0$ for all $\mathbf{z} \in \mathrm{D}$ with $\delta > 0$.

Conversely, if $\delta \nabla f(\mathbf{x})^T(\mathbf{z}-\mathbf{x}) \geq 0$ for all $\mathbf{z} \in \mathrm{D}$, then $(\mathbf{x}-\nabla f(\mathbf{x})-\mathbf{x})^T(\mathbf{z}-\mathbf{x}) \leq 0$, from which we have $T(\mathbf{x}) = \mathbf{x}$. The assertion for a convex function follows from (ii) in Theorem B.30.

Now let $\{\mathbf{x}(n)\}$ be the sequence generated by (B.23). Provided that $0 < \delta < 2/M$, it follows from (i) that $\{f(\mathbf{x}(n))\}$ is nonincreasing, and therefore this sequence converges since f is bounded below on D. From this, the left-hand side of

$$f(\mathbf{x}(n+1)) - f(\mathbf{x}(n)) \leq \Big(\frac{M}{2} - \frac{1}{\delta}\Big)\|T(\mathbf{x}(n)) - \mathbf{x}(n)\|_2^2$$

tends to zero, whereas the right-hand side is nonpositive if $0 < \delta < 2/M$ so that $\|T(\mathbf{x}(n)) - \mathbf{x}(n)\|_2^2$ must tend to zero as well. Hence, if $\mathbf{x}^*$ is a limit point of $\{\mathbf{x}(n)\}$, the sequence $T(\mathbf{x}(n))$ converges to $\mathbf{x}^*$. Moreover, by continuity of T (T is continuous since it is a concatenation of continuous maps), we must have $T(\mathbf{x}^*) = \mathbf{x}^*$. Finally, if f is convex, then the sequence $\{\mathbf{x}(n)\}$ generated by (B.23) converges to some $\mathbf{x}^*$ that minimizes f over D. We summarize this in a theorem.

Theorem B.36. *Suppose that the conditions of Theorem B.35 are satisfied. Then, provided that* $0 < \delta < 2/M$, *the sequence* $\{\mathbf{x}(n)\}$ *generated by (B.23) converges to some* $\mathbf{x}^*$ *satisfying* $(\mathbf{z}-\mathbf{x}^*)\nabla f(\mathbf{x}^*) \geq 0$ *for all* $\mathbf{z} \in \mathrm{D}$. *If* $f : \mathrm{D} \to \mathbb{R}$ *is a convex function, then* $\mathbf{x}^*$ *minimizes* f *over* D.

Finally we point out that if $f : \mathrm{D} \to \mathbb{R}$ is strongly convex (see Sect. B.2.1), then the rate of convergence is geometric (Definition 6.27). The proof can be found in [88].

References

1. Bertsekas DP, Gallager RG. Data Networks. Prentice Hall, Englewood Cliffs; 1992.
2. Seneta E. Non-Negative Matrices and Markov Chains. Springer, Berlin; 1981.
3. Horn RA, Johnson CR. Matrix Analysis. Cambridge University Press; 1985.
4. Gantmacher FR. Matritzentheorie. Springer, Berlin; 1986. (German translation of the Russian original).
5. Serre D. Matrices: Theory and Applications. Springer, Berlin; 2001.
6. Arnold L, Gundlach V, Demetrius L. Evolutionary Formalism for Products of positive Random Matrices. Ann Appl Probab. 1994;4(3):859–901.
7. Horn RA, Johnson CR. Topics in Matrix Analysis. Cambridge University Press; 1985.
8. Ahlswede R, Gacs P. Spreading of sets in product spaces and hypercontraction of the Markov operator. Ann Prob. 1976;4(6):925–939.
9. Cover TM, Thomas JA. Elements of Information Theory. Wiley Series in Telecommunications. John Wiley, New York; 1991.
10. Lesniewski A, Ruskai MB. Monotone Riemannian Metrics and Relative Entropy on Non-Commutative Probability Spaces. J Math Phys. 1999;40:5702–5724.
11. Boyd S, Vandenberghe L. Convex Optimization. Cambridge University Press; 2004.
12. Meyer CD. Matrix Analysis and Applied Linear Algebra. SIAM, Philadelphia; 2000.
13. Boche H, Stanczak S. On Systems of Linear Equations with Nonnegative Coefficients. Appl Alg Eng Comm Comp. 2004;14(6):397–414.
14. Boche H, Stanczak S. Convexity of Some Feasible QoS Regions and Asymptotic Behavior of the Minimum Total Power in CDMA Systems. IEEE Trans Commun. 2004;52(12):2190–2197.
15. Boche H, Stanczak S. Log-Convexity of the Minimum Total Power in CDMA Systems with Certain Quality-Of-Service Guaranteed. IEEE Trans Inform Theory. 2005;51(1):374–381.
16. Stanczak S, Boche H. The Infeasible SIR Region Is Not a Convex Set. IEEE Trans Commun. 2006;To appear.
17. Stanczak S, Boche H. Towards a better understanding of the QoS tradeoff in multiuser multiple antenna systems. In: Smart Antennas - State-of-the-Art. EURASIP Book Series on Signal Processing and Communications. Hindawi Publishing Corporation; 2005. p. 521–543.

18. Stanczak S, Boche H, Wiczanowski M. Towards a Better Understanding of Medium Access Control for Multiuser Beamforming Systems. In: Proc. IEEE Wireless Communications and Networking Conference (WCNC), New Orleans, LA USA; 2005. .
19. Boche H, Wiczanowski M, Stanczak S. Unifying View on Min-Max Fairness, Max-Min Fairness, and Utility Optimization in Cellular Networks. EURASIP J Wireless Commun and Net. 2006;To appear.
20. Imhof L, Mathar R. Capacity Regions and Optimal Power Allocation for CDMA Cellular Radio. IEEE Trans Inform Theory. 2005;51(6):2011–2019.
21. Imhof L, Mathar R. The Geometry of the Capacity Region for CDMA Systems with General Power Constraints. IEEE Trans Wireless Commun. 2005;4(5).
22. Stanczak S, Boche H. Strict Log-Convexity of the Minimum Power Vector. In: Proc. IEEE International Symposium on Information Theory (ISIT), Seattle, WA, USA; 2006. .
23. Kirkland SJ, Neumann M, Ormes N, Xu J. On the Elasticity of the Perron Root of a Nonnegative Matrix. SIAM J Matrix Anal Appl. 2002;24(2):454–464.
24. Friedland S, Karlin S. Some Inequalities for the Spectral Radius of Non-Negative Matrices and Applications. Duke Math J. 1975;42(3):459–490.
25. Kingman JFC. A Convexity Property of Positive Matrices. Quart, J Math Oxford Ser. 1961;12(2):283–284.
26. Catrein D, Imhof L, Mathar R. Power Control, Capacity, and Duality of Up- and Downlink in Cellular CDMA Systems. IEEE Trans Commun. 2004;52(10):1777–1785.
27. Cohen JE, Friedland S, Kato T, Kelly FP. Eigenvalue Inequalities for Products of Matrix Exponentials. Linear Algebra Appl. 1982;45:55–95.
28. Friedland S. Convex Spectral Functions. Linear Multilin Alg. 1981;9:293–316.
29. Cohen JE. Random Evolutions and the Spectral Radius of a Non-Negative matrix. Math Proc Cambridge Philos Soc. 1979;86:345–350.
30. Deutsch E, Neumann M. Derivatives of the Perron Root at an Essentially Nonnegative Matrix and the Group Inverse of an M-Matrix. J Math Anal Appl. 1984;I-29(102):1–29.
31. Elsner L. On Convexity Properties of the Spectral Radius of Nonnegative Matrices. Linear Algebra Appl. 1984;61:31–35.
32. Elsner L. Über Eigenwereinschliessungen mit Hilfe von Gerschgorin-Kreisen. Z Angew Math Mech. 1970;50:381–384.
33. Goldsmith AJ, Wicker SB. Design Challenges for Energy-Constrained Ad-Hoc Wireless Networks. IEEE Wireless Commun Mag. 2002;9:8–27.
34. Zander J. Distributed Cochannel Interference Control in Cellular Radio Systems. IEEE Trans Veh Technol. 1992;41:305–311.
35. Zander J. Performance of Optimum Transmitter Power Control in Cellular Radio Systems. IEEE Trans Veh Technol. 1992;41(1):57–62.
36. Foschini GJ, Miljanic Z. A Simple Distributed Autonomous Power Control Algorithm and its Convergence. IEEE Trans on Vehicul Technol. 1993;42(4):641–646.
37. Yates RD. A Framework for Uplink Power Control in Cellular Radio Systems. IEEE J Select Areas Commun. 1995;13(7):1341–1347.
38. Yates RD, Huang CY. Integrated Power Control and Base Station Assignment. IEEE Trans Veh Technol. 1995;44(3):638–644.

39. Bambos N. Toward Power-Sensitive Network Architectures in Wireless COmmunications: Concepts, Issues, and Design Aspects. IEEE Personal Commun Mag. 1998;5:50–59.
40. ElBatt T, Ephremides A. Joint Scheduling and Power Control for Wireless Ad Hoc Networks. IEEE Trans Wireless Commun. 2004;3(1):74–85.
41. Wu Q. Optimum Transmitter Power Control in Cellular Systems with Heterogeneous SIR Thresholds. IEEE Trans Veh Technol. 2000;49(4):1424–1429.
42. Bambos N, Chen SC, Pottie GJ. Channel Access Algorithms with Active Link Protection for Wireless Communication Networks with Power Control. IEEE/ACM Trans Networking. 2000;8(5):583–597.
43. Feiten A, Mathar R. Optimal Power Control for Multiuser CDMA Channels. In: Proc. 2005 IEEE International Symposium on Information Theory (ISIT), Adelaide, Australia; 2005. .
44. Goodman D, Mandayam N. Power Control for Wireless Data. IEEE Personal Commun Mag. 2000;7:48–54.
45. Saraydar CU, Mandayam NB, Goodman DJ. Efficient Power Control via Pricing in Wireless Data Networks. IEEE Trans Commun. 2002;50(2):291–303.
46. Xiao M, Schroff NB, Chong EKP. A Utility-Based Power Control Scheme in Wireless Cellular Systems. IEEE/ACM Trans Networking. 2003;11(2):210–221.
47. Johansson M, Xiao L, Boyd S. Simultaneous routing and resource allocation in CDMA wireless data networks. In: Proc. IEEE International Conference on Communications, Anchorage, Alaska; 2003. .
48. ONeill D, Julian D, Boyd S. Seeking Foschini's Genie: Optimal Rates and Powers in Wireless Networks. IEEE Trans Veh Technol. 2003 (submitted);Available from http://www.stanford.edu/ boyd/.
49. Chiang M. To Layer or Not To Layer: Balancing Transport and Physical Layers in Wireless Multihop Networks. In: Proc. 23rd IEEE Conference on Computer Communications (INFOCOM), Hong Kong; 2004. .
50. Chiang M. Balancing Transport and Physical Layers in Wireless Multihop Networks: Jointly Optimal Congestion Control and Power Control. IEEE J Select Areas Commun. 2005;23(1):104–116.
51. Huang J, Berry R, Honig ML. A Game Theoretic Analysis of Distributed Power Control for Spread Spectrum Ad Hoc Networks. In: Proc. IEEE International Symposium on Information Theory (ISIT), Adelaide, Australia; 2005. .
52. Huang J, Berry R, Honig ML. Distributed Interference Compensation for Multi-Channel Wireless Networks. In: Proc. 43nd Annual Allerton Conference on Communication, Control and Computing, Monticello, IL, USA; 2005. .
53. Stanczak S, Wiczanowski M. Distributed Fair Power Control for Wireless Networks: Objectives and Algorithms. In: Proc. the 43rd Annual Allerton Conference on Communications, Control, and Computing; 2005. Invited.
54. Stanczak S, Wiczanowski M, Boche H. Distributed Power Control for Optimizing a Weighted Sum of QoS Parameter Values. In: Proc. IEEE Global Telecommunications Conference (GLOBECOM), St. Louis, MO, USA; 2005. .
55. Wiczanowski M, Stanczak S, Boche H. Distributed Optimization and Duality in QoS Control for Wireless Best-Effort Traffic. In: Proc. 39th Asilomar Conference on Signals, Systems and Computers, Monterey, CA, USA; 2005. .
56. Kelly FP, Maulloo AK, Tan DKH. Rate Control for Communication Networks: Shadow Prices, Proportional Fairness and Stability. J Oper Res Soc. 1998;49(3):237–252.

57. Mo J, Walrand J. Fair End-to-End Window-Based Congestion Control. IEEE/ACM Trans on Networking. 2000;8(5):556–567.
58. Tang A, Wang J, Low S. Is Fair Allocation Always Inefficient? In: Proc. 23rd IEEE Conference on Computer Communications (INFOCOM), Hong Kong; 2004.
59. Helleseth T, Kumar PV. Sequences with Low Correlation. In: Handbook of Coding Theory. vol. 2. Elsevier Science, Amsterdam; 1998. p. 1765–1853.
60. Verdu S. Multiuser Detection. Cambridge University Press; 1998.
61. Proakis JG. Digital Communications. 3rd ed. McGraw Hill, New York; 1995.
62. Tassiulas L, Sarkar S. Maxmin Fair Scheduling in Wireless Ad Hoc Networks. IEEE J Select Areas Commun. 2005;23(1):163–173.
63. Schubert M, Boche H. Iterative Multiuser Uplink and Downlink Beamforming under SINR Constraints. IEEE Trans Signal Processing. 2005;53(7):2324– 2334.
64. Boche H, Schubert M. Duality Theory for Uplink and Downlink Multiuser Beamforming. In: Smart Antennas–State-of-the-Art. EURASIP Book Series on Signal Processing and Communications. Hindawi Publishing Corporation; 2005. p. 545–575.
65. Massoulie L, Roberts J. Bandwidth Sharing: Objectives and Algorithms. IEEE/ACM Trans on Networking. 2002;10(3):320–328.
66. Hahne E. Round-Robin Scheduling for Fair Flow Control in Data Communication Networks. MIT, Deptartment of Electrical Engineering and Computer Science, Cambridge, MA; 1986.
67. Hahne E. Round-robin Scheduling for Max-Min Fairness in Data Networks. IEEE J Select Areas Commun. 1991 Sep;9(7):1024–1039.
68. Charny A, Clark D, Jain R. Congestion Control with Explicit Rate Indication. In: Proc. IEEE International Conference on Communications (ICC); 1995. .
69. Ji H, Huang CY. Non-Cooperative Uplink Power Control in Cellular Radio Systems. Wireless Networks. 1998;4:233–240.
70. Kozat UC, Koutsopoulos I, Tassiulas L. A Framework for Cross-layer Design of Energy-efficient Communication with QoS Provisioning in Multi-hop Wireless Networks. In: Proc. 23rd IEEE Conference on Computer Communications (INFOCOM), Hong Kong; 2004. .
71. Neely MJ, Modiano E, Rohrs CE. Dynamic Power Allocation and Routing for Time Varying Wireless Networks. In: Proc. 22nd IEEE Conference on Computer Communications (INFOCOM), San Francisco, CA, USA; 2003. .
72. Neely MJ. Dynamic Power Allocation and Routing for Satellite and Wireless Networks with Time Varying Channels. Massachusetts Institute of Technology, LIDS; 2003.
73. Low S, Peterson L, Wang L. Understanding TCP Vegas: A Duality Model. Journal of the ACM (JACM). 2002;49(2):207–235.
74. Yi Y, Shakkottai S. Hop-by-hop Congestion Control over a Wireless Multi-hop Network. In: Proc. 23rd IEEE Conference on Computer Communications (INFOCOM), Hong Kong; 2004. .
75. Sarkar S, Tassiulas L. End-to-End Bandwidth Guarantees Through Fair Local Spectrum Share in Wireless Ad Hoc Networks. IEEE Trans Automat Contr. 2005;50(9):1246–1259.
76. Nandagopal T, Kim T, Gao X, Bharghavan V. Achieving MAC Layer Fairness in Wireless Packet Networks. In: Proc. ACM Mobicom, Boston, MA, USA; 2000. p. 87–98.

77. Tassiulas L, Ephremides A. Stability Properties of Constrained Queueing Systems and Scheduling Policies for Maximum Throughput in Multihop Radio Networks. IEEE Trans Automat Contr. 1992;37(12):1936–1948.
78. Tassiulas L, Ephremides A. Jointly optimal routing and scheduling in packet radio networks. IEEE Trans Inform Theory. 1992;38(1):165–168.
79. Tse D, Hanly S. Linear Multiuser Receivers: Effective Interference, Effective Bandwidth and User Capacity. IEEE Trans Inform Theory. 1999;45(2):641–657.
80. Verdu S. On Channel Capacity per Unit Cost. IEEE Trans Inform Theory. 1990;36(5):1019–1030.
81. Verdu S. Recent Results on the Capacity of Wideband Channels in the Low-Power Regime. IEEE Wireless Commun Mag. 2002;9:40–45.
82. Schubert M, Boche H. QoS-Based Resource Allocation and Transceiver Optimization. Foundation and Trends in Communications and Information Theory. 2006;2(6).
83. Bertsekas DP, Tsitsiklis JN. Parallel and Distributed Computation. Prentice Hall, Englewood Cliffs; 1989.
84. Hardy G, Littlewood JE, Polya G. Inequalities. 2nd ed. Cambridge University Press; 1952.
85. Boche H, Stanczak S. Log-Concavity of SIR and Characterization of the Feasible SIR Region for CDMA Channels. In: Proc. 37th Asilomar Conference on Signals, Systems, and Computers, Monterey, CA, USA; 2003. .
86. Boche H, Wiczanowski M, Stanczak S. Characterization of Optimal Resource Allocation in Cellular Networks. In: Proc. 5th IEEE Workshop on Signal Processing Advances in Wireless Communications, Lisboa, Portugal; 2004. .
87. Rudin W. Functional Analysis. 2nd ed. McGraw Hill, New York; 1991.
88. Bertsekas DP. Nonlinear Programming. Athena Scientific, Belmont, Massachusetts; 1995.
89. Ortega JM, Rheinboldt WC. Iterative Solution of Nonlinear Equations in Several Variables. Classics in Applied Mathematics 30, SIAM, Philadelphia; 2000.
90. Giannakis GB, Hua Y, Stoica P, Tong L, editors. Signal Processing Advances in Wireless and Mobile Communications. Prentice Hall, Englewood Cliffs; 2000.
91. Poor HV. An Introduction to Signal Detection and Estimation. Springer, Berlin; 1998.
92. Kushner HJ. Stochastic Approximation and Recursive Algorithms and Applications. Springer, Berlin; 2003.
93. Robinson H, Monro S. A Stochastic Approximation Method. Ann Math Statist. 1951;22:400–407.
94. Ljung L. Analysis of Recursive Stochastic Algorithms. IEEE Trans Automat Contr. 1977;AC-22(4):551–575.
95. Polyak BT, Juditsky AB. Acceleration of Stochastic Approximation by Averaging. SIAM J Control Optim. 1992;30:838–855.
96. Williams D. Probability with Martingales. Cambridge University Press; 1991.
97. Mitrinovic DS. Analytic Inequalities. Springer, Berlin; 1970.
98. Rudin W. Principles of Mathematical Analysis. McGraw-Hill, New York; 1976.

Printing: Mercedes-Druck, Berlin
Binding: Stein + Lehmann, Berlin

Lecture Notes in Computer Science

For information about Vols. 1–4188

please contact your bookseller or Springer

Vol. 4272: P. Havinga, M. Lijding, N. Meratnia, M. Wegdam (Eds.), Smart Sensing and Context. XI, 267 pages. 2006.

Vol. 4270: H. Zha, Z. Pan, H. Thwaites, A.C. Addison, M. Forte (Eds.), Interactive Technologies and Sociotechnical Systems. XVI, 547 pages. 2006.

Vol. 4269: R. State, S. van der Meer, D. O'Sullivan, T. Pfeifer (Eds.), Large Scale Management of Distributed Systems. XIII, 282 pages. 2006.

Vol. 4268: G. Parr, D. Malone, M. Ó Foghlú (Eds.), Autonomic Principles of IP Operations and Management. XIII, 237 pages. 2006.

Vol. 4267: A. Helmy, B. Jennings, L. Murphy, T. Pfeifer (Eds.), Autonomic Management of Mobile Multimedia Services. XIII, 257 pages. 2006.

Vol. 4266: H. Yoshiura, K. Sakurai, K. Rannenberg, Y. Murayama, S. Kawamura (Eds.), Advances in Information and Computer Security. XIII, 438 pages. 2006.

Vol. 4265: N. Lavrač, L. Todorovski, K.P. Jantke (Eds.), Discovery Science. XIV, 384 pages. 2006. (Sublibrary LNAI).

Vol. 4264: J.L. Balcázar, P.M. Long, F. Stephan (Eds.), Algorithmic Learning Theory. XIII, 393 pages. 2006. (Sublibrary LNAI).

Vol. 4263: A. Levi, E. Savas, H. Yenigün, S. Balcisoy, Y. Saygin (Eds.), Computer and Information Sciences – ISCIS 2006. XXIII, 1084 pages. 2006.

Vol. 4260: Z. Liu, J. He (Eds.), Formal Methods and Software Engineering. XII, 778 pages. 2006.

Vol. 4257: I. Richardson, P. Runeson, R. Messnarz (Eds.), Software Process Improvement. XI, 219 pages. 2006.

Vol. 4256: L. Feng, G. Wang, C. Zeng, R. Huang (Eds.), Web Information Systems Engineering – WISE 2006 Workshops. XIV, 320 pages. 2006.

Vol. 4255: K. Aberer, Z. Peng, E.A. Rundensteiner, Y. Zhang, X. Li (Eds.), Web Information Systems Engineering – WISE 2006. XIV, 563 pages. 2006.

Vol. 4254: T. Grust, H. Höpfner, A. Illarramendi, S. Jablonski, M. Mesiti, S. Müller, P.-L. Patranjan, K.-U. Sattler, M. Spiliopoulou (Eds.), Current Trends in Database Technology – EDBT 2006. XXXI, 932 pages. 2006.

Vol. 4253: B. Gabrys, R.J. Howlett, L.C. Jain (Eds.), Knowledge-Based Intelligent Information and Engineering Systems, Part III. XXXII, 1301 pages. 2006. (Sublibrary LNAI).

Vol. 4252: B. Gabrys, R.J. Howlett, L.C. Jain (Eds.), Knowledge-Based Intelligent Information and Engineering Systems, Part II. XXXIII, 1335 pages. 2006. (Sublibrary LNAI).

Vol. 4251: B. Gabrys, R.J. Howlett, L.C. Jain (Eds.), Knowledge-Based Intelligent Information and Engineering Systems, Part I. LXVI, 1297 pages. 2006. (Sublibrary LNAI).

Vol. 4249: L. Goubin, M. Matsui (Eds.), Cryptographic Hardware and Embedded Systems - CHES 2006. XII, 462 pages. 2006.

Vol. 4248: S. Staab, V. Svátek (Eds.), Engineering Knowledge in the Age of the Semantic Web. XIV, 400 pages. 2006. (Sublibrary LNAI).

Vol. 4247: T.-D. Wang, X. Li, S.-H. Chen, X. Wang, H. Abbass, H. Iba, G. Chen, X. Yao (Eds.), Simulated Evolution and Learning. XXI, 940 pages. 2006.

Vol. 4245: A. Kuba, L.G. Nyúl, K. Palágyi (Eds.), Discrete Geometry for Computer Imagery. XIII, 688 pages. 2006.

Vol. 4244: S. Spaccapietra (Ed.), Journal on Data Semantics VII. XI, 267 pages. 2006.

Vol. 4243: T. Yakhno, E.J. Neuhold (Eds.), Advances in Information Systems. XIII, 420 pages. 2006.

Vol. 4241: R.R. Beichel, M. Sonka (Eds.), Computer Vision Approaches to Medical Image Analysis. XI, 262 pages. 2006.

Vol. 4239: H.Y. Youn, M. Kim, H. Morikawa (Eds.), Ubiquitous Computing Systems. XVI, 548 pages. 2006.

Vol. 4238: Y.-T. Kim, M. Takano (Eds.), Management of Convergence Networks and Services. XVIII, 605 pages. 2006.

Vol. 4237: H. Leitold, E. Markatos (Eds.), Communications and Multimedia Security. XII, 253 pages. 2006.

Vol. 4236: L. Breveglieri, I. Koren, D. Naccache, J.-P. Seifert (Eds.), Fault Diagnosis and Tolerance in Cryptography. XIII, 253 pages. 2006.

Vol. 4234: I. King, J. Wang, L. Chan, D. Wang (Eds.), Neural Information Processing, Part III. XXII, 1227 pages. 2006.

Vol. 4233: I. King, J. Wang, L. Chan, D. Wang (Eds.), Neural Information Processing, Part II. XXII, 1203 pages. 2006.

Vol. 4232: I. King, J. Wang, L. Chan, D. Wang (Eds.), Neural Information Processing, Part I. XLVI, 1153 pages. 2006.

Vol. 4231: J. F. Roddick, R. Benjamins, S.S.-S. Cherfi, R. Chiang, C. Claramunt, R. Elmasri, F. Grandi, H. Han, M. Hepp, M. Hepp, M. Lytras, V.B. Mišić, G. Poels, I.-Y. Song, J. Trujillo, C. Vangenot (Eds.), Advances in Conceptual Modeling - Theory and Practice. XXII, 456 pages. 2006.

Vol. 4229: E. Najm, J.F. Pradat-Peyre, V.V. Donzeau-Gouge (Eds.), Formal Techniques for Networked and Distributed Systems - FORTE 2006. X, 486 pages. 2006.

Vol. 4228: D.E. Lightfoot, C.A. Szyperski (Eds.), Modular Programming Languages. X, 415 pages. 2006.

Vol. 4227: W. Nejdl, K. Tochtermann (Eds.), Innovative Approaches for Learning and Knowledge Sharing. XVII, 721 pages. 2006.

Vol. 4225: J.F. Martínez-Trinidad, J.A. Carrasco Ochoa, J. Kittler (Eds.), Progress in Pattern Recognition, Image Analysis and Applications. XIX, 995 pages. 2006.

Vol. 4224: E. Corchado, H. Yin, V. Botti, C. Fyfe (Eds.), Intelligent Data Engineering and Automated Learning – IDEAL 2006. XXVII, 1447 pages. 2006.

Vol. 4223: L. Wang, L. Jiao, G. Shi, X. Li, J. Liu (Eds.), Fuzzy Systems and Knowledge Discovery. XXVIII, 1335 pages. 2006. (Sublibrary LNAI).

Vol. 4222: L. Jiao, L. Wang, X. Gao, J. Liu, F. Wu (Eds.), Advances in Natural Computation, Part II. XLII, 998 pages. 2006.

Vol. 4221: L. Jiao, L. Wang, X. Gao, J. Liu, F. Wu (Eds.), Advances in Natural Computation, Part I. XLI, 992 pages. 2006.

Vol. 4219: D. Zamboni, C. Kruegel (Eds.), Recent Advances in Intrusion Detection. XII, 331 pages. 2006.

Vol. 4218: S. Graf, W. Zhang (Eds.), Automated Technology for Verification and Analysis. XIV, 540 pages. 2006.

Vol. 4217: P. Cuenca, L. Orozco-Barbosa (Eds.), Personal Wireless Communications. XV, 532 pages. 2006.

Vol. 4216: M.R. Berthold, R. Glen, I. Fischer (Eds.), Computational Life Sciences II. XIII, 269 pages. 2006. (Sublibrary LNBI).

Vol. 4215: D.W. Embley, A. Olivé, S. Ram (Eds.), Conceptual Modeling - ER 2006. XVI, 590 pages. 2006.

Vol. 4213: J. Fürnkranz, T. Scheffer, M. Spiliopoulou (Eds.), Knowledge Discovery in Databases: PKDD 2006. XXII, 660 pages. 2006. (Sublibrary LNAI).

Vol. 4212: J. Fürnkranz, T. Scheffer, M. Spiliopoulou (Eds.), Machine Learning: ECML 2006. XXIII, 851 pages. 2006. (Sublibrary LNAI).

Vol. 4211: P. Vogt, Y. Sugita, E. Tuci, C. Nehaniv (Eds.), Symbol Grounding and Beyond. VIII, 237 pages. 2006. (Sublibrary LNAI).

Vol. 4210: C. Priami (Ed.), Computational Methods in Systems Biology. X, 323 pages. 2006. (Sublibrary LNBI).

Vol. 4209: F. Crestani, P. Ferragina, M. Sanderson (Eds.), String Processing and Information Retrieval. XIV, 367 pages. 2006.

Vol. 4208: M. Gerndt, D. Kranzlmüller (Eds.), High Performance Computing and Communications. XXII, 938 pages. 2006.

Vol. 4207: Z. Ésik (Ed.), Computer Science Logic. XII, 627 pages. 2006.

Vol. 4206: P. Dourish, A. Friday (Eds.), UbiComp 2006: Ubiquitous Computing. XIX, 526 pages. 2006.

Vol. 4205: G. Bourque, N. El-Mabrouk (Eds.), Comparative Genomics. X, 231 pages. 2006. (Sublibrary LNBI).

Vol. 4204: F. Benhamou (Ed.), Principles and Practice of Constraint Programming - CP 2006. XVIII, 774 pages. 2006.

Vol. 4203: F. Esposito, Z.W. Raś, D. Malerba, G. Semeraro (Eds.), Foundations of Intelligent Systems. XVIII, 767 pages. 2006. (Sublibrary LNAI).

Vol. 4202: E. Asarin, P. Bouyer (Eds.), Formal Modeling and Analysis of Timed Systems. XI, 369 pages. 2006.

Vol. 4201: Y. Sakakibara, S. Kobayashi, K. Sato, T. Nishino, E. Tomita (Eds.), Grammatical Inference: Algorithms and Applications. XII, 359 pages. 2006. (Sublibrary LNAI).

Vol. 4200: I.F.C. Smith (Ed.), Intelligent Computing in Engineering and Architecture. XIII, 692 pages. 2006. (Sublibrary LNAI).

Vol. 4199: O. Nierstrasz, J. Whittle, D. Harel, G. Reggio (Eds.), Model Driven Engineering Languages and Systems. XVI, 798 pages. 2006.

Vol. 4198: O. Nasraoui, O. Zaiane, M. Spiliopoulou, B. Mobasher, B. Masand, P. Yu (Eds.), Advances in Web Minding and Web Usage Analysis. IX, 177 pages. 2006. (Sublibrary LNAI).

Vol. 4197: M. Raubal, H.J. Miller, A.U. Frank, M.F. Goodchild (Eds.), Geographic, Information Science. XIII, 419 pages. 2006.

Vol. 4196: K. Fischer, I.J. Timm, E. André, N. Zhong (Eds.), Multiagent System Technologies. X, 185 pages. 2006. (Sublibrary LNAI).

Vol. 4195: D. Gaiti, G. Pujolle, E. Al-Shaer, K. Calvert, S. Dobson, G. Leduc, O. Martikainen (Eds.), Autonomic Networking. IX, 316 pages. 2006.

Vol. 4194: V.G. Ganzha, E.W. Mayr, E.V. Vorozhtsov (Eds.), Computer Algebra in Scientific Computing. XI, 313 pages. 2006.

Vol. 4193: T.P. Runarsson, H.-G. Beyer, E. Burke, J.J. Merelo-Guervós, L.D. Whitley, X. Yao (Eds.), Parallel Problem Solving from Nature - PPSN IX. XIX, 1061 pages. 2006.

Vol. 4192: B. Mohr, J.L. Träff, J. Worringen, J. Dongarra (Eds.), Recent Advances in Parallel Virtual Machine and Message Passing Interface. XVI, 414 pages. 2006.

Vol. 4191: R. Larsen, M. Nielsen, J. Sporring (Eds.), Medical Image Computing and Computer-Assisted Intervention – MICCAI 2006, Part II. XXXVIII, 981 pages. 2006.

Vol. 4190: R. Larsen, M. Nielsen, J. Sporring (Eds.), Medical Image Computing and Computer-Assisted Intervention – MICCAI 2006, Part I. XXXVVIII, 949 pages. 2006.

Vol. 4189: D. Gollmann, J. Meier, A. Sabelfeld (Eds.), Computer Security – ESORICS 2006. XI, 548 pages. 2006.